MATH 87

An Incremental Development

MATH 87

An Incremental Development

Stephen Hake
John Saxon

SAXON PUBLISHERS, INC.

Math 87: An Incremental Development

Printed in the United States of America

ISBN: 0-939798-54-9

Editor: Nancy Warren
Production supervisors: Joan Coleman and David Pond
Graphic artists: Scott Kirby, John Chitwood, and David Pond

Sixth printing: March 1995

Saxon Publishers, Inc.
1320 W. Lindsey, Suite 100
Norman, Oklahoma   73069

Contents

		Preface		**xi**
LESSON	A	Operations of Arithmetic • Parentheses • Arithmetic with Whole Numbers		1
LESSON	B	Operations of Arithmetic with Money		6
LESSON	1	Missing Numbers in Addition, Subtraction, and Multiplication		11
LESSON	2	Number Line • Positive and Negative Numbers • Ordering and Comparing Numbers		15
LESSON	3	Adding and Subtracting on the Number Line		20
LESSON	4	Place Value through Hundred Trillions • Reading and Writing Whole Numbers		23
LESSON	5	Factors • Divisibility		27
LESSON	6	Lines, Rays, and Segments		32
LESSON	7	Fractions and Mixed Numbers		36
LESSON	8	Adding, Subtracting, and Multiplying Fractions		41
LESSON	9	"Some and Some More" Word Problems		44
LESSON	10	"Some Went Away" Word Problems		49
LESSON	11	"Larger-Smaller-Difference" Word Problems • Time Problems		54
LESSON	12	Equal Groups Word Problems		59
LESSON	13	Part-Part-Whole Word Problems		63
LESSON	14	Fractions Equal to 1 • Improper Fractions		69
LESSON	15	Equivalent Fractions		73
LESSON	16	Reducing Fractions, Part 1		77
LESSON	17	Linear Measure		82
LESSON	18	Pairs of Lines • Angles		87
LESSON	19	Polygons		92

LESSON	*20*	Perimeter, Part 1	**96**
LESSON	*21*	Solving Equations	**100**
LESSON	*22*	Prime and Composite Numbers • Prime Factorization	**104**
LESSON	*23*	Simplifying Fractions and Mixed Numbers	**110**
LESSON	*24*	Writing Mixed Numbers and Whole Numbers as Fractions	**114**
LESSON	*25*	Fraction-of-a-Group Problems	**117**
LESSON	*26*	Adding and Subtracting Mixed Numbers	**121**
LESSON	*27*	Reciprocals	**127**
LESSON	*28*	Reducing Fractions, Part 2	**131**
LESSON	*29*	Dividing Fractions	**135**
LESSON	*30*	Multiplying and Dividing Mixed Numbers	**140**
LESSON	*31*	Multiples • Least Common Multiple	**143**
LESSON	*32*	Two-Step Word Problems	**147**
LESSON	*33*	Average, Part 1	**151**
LESSON	*34*	Rounding Whole Numbers • Estimating Answers	**155**
LESSON	*35*	Common Denominators • Adding and Subtracting Fractions with Different Denominators	**160**
LESSON	*36*	Decimal Fractions • Decimal Place Value	**166**
LESSON	*37*	Reading and Writing Decimal Numbers • Comparing Decimal Numbers	**171**
LESSON	*38*	Rounding Decimal Numbers	**176**
LESSON	*39*	Decimal Numbers on the Number Line	**180**
LESSON	*40*	Adding and Subtracting Decimal Numbers	**184**
LESSON	*41*	Ratio	**188**
LESSON	*42*	Perimeter, Part 2	**193**
LESSON	*43*	Graphs	**197**
LESSON	*44*	Proportions	**203**
LESSON	*45*	Multiplying Decimal Numbers	**207**
LESSON	*46*	Dividing a Decimal Number by a Whole Number	**211**
LESSON	*47*	Repeating Digits	**215**
LESSON	*48*	Decimals to Fractions and Fractions to Decimals	**220**
LESSON	*49*	Division Answers	**225**

LESSON	**50**	Dividing by a Decimal Number	**228**
LESSON	**51**	Unit Price	**233**
LESSON	**52**	Exponents	**236**
LESSON	**53**	Powers of 10	**240**
LESSON	**54**	Rectangular Area, Part 1	**244**
LESSON	**55**	Square Root	**249**
LESSON	**56**	Rates	**253**
LESSON	**57**	Percent	**256**
LESSON	**58**	Mixed Measures • Adding Mixed Measures	**262**
LESSON	**59**	Multiplying Rates	**266**
LESSON	**60**	Rectangular Area, Part 2	**271**
LESSON	**61**	Scientific Notation for Large Numbers	**275**
LESSON	**62**	Order of Operations	**279**
LESSON	**63**	Unit Multipliers • Unit Conversion	**283**
LESSON	**64**	Ratio Word Problems	**288**
LESSON	**65**	Average, Part 2	**292**
LESSON	**66**	Subtracting Mixed Measures	**296**
LESSON	**67**	Liquid Measure	**299**
LESSON	**68**	Scientific Notation for Small Numbers	**304**
LESSON	**69**	Classifying Quadrilaterals	**307**
LESSON	**70**	Area of a Parallelogram	**312**
LESSON	**71**	Fraction-Decimal-Percent Equivalents	**316**
LESSON	**72**	Sequences • Functions, Part 1	**320**
LESSON	**73**	Adding Integers on the Number Line	**324**
LESSON	**74**	Fractional Part of a Number and Decimal Part of a Number, Part 1	**330**
LESSON	**75**	Variables and Evaluation	**334**
LESSON	**76**	Classifying Triangles	**338**
LESSON	**77**	Symbols of Inclusion	**343**
LESSON	**78**	Adding Signed Numbers	**347**
LESSON	**79**	Area of a Triangle	**352**
LESSON	**80**	Percent of a Number	**357**
LESSON	**81**	Ratio Problems Involving Totals	**361**

LESSON 82	Geometric Solids	366
LESSON 83	Weight	370
LESSON 84	Circles • Investigating Circumference	375
LESSON 85	Circumference and Pi	379
LESSON 86	The Opposite of the Opposite • Algebraic Addition	384
LESSON 87	Operations with Fractions and Decimals	388
LESSON 88	The Addition Rule for Equations	393
LESSON 89	More on Scientific Notation	399
LESSON 90	Multiplication Rule for Equations	403
LESSON 91	Volume	409
LESSON 92	Finding the Whole Group When a Fraction Is Known	414
LESSON 93	Implied Ratios	417
LESSON 94	Fractional Part of a Number and Decimal Part of a Number, Part 2	422
LESSON 95	Multiplying and Dividing Signed Numbers	427
LESSON 96	Area of a Complex Figure • Area of a Trapezoid	432
LESSON 97	Inverting the Divisor	437
LESSON 98	More on Percent	442
LESSON 99	Graphing Inequalities	447
LESSON 100	Insufficient Information • Quantitative Comparisons	451
LESSON 101	Measuring Angles with a Protractor	456
LESSON 102	Using Proportions to Solve Percent Problems	461
LESSON 103	Area of a Circle	467
LESSON 104	Multiplying Powers of 10 • Multiplying Numbers in Scientific Notation	472
LESSON 105	Mean, Median, Mode, and Range	477
LESSON 106	Order of Operations with Signed Numbers • Functions, Part 2	482
LESSON 107	Number Families	488
LESSON 108	Memorizing Common Fraction-Percent Equivalents	494
LESSON 109	Multiple Unit Multipliers • Conversion of Units of Area	498
LESSON 110	Sum of the Angle Measures of a Triangle • Straight Angles	504

LESSON 111	Equations with Mixed Numbers	**510**
LESSON 112	Evaluations with Signed Numbers • Signed Numbers without Parentheses	**514**
LESSON 113	Sales Tax	**519**
LESSON 114	Percents Greater than 100, Part 1	**523**
LESSON 115	Percents Greater than 100, Part 2	**528**
LESSON 116	Solving Two-Step Equations	**533**
LESSON 117	Simple Probability	**538**
LESSON 118	Volume of a Right Solid	**544**
LESSON 119	Rectangular Coordinates	**548**
LESSON 120	Estimating Angle Measures	**554**
LESSON 121	Similar Triangles	**561**
LESSON 122	Scale and Scale Factor	**567**
LESSON 123	Pythagorean Theorem	**572**
LESSON 124	Estimating Square Roots • Special Angles	**578**
LESSON 125	Multiplying Three or More Signed Numbers • Powers of Negative Numbers	**584**
LESSON 126	Semicircles	**588**
LESSON 127	Surface Area	**592**
LESSON 128	Solving Literal Equations • Transforming Formulas	**596**
LESSON 129	Graphing Functions	**600**
LESSON 130	Formulas and Substitution	**605**
LESSON 131	Simple Interest	**609**
LESSON 132	Compound Probability	**613**
LESSON 133	Volume of a Pyramid and a Cone	**618**
LESSON 134	Probability, Chance, and Odds	**622**
LESSON 135	Volume, Capacity, and Weight in the Metric System	**627**
APPENDIX	Supplemental Practice Problems for Selected Lessons	**633**
	Glossary	**647**
	Index	**661**

Preface

To the Student

We study mathematics because it is an important part of our daily lives. Our school schedule, our trip to the store, the preparation of our meals, and many of the games we play all involve mathematics. Many of the word problems you will see in this book are drawn from our daily experiences.

Mathematics is even more important in the adult world. In fact, your personal future in the adult world may depend in part upon the mathematics you have learned. This book was written with the hope that more students will learn mathematics and learn it well. For this to happen, you must use this book properly. As you work through the pages of this book, you will find similar problems presented over and over again. **Solving these problems day after day is the secret to success. Work every problem in every practice set and in every problem set. Do not skip problems. With honest effort you will experience success and true learning which will stay with you and serve you well in the future.**

Acknowledgments

We thank Shirley McQuade Davis for her ideas on teaching word problem thinking patterns.

Stephen Hake
Temple City, California

John Saxon
Norman, Oklahoma

Operations of Arithmetic • Parentheses • Arithmetic with Whole Numbers

Operations of arithmetic

The operations of arithmetic are addition, subtraction, multiplication, and division. Each of these operations is a **binary operation**. The English word *binary* comes from the Latin word *binarius,* which means "in two parts." We say that addition is a binary operation because we can add only two numbers in one step. If we wish to add

$$2 + 3 + 4$$

we can add two of the numbers and then add the other number. We can add 2 and 3 to get 5 and then add 5 and 4 to get 9.

$$2 + 3 + 4 \qquad \text{problem}$$
$$5 + 4 \qquad \text{added 2 and 3}$$
$$9 \qquad \text{added 5 and 4}$$

The two numbers we add are called **addends,** and the result is called the **sum**. Subtraction is also a binary operation. We subtract the **subtrahend** from the **minuend** and call the result the **difference.**

ADDITION		
2	←	addend
+3	←	addend
5	←	sum

SUBTRACTION		
9	←	minuend
− 7	←	subtrahend
2	←	difference

The other two fundamental operations of arithmetic are multiplication and division. These operations are also binary operations. The numbers that are multiplied are called **factors,** and the result is called the **product**. When we divide, we divide the **dividend** by the **divisor,** and the result is called the **quotient**.

MULTIPLICATION		
5	←	factor
× 4	←	factor
20	←	product

DIVISION

$$\text{divisor} \rightarrow 5\overline{)20} \quad \begin{array}{l} \leftarrow \text{quotient} \\ \leftarrow \text{dividend} \end{array}$$

1

To indicate multiplication, we can use a multiplication symbol shaped like the letter X. We can also use a center dot. We can also use symbols of inclusion such as parentheses.

$$4 \times 5 \qquad \text{multiplication symbol}$$

$$4 \cdot 5 \qquad \text{center dot}$$

$$4(5) \qquad \text{parentheses}$$

Many computer programs use an asterisk to indicate multiplication.

$$4 * 5 \qquad \text{asterisk (computers)}$$

There are also several ways to designate division. Each of the following indicates that 12 is to be divided by 3.

$$3\overline{)12} \qquad \text{division box}$$

$$12 \div 3 \qquad \text{division symbol}$$

$$\frac{12}{3} \qquad \text{division bar or fraction line}$$

Computers often use a slanted fraction line so that everything can be written on one line.

$$12/3 \qquad \text{slanted fraction line (computers)}$$

Example 1   When the sum of 3 and 4 is subtracted from the product of 3 and 4, what is the difference?

Solution   We add to find a sum, so the sum of 3 and 4 is 7. We multiply to find a product, so the product of 3 and 4 is 12. We subtract to find the difference, so the difference of 12 and 7 is **5.**

$$4 + 3 = 7 \qquad \text{sum}$$

$$4 \cdot 3 = 12 \qquad \text{product}$$

$$12 - 7 = \mathbf{5} \qquad \text{difference}$$

Parentheses   The numbers inside parentheses are considered to be a single quantity. **If an expression contains parentheses, the first step is to simplify within the parentheses.**

Example 2   Simplify: $8 - (4 + 2)$

Solution   We simplify within the parentheses as the first step. Then we subtract.

$8 - (4 + 2)$	problem
$8 - 6$	simplified
2	subtracted

Arithmetic with whole numbers

The **counting numbers** are the numbers we use to count. They are

$$1, 2, 3, 4, 5, 6, 7, \ldots$$

The three dots, called an *ellipsis,* means the list goes on and on and never ends. Every counting number is also a **whole number**. The **number zero** is also a whole number. To make a list of the whole numbers, we write

$$0, 1, 2, 3, 4, 5, \ldots$$

In the following examples we review the procedures for adding, subtracting, multiplying, and dividing whole numbers.

Example 3   Add: $36 + 472 + 3614$

Solution   To add whole numbers, we align the ones' digits so that we add digits with the same place value. We add in columns from right to left. We may add the digits in any order. Looking for combinations of digits that total 10 may speed our work.

$$
\begin{array}{r}
111 \\
36 \\
472 \\
+\ 3614 \\
\hline
\mathbf{4122}
\end{array}
$$

Example 4   Subtract: $5207 - 948$

Solution   To subtract whole numbers, we also align the ones' digits. There is an order to subtraction. We write the first number above the second number. Then we subtract.

$$
\begin{array}{r}
\overset{4\ 11\ 9}{\cancel{5}\,\cancel{2}\,\cancel{0}\,^{1}7} \\
-\ \ 9\,4\,8 \\
\hline
\mathbf{4\,2\,5\,9}
\end{array}
$$

Example 5 Multiply: 164 × 23

Solution In a multiplication problem, either number can go on top. We usually put the number with the most digits on top. We begin by multiplying 164 × 3. Then we multiply 164 by 2 and write this result below the result of the first multiplication. Since this 2 is really 20, we either write a zero at the end, or we leave the place empty. Then we add the partial products to find the final product.

$$
\begin{array}{r}
164 \\
\times\ 23 \\
\hline
492 \\
328\underline{0} \\
\hline
\mathbf{3772}
\end{array}
$$

← This zero is optional.

Example 6 Multiply: 468 × 200

Solution We write the 468 on top. We may let the zeros in 200 "hang out" to the right. Then we write zeros below the line. Then we multiply 468 by 2.

$$
\begin{array}{r}
468 \\
\times\ \ 200 \\
\hline
00
\end{array}
\qquad
\begin{array}{r}
468 \\
\times\ \ 200 \\
\hline
\mathbf{93,600}
\end{array}
$$

Example 7 Divide: $\dfrac{234}{6}$

Solution We rewrite the problem using a division box. We put the top number inside the box. This division comes out even. There is no remainder.

$$
\begin{array}{r}
\mathbf{39} \\
6\overline{)234} \\
\underline{18}\ \ \\
54 \\
\underline{54} \\
0
\end{array}
$$

Example 8 Divide: 1234 ÷ 56

Solution This division does not come out even. There is a remainder. Note how we write the answer. Other methods for dealing with a remainder will be considered later.

$$
\begin{array}{r}
\mathbf{22\ r\ 2} \\
56\overline{)1234} \\
\underline{112}\ \ \\
114 \\
\underline{112} \\
2
\end{array}
$$

Practice **a.** When the product of 4 and 4 is divided by the sum of 4 and 4, what is the quotient?

Add, subtract, multiply, or divide, as indicated:

b. $12 \div (6 \div 2)$ **c.** $42 + 1250 + 568$

d. $3014 - 426$ **e.** 42×367

f. $365 \div 12$ **g.** $\dfrac{234}{18}$

Problem set A

1. When the sum of 5 and 6 is subtracted from the product of 5 and 6, what is the difference?

2. If the subtrahend is 9 and the difference is 8, what is the minuend?

3. If the divisor is 4 and the quotient is 8, what is the dividend?

4. When the product of 6 and 6 is divided by the sum of 6 and 6, what is the quotient?

5. Name the four fundamental operations of arithmetic.

6. If the sum is 12 and one addend is 4, what is the other addend?

Add, subtract, multiply, or divide, as indicated:

7. $\dfrac{1000}{8}$

8. $\begin{array}{r} 4374 \\ -\ 1659 \end{array}$

9. $\begin{array}{r} 64 \\ \times\ 37 \end{array}$

10. $\begin{array}{r} 7 \\ 8 \\ 4 \\ 6 \\ 9 \\ 3 \\ 5 \\ +\ 7 \end{array}$

11. $364 + 52 + 867 + 9$ **12.** $4000 - 3625$

13. 316×18 **14.** $4360 \div 20$

15. $25\overline{)767}$ **16.** $64 \div (8 \div 4)$

17. $(64 \div 8) \div 4$ **18.** $8 \cdot 64$

19. $76 - (37 + 16)$

20. $(76 - 37) + 16$ **21.** $3708 \div 12$

22. $431 + 562 + 54 + 29 + 8$

23. $3000 - (1200 - 457)$

24. 365×20

25. $30(40)$ **26.** $1010 - 234$ **27.** $\dfrac{560}{14}$

28.
$$\begin{array}{r} 4017 \\ -\ 3952 \end{array}$$

29.
$$\begin{array}{r} 250 \\ \times\ 80 \end{array}$$

30.
$$\begin{array}{r} 3634 \\ +\ 2957 \end{array}$$

REVIEW LESSON B

Operations of Arithmetic with Money

Two forms of writing money   Amounts of money in the United States are measured in dollars and cents. To indicate an amount of money, we may use the dollar sign ($) or the cent sign (¢), but not both.

Since 1 dollar equals 100 cents, we may write 1 dollar either of these two ways:

$$\$1.00 \qquad \text{or} \qquad 100¢$$

A **decimal point** is used with a dollar sign. Numbers to the left of the decimal point name whole dollars. Numbers to the right of the decimal point name hundredths of a dollar, that is, cents. If a dollar value is even, that is, with no cents, it may be written without the two zeros following the decimal point. For example, $1.00 may be written $1. A cent sign means the value is in pennies, not dollars. So a decimal point is used with a cent sign only when naming part of a cent.

Example 1 Write twenty-five cents with (a) a cent sign and (b) a dollar sign.

Solution (a) **25¢**

(b) **$0.25**

In (b) we wrote a zero to the left of the decimal point to show that there are no dollars. This zero is not necessary, but it is customary to use it.

Example 2 Occasionally we will see the dollar sign and the cent sign used incorrectly. This sign is incorrect.

> Soft Drinks
> **0.50¢ each**

Show two ways to correct this sign.

Solution We assume that the sign means that soft drinks cost 50 cents each. Fifty cents may be written with a dollar sign, **$0.50** (no dollars and 50 cents), or with a cent sign, **50¢** (50 pennies). The sign is incorrect because it used a decimal point with the cent sign. The incorrect sign literally means that soft drinks cost not half a dollar but half a cent! Both of the following signs are correct.

> Soft Drinks
> **50¢ each**

> Soft Drinks
> **$0.50 each**

Arithmetic with money   The following examples provide a review of arithmetic with money.

Example 3   Add: $1.45 + $6 + 8¢

Solution   We begin by rewriting the amounts of money so that each is written with a dollar sign and has two digits to the right of the decimal point.

$$\$1.45 + \$6.00 + \$0.08$$

Next we write the numbers in columns and we **align the decimal points**. Then we add. The total is written with a dollar sign and with a decimal point in line with the other decimal points.

$$
\begin{array}{r}
\$1.45 \\
6.00 \\
+\ 0.08 \\
\hline
\$7.53
\end{array}
$$

Example 4   Subtract: $5 − 25¢

Solution   We rewrite the amounts so that each amount is written with a dollar sign and with two digits after the decimal point to show cents.

$$\$5.00 - \$0.25$$

We write the first number on top. We are careful to line up the decimal points. The decimal point in the answer is in line with the other two decimal points.

$$
\begin{array}{r}
\$5.00 \\
-\ 0.25 \\
\hline
\$4.75
\end{array}
$$

Example 5   Multiply: (a) $1.45 × 6   (b) 29¢ × 5

Solution   (a)   We set up the multiplication the same way as we do when we multiply whole numbers. We write the product with a dollar sign and have two digits to the right of the decimal point.

$$
\begin{array}{r}
\$1.45 \\
\times\ \ \ \ 6 \\
\hline
\$8.70
\end{array}
$$

(b)   First we multiply 29¢ × 5. The answer is greater than $1, so we use a dollar sign and a decimal point to write the answer.

$$
\begin{array}{r}
29¢ \\
\times\ 5 \\
\hline
145¢
\end{array} = \mathbf{\$1.45}
$$

Example 6   Divide: $12.60 ÷ 5

Solution   We divide the same way we divide whole numbers. We write the quotient with a dollar sign. We place the decimal point in the answer directly above the decimal point in the dividend.

$$\begin{array}{r} \$2.52 \\ 5\overline{)\$12.60} \\ \underline{10} \\ 26 \\ \underline{25} \\ 10 \\ \underline{10} \\ 0 \end{array}$$

Practice   **a.**   Write five cents with (a) a cent sign and (b) a dollar sign.

b.   The sign shown is incorrect. Show two ways to correct this sign.

> Lemonade
> **0.45¢ per cup**

Add, subtract, multiply, or divide, as indicated:

c. $1.75 + 60¢ + $3

d. $2 − 47¢

e.   65¢ × 8

f. $24.00 ÷ 5

Problem set B

1.   When the product of 2 and 3 is subtracted from the sum of 4 and 5, what is the difference?

2.   Write four cents (a) with a cent sign and (b) with a dollar sign.

3.   The sign shown is incorrect. Show two ways to correct this sign.

> Orange Juice
> **0.75¢ per cup**

4.   Name the four fundamental operations of arithmetic.

5.   If the dividend is 60 and the divisor is 4, what is the quotient?

6. If the product is 12 and one factor is 4, what is the other factor?

Add, subtract, multiply, or divide, as indicated:

7. $42.47
+ 63.89

8. $20.00
− 14.79

9. $1.54
×   7

10. $\dfrac{\$30.00}{8}$

11. $4.36 + 75¢ + $12 + 6¢

12. $10.00 − ($4.89 + 74¢)

13. 8
5
4
6
5
4
3
7
2
4
1
+ 8

14. 3105 ÷ 15

15. 40)‾1630

16. 81 ÷ (9 ÷ 3)

17. (81 ÷ 9) ÷ 3

18. (10)($3.75)

19. 3167 − (450 − 78)

20. (3167 − 450) − 78

21. $20.00 ÷ 16

22. 70 · 800

23. $10 − $8.45

24. 3714 + 268 + 47 + 9

25. $51.71 + $6.49 + 79¢ + $15

26. $20 − ($1.47 + $8)

27. $75.00 ÷ 12

28. $0.45
×   30

29. 45¢
× 72

30. $\dfrac{6324}{31}$

LESSON 1

Missing Numbers in Addition, Subtraction, and Multiplication

Missing numbers in addition

In every addition problem there are at least two addends. There is only one final sum in an addition problem. Sometimes we encounter addition problems in which the sum is missing. Sometimes we encounter addition problems in which an addend is missing. Our job is to find the missing number. We can use a letter to represent a missing number.

In the problem on the left, we use the letter N to represent the missing sum.

MISSING SUM	MISSING ADDEND	MISSING ADDEND
2	2	B
$+\,3$	$+\,A$	$+\,3$
N	5	5

In the problem in the center, we use the letter A to represent the missing addend. In the problem on the right, we use the letter B to represent the missing addend.

Example 1 Find each missing number:

(a)
$$\begin{array}{r} N \\ +\,53 \\ \hline 75 \end{array}$$

(b)
$$\begin{array}{r} 26 \\ +\,A \\ \hline 61 \end{array}$$

Solution In both (a) and (b) the missing number is an addend. We can find each missing addend by subtracting the known addend from the sum. Then we check.

(a) Subtract.　　Try it.
$$\begin{array}{r} 75 \\ -\,53 \\ \hline 22 \end{array} \qquad \begin{array}{r} 22 \\ +\,53 \\ \hline 75 \end{array} \text{ check}$$

So the missing number in (a) is **22.**

(b) Subtract.　　Try it.
$$\begin{array}{r} 61 \\ -\,26 \\ \hline 35 \end{array} \qquad \begin{array}{r} 26 \\ +\,35 \\ \hline 61 \end{array} \text{ check}$$

So the missing number in (b) is **35.**

Missing numbers in subtraction

There are three numbers in a subtraction problem. If any one of the three numbers is missing, our job is to find the missing number.

Missing Minuend	Missing Subtrahend	Missing Difference
N	5	5
$-\ 3$	$-\ X$	$-\ 3$
2	2	M

To find a missing minuend (top number), we add the other two numbers. To find a missing subtrahend or difference, we subtract.

Example 2 Find each missing number: (a) $\quad P$ $\quad$ (b) $\quad$ 32

$$\begin{array}{r} P \\ -\ 24 \\ \hline 17 \end{array} \qquad \begin{array}{r} 32 \\ -\ X \\ \hline 14 \end{array}$$

Solution

(a) To find the top number, we add the bottom two numbers. We find that the missing number in (a) is **41**.

Add.	Try it.
17	41
$+\ 24$	$-\ 24$
41	17 $\quad$ check

(b) To find one of the two lower numbers, we subtract the smaller given number from the larger. So the missing number in (b) is **18**.

Subtract.	Try it.
32	32
$-\ 14$	$-\ 18$
18	14 $\quad$ check

Missing numbers in multiplication

There are three numbers in a multiplication problem. Two of the numbers are factors, and the third number is the product. If any one of the three numbers is missing, we can figure out what it is.

Missing Product	Missing Factor	Missing Factor
3	3	R
$\times\ 2$	$\times\ F$	$\times\ 2$
P	6	6

To find a missing product, we multiply the factors. To find a missing factor, we divide the product by the known factor.

Example 3 Find each missing number:
(a) $\begin{array}{r} 12 \\ \times\ N \\ \hline 168 \end{array}$
(b) $\begin{array}{r} K \\ \times\ 7 \\ \hline 105 \end{array}$

Solution In both (a) and (b) the missing number is a factor. We can find a missing factor by dividing the product by the known factor.

(a) $\begin{array}{r} 14 \\ 12\overline{)168} \\ \underline{12} \\ 48 \\ \underline{48} \\ 0 \end{array}$
Try it.
$\begin{array}{r} 12 \\ \times\ 14 \\ \hline 48 \\ \underline{12\ \ } \\ 168 \end{array}$ check

(b) $\begin{array}{r} 15 \\ 7\overline{)105} \\ \underline{7} \\ 35 \\ \underline{35} \\ 0 \end{array}$
Try it.
$\begin{array}{r} 15 \\ \times\ 7 \\ \hline 105 \end{array}$ check

So the missing number in (a) is **14**.

So the missing number in (b) is **15**.

Practice Find each missing number. Pay close attention to the sign so you know what type of problem it is.

a. $\begin{array}{r} A \\ +\ 12 \\ \hline 31 \end{array}$

b. $\begin{array}{r} B \\ -\ 24 \\ \hline 15 \end{array}$

c. $\begin{array}{r} C \\ \times\ 15 \\ \hline 180 \end{array}$

d. $\begin{array}{r} 14 \\ \times\ D \\ \hline 420 \end{array}$

e. $\begin{array}{r} 26 \\ +\ E \\ \hline 43 \end{array}$

f. $\begin{array}{r} 51 \\ -\ F \\ \hline 20 \end{array}$

Problem set 1

1. When the product of 4 and 4 is divided by the sum of 4 and 4, what is the quotient?

2. If the subtrahend is 9 and the difference is 9, what is the minuend?

3. Write the value of 5 quarters (a) with a cent sign and (b) with a dollar sign.

4. If one addend is 7 and the sum is 21, what is the other addend?

5. A center dot may be used to indicate what operation of arithmetic?

6. If the product of two identical factors is 36, what is each factor?

Find each missing number:

7. X
      $+\ 83$
      $\overline{112}$

8.   96
      $-\ R$
      $\overline{27}$

9. K
      $\times\ 7$
      $\overline{119}$

10.   127
      $+\ Z$
      $\overline{300}$

11. M
      $-\ 137$
      $\overline{731}$

12.   25
      $\times\ N$
      $\overline{400}$

Add, subtract, multiply, or divide, as indicated:

13. $3517 \div 14$

14. $60(700)$

15. $\begin{array}{r} 8 \\ 5 \\ 6 \\ 1 \\ 8 \\ 7 \\ 4 \\ 3 \\ 5 \\ 8 \\ 5 \\ +\ 3 \end{array}$

16. $96 \div (16 \div 2)$

17. $(96 \div 16) \div 2$

18. $35\overline{)2104}$

19. $\$16.47 + \$15 + 63¢$

20. $\$50.00 - (\$6.48 + \$31.75)$

21.   47
      $\times\ 39$

22. $\$8.79$
      $\times\ \ \ 80$

23. $\dfrac{4740}{30}$

24. $1100 - (374 - 87)$

25. $(1100 - 374) - 87$

26. $4736 + 271 + 9 + 88$

27. 30,145 − 4,299

28. $0.48
       ×   40

29. $\frac{\$40.00}{32}$

30. 32¢
       × 48

LESSON 2

Number Line • Positive and Negative Numbers • Ordering and Comparing Numbers

Number line A **number line** can be used to help us arrange numbers in order. To construct a number line, we draw a straight line with a straightedge. Then we put marks on the line. The distance between any two adjacent marks is the same.

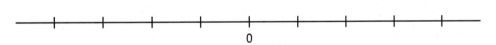

Then we choose one of the marks on the line and write a zero below the mark. We call this mark the **origin.**

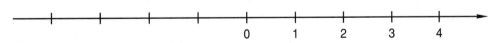

Then below the marks to the right of the origin we write the counting numbers in order beginning with the number 1 as we show here.

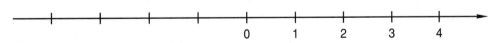

The arrowhead reminds us that we have just drawn a part of the number line. The number line goes on and on and does not end.

Positive and negative numbers

The numbers to the right of the origin are called the **positive numbers,** and they are all **greater than zero.** Every positive number has an **opposite** that is the same distance to the left of the origin. The numbers to the left of the origin are called **negative numbers**. The negative numbers are all **less than zero**. We always use a negative sign when we write a negative number, as we show on the number line below. If the number does not have a sign, we know that the number is a positive number. A number is **greater than** another number if it is farther to the right on the number line.

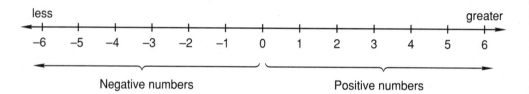

We note that each number and its opposite are the same distance from zero on the number line. Note that 5 is 5 to the right of zero and -5 is 5 to the left of zero. We read -5 by saying "negative five."

Ordering and comparing numbers

The number line lets us see how numbers can be arranged in order. As we move to the right on the number line, the numbers become greater and greater. As we move to the left on the number line, the numbers become less and less.

We **compare** two numbers by determining whether one number is greater than another number or whether the two numbers are equal. We place a **comparison symbol** between two numbers to show the comparison. The comparison symbols are the equals sign (=) and the greater than/less than symbol (> or <). The greater than/less than symbol may point in either direction. We write this symbol so that the smaller end (the point) points to the "smaller" number.

We may read a comparison from left to right or from right to left. When we read the greater than/less than symbol, we say the end of the symbol we get to first. We read the pointed end by saying "is less than." We read the open end by saying "is greater than."

Below we show three comparisons.

$$-5 < 4 \qquad 3 + 2 = 5 \qquad 5 > -6$$

First we will read these expressions from left to right.

$$-5 < 4 \qquad 3 + 2 = 5 \qquad 5 > -6$$

−5 is less than 4   3 plus 2 equals 5   5 is greater than −6

Now we will read the same comparisons by reading the right side first and then the left side.

$$-5 < 4 \qquad 3 + 2 = 5 \qquad 5 > -6$$

4 is greater than −5   5 equals 3 plus 2   −6 is less than 5

Although comparison expressions may be read in either direction, for the exercises in this book we will assume the left-to-right direction is intended.

Example 1   Arrange these numbers in order from least to greatest:

$$0, 1, -2$$

Solution   We arrange the numbers in the order in which they appear on the number line.

$$-2, 0, 1$$

Example 2   Rewrite each expression by replacing the circle with the correct comparison symbol. Then use words to write the comparison.

(a) $-5 \bigcirc 4$ (b) $1 + 1 \bigcirc 1 \times 1$

Solution   (a)   Since −5 is less than 4, we write

$$-5 < 4$$

negative 5 is less than 4

(b)   First we mentally simplify each expression.

$$1 + 1 = 2 \quad \text{and} \quad 1 \times 1 = 1$$

Since 2 is greater than 1, we write

$$1 + 1 > 1 \times 1$$

1 plus 1 is greater than 1 times 1

Practice   **a.** Copy the number line in this lesson.

b. Arrange these numbers in order from least to greatest:

$$0, -1, 2, -3$$

c. Use digits and a comparison symbol to write "The sum of 2 and 3 is less than the product of 2 and 3."

Replace each circle with the proper comparison symbol.

d. $3 \bigcirc -4$

e. $312 \bigcirc 321$

f. $2 \cdot 2 \bigcirc 2 + 2$

Problem set 2

1. Find the quotient when the product of 10 and 10 is divided by the sum of 10 and 10.

2. Use digits and symbols to write "The sum of 3 and 3 is less than the product of 3 and 3."

3. Write the value of a nickel and 3 pennies with a dollar sign.

4. Arrange these numbers in order from least to greatest:

$$-1, 1, -3, 3$$

5. If the sum of two identical addends is 16, what are the addends?

6. Replace each circle with the proper comparison symbol.

(a) $-2 \bigcirc 2$ (b) $432 \bigcirc 423$

7. What is the name for numbers that are less than zero?

8. How many units is it from -2 to 3 on the number line?

Find each missing number:

9.   367
   + N
   432

10.   P
  − 216
   198

11.   4
   8
   6
   2
   1
   5
   3
   8
   2
   7
   4
 + 6
   N

12.   24
  × K
  576

13.   M
 + 329
  673

14.   888
  − X
  147

15.   Y
  × 8
  624

Add, subtract, multiply, or divide, as indicated:

16. 68×706

17. $\$31.50 \div 25$

18. $90(700)$

19. $\$10.00 - (\$1.45 + 89¢)$

20. $(\$10.00 - \$1.45) + 89¢$

21. $\$3.75 + \$24 + \$76.38$

22. $348 + 76 + 3859 + 7 + 15$

23. $200 \div (10 \div 5)$

24. $43\overline{)974}$

25. $3 \cdot 4 \cdot 5$

26. $10(20)(30)$

27. $\$64.48 \div 16$

28. $\$0.75$
    $\times\;\;\;36$

29. $\dfrac{4800}{60}$

30. $15¢$
    $\times 78$

LESSON 3

Adding and Subtracting on the Number Line

We can use the number line to help us add and subtract. In this lesson we will sketch number lines and use arrows to show addition and subtraction. To add, we let the arrow point to the right. To subtract, we let the arrow point to the left.

Example 1 Show this addition problem on a number line: 3 + 2

Solution First we sketch a number line. **Next we start at the origin** (at zero) and draw an arrow that points to the right that is 3 units long. **From this arrowhead** we draw a second arrow that points to the right that is 2 units long.

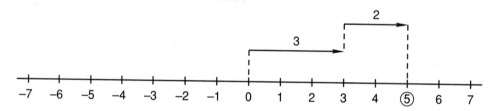

The second arrow ends at 5. This shows that 3 + 2 = **5**.

Example 2 Show this subtraction problem on a number line: 5 − 3

Solution We sketch a number line. **Then, starting at the origin,** we draw an arrow 5 units long that points to the right. Now, to subtract, we draw the second arrow that points **to the left** that is 3 units long. Remember to draw the second arrow from the first arrowhead.

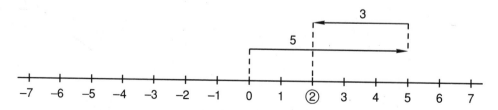

The second arrow ends at 2. This shows that 5 − 3 = **2**.

Example 3 Show this subtraction problem on a number line: 3 − 5

Solution We take the numbers in the order given. We always begin at the origin. Starting from the origin, we draw an arrow 3 units long that points to the right. From this arrowhead we draw a second arrow 5 units long that points to the **left**.

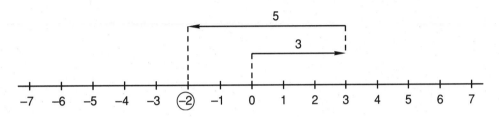

The second arrow ends at −2. This shows that 3 − 5 = **−2**.

Practice Use arrows to show each addition or subtraction problem on a number line.

    a. 4 + 2   **b.** 4 − 2   **c.** 2 − 4

Problem set 3

1. What is the difference when the sum of 5 and 4 is subtracted from the product of 3 and 3?

2. If the minuend is 27 and the difference is 9, what is the subtrahend?

3. What is the name for numbers that are greater than zero?

4. The sign shown is incorrect. Show two ways to correct this sign.

> *Grapefruit*
> **0.15¢ each**

5. Use digits and other symbols to write "The product of 5 and 2 is greater than the sum of 5 and 2."

6. Arrange these numbers in order from least to greatest:

$$-2, 1, 0, -1$$

7. Replace each circle with the proper comparison symbol.

(a) $3 \cdot 4 \bigcirc 2(6)$ (b) $-3 \bigcirc -2$

8. Show this addition problem on a number line: $2 + 3$

9. Show this subtraction problem on a number line: $2 - 3$

Find each missing number:

10.
$$\begin{array}{r} X \\ \times\ 12 \\ \hline 372 \end{array}$$

11.
$$\begin{array}{r} 439 \\ +\ \ Y \\ \hline 512 \end{array}$$

12.
$$\begin{array}{r} 8 \\ 7 \\ 4 \\ 2 \\ 3 \\ 6 \\ 5 \\ 8 \\ 7 \\ 5 \\ 2 \\ +\ 9 \\ \hline N \end{array}$$

13.
$$\begin{array}{r} Z \\ -\ 123 \\ \hline 654 \end{array}$$

14.
$$\begin{array}{r} 1000 \\ -\ \ M \\ \hline 101 \end{array}$$

15.
$$\begin{array}{r} P \\ +\ \$1.45 \\ \hline \$4.95 \end{array}$$

16.
$$\begin{array}{r} 32 \\ \times\ \ K \\ \hline 224 \end{array}$$

Add, subtract, multiply, or divide, as indicated:

17. $\$3.67 + 14¢ + \$52.75$

18. $\$100.00 - \$36.49$

19. $47(31)$

20. $5 \cdot 6 \cdot 7$

21. $9900 \div 18$

22. $30(20)(40)$

23. $(130 - 57) + 9$

24. $2014 - 1987$

25. $\$68.60 \div 7$

26. $46¢ + 64¢$

27. $21\overline{)6414}$

28.
$$\begin{array}{r} \$3.75 \\ \times\ \ \ \ 30 \\ \hline \end{array}$$

29. $\dfrac{4640}{80}$

30.
$$\begin{array}{r} 36¢ \\ \times 48 \\ \hline \end{array}$$

LESSON 4

Place Value through Hundred Trillions • Reading and Writing Whole Numbers

Place value   In our number system the value of a digit depends upon its position within a number. The value of each position is its **place value**. The chart below shows place values from the ones' place to the hundred trillions' place.

WHOLE NUMBER PLACE VALUES

hundred trillions	ten trillions	trillions	hundred billions	ten billions	billions	hundred millions	ten millions	millions	hundred thousands	ten thousands	thousands	hundreds	tens	ones	decimal point

— — — , — — — , — — — , — — — , — — — •

Example 1   (a)   Which digit is in the trillions' place in 32,567,890,000,000?

(b)   In 12,457,697,380,000 what is the place value of the digit 4?

Solution   (a)   The digit in the trillions' place is **2**.

(b)   The place value of the digit 4 is **hundred billions**.

Reading and writing whole numbers   Whole numbers with more than three digits are often written with commas to make the numbers easier to read. Commas help us read large numbers by marking the end of the trillions', billions', millions', and thousands' places. We need only to read the three-digit number in front of each comma and then say "trillion" or "billion" or "million" or "thousand" when we reach the comma.

— — — , — — — , — — — , — — — , — — —

trillion   billion   million   thousand

We will use the following guidelines when writing out numbers.

1. Put commas after the words trillion, billion, million, and thousand.

2. Hyphenate numbers between 20 and 100 that do not end in zero. For example, 52, 76, and 95 are written fifty-two, seventy-six, and ninety-five.

Example 2 Use words to write 1,380,000,050,200.

Solution **One trillion, three hundred eighty billion, fifty thousand, two hundred.**

Note: Since there are no millions, we do not read the millions' comma.

Example 3 Use words to write 3406521.

Solution First we start on the right and insert commas every three places as we move to the left.

$$3,406,521$$

Three million, four hundred six thousand, five hundred twenty-one.

Example 4 Use digits to write twenty trillion, five hundred ten million, seventy-eight thousand.

Solution It may be helpful to draw a "skeleton" of the number. We see that the number is more than one trillion, so we draw this skeleton.

$$\underline{}\,,\,\underline{}\,,\,\underline{}\,,\,\underline{}\,,\,\underline{}$$
$$\text{T} \qquad \text{B} \qquad \text{M} \qquad \text{T}$$

The letters below the commas stand for trillion, billion, million, and thousand. We will read to a comma, then pause to write what we have read. We read "twenty trillion." We write:

$$\underline{2\,0}\,,\,\underline{}\,,\,\underline{}\,,\,\underline{}\,,\,\underline{}$$
$$\text{T} \qquad \text{B} \qquad \text{M} \qquad \text{T}$$

Next we read "five hundred ten million." We write 510 before the **millions'** comma.

$$\underline{2\ 0}\ ,\ \_\,\_\,\_\ ,\ \underline{5\ 1\ 0}\ ,\ \_\,\_\,\_\ ,\ \_\,\_\,\_$$
T   B   M   T

Since there are **no billions**, we write zeros in the three places before the billions' comma.

$$\underline{2\ 0}\ ,\ \underline{0\ 0\ 0}\ ,\ \underline{5\ 1\ 0}\ ,\ \_\,\_\,\_\ ,\ \_\,\_\,\_$$
T   B   M   T

Then we read "seventy-eight thousand." We write 78 just before the thousands' comma. Since 78 is only two digits, we write a zero in front.

$$\underline{2\ 0}\ ,\ \underline{0\ 0\ 0}\ ,\ \underline{5\ 1\ 0}\ ,\ \underline{0\ 7\ 8}\ ,\ \_\,\_\,\_$$
T   B   M   T

Since there are no hundreds, tens, or ones, we write zeros in the last three places. Now we omit the dashes and write the number.

20,000,510,078,000

Practice   **a.** In 217,534,896,000,000, which digit is in the ten billions' place?

   b. In 9,876,543,210,000, what is the place value of the digit 6?

Use words to write each number.

   c. 36427580

   d. 40302010

Use digits to write each number.

   e. Twenty-five million, two hundred six thousand, forty

   f. Fifty billion, four hundred two million, one hundred thousand

**Problem set
4**

1. What is the sum of six hundred seven and two thousand, three hundred ninety-three?

2. Use digits and symbols to write "One hundred one thousand is greater than one thousand, one hundred."

3. Use words to write 50,574,006.

4. Which digit is in the trillions' place in 12,345,678,900,000?

5. Use digits to write two hundred fifty million, five thousand, seventy.

6. Replace the circle with the proper comparison symbol. Then use words to write the comparison.

$$-12 \bigcirc -15$$

7. Arrange these numbers in order from least to greatest:

$$-1, 4, -7, 0, 5, 7$$

8. Show this subtraction problem on a number line: $5 - 4$

9. How many units is it from negative 5 to positive 2 on the number line?

Find each missing number:

10.
$$\begin{array}{r} N \\ \times\ 30 \\ \hline 960 \end{array}$$

11.
$$\begin{array}{r} A \\ -\ 1367 \\ \hline 2500 \end{array}$$

12.
$$\begin{array}{r} 4 \\ 3 \\ 1 \\ 2 \\ 5 \\ 7 \\ 2 \\ 3 \\ 8 \\ 5 \\ 4 \\ +\ 9 \\ \hline N \end{array}$$

13.
$$\begin{array}{r} B \\ +\ 571 \\ \hline 3142 \end{array}$$

14.
$$\begin{array}{r} \$25.00 \\ -\ \qquad K \\ \hline \$18.70 \end{array}$$

15.
$$\begin{array}{r} 6400 \\ +\ \qquad D \\ \hline 10{,}000 \end{array}$$

16.
$$\begin{array}{r} 26 \\ \times\ E \\ \hline 624 \end{array}$$

Add, subtract, multiply, or divide, as indicated:

17. 37,428
+ 59,775

18. 31,014
− 24,767

19. 45 + 362 + 7 + 4319

20. \$64.59 + \$124 + \$6.30 + 37¢

21. 144 ÷ (12 ÷ 3)

22. (144 ÷ 12) ÷ 3

23. 40(500)

24. $20\overline{)1000}$

25. 6 · 5 · 4

26. \$10 − (\$4.60 − 39¢)

27. (\$10 − \$4.60) − 39¢

28. 29¢
× 36

29. $\dfrac{8505}{21}$

30. \$12.47
×   10

LESSON 5

Factors • Divisibility

Factors Recall that a factor is one of the numbers multiplied to form a product.

In 3 × 5 = 15,   the factors are 3 and 5, so both 3 and 5 are factors of 15.

In 1 × 15 = 15,   the factors are 1 and 15, so both 1 and 15 are factors of 15.

Therefore, any of the numbers 1, 3, 5, and 15 can serve as a factor of 15.

Notice that 15 can be divided by 1, 3, 5, and 15 without a remainder. This leads us to another definition of **factor**.

> A *factor* is a whole number that divides another whole number without a remainder.

For exam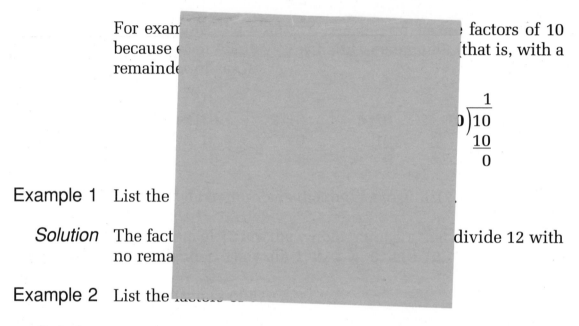 factors of 10 because e   that is, with a remainde

$$\begin{array}{r} 1 \\ 0\overline{)10} \\ \underline{10} \\ 0 \end{array}$$

Example 1 List the

Solution The fact   divide 12 with no rema

Example 2 List the factors of 51

Solution As we try to think of whole numbers that divide 51 with no remainder, we may think that 51 has only two factors, 1 and 51. However, there are actually four factors of 51. Notice that 3 and 17 are factors of 51.

$$\text{3 is a factor of 51} \quad \rightarrow \quad 3\overline{)51} \quad \leftarrow \quad \text{17 is a factor of 51}$$

Since 3 · 17 equals 51, both 3 and 17 are factors of 51. Thus, the four factors of 51 are **1, 3, 17,** and **51**.

Divisibility As we saw in Example 2, 51 **can be divided** by 1, 3, 17, and 51, and the remainder is zero. The capability of a whole number to be divided by another whole number with no remainder is called **divisibility**. Thus, 51 is **divisible** by 1, 3, 17, and 51.

There are several methods for testing the divisibility of a number without actually performing the division. For example, if a number is divisible by 5, its last digit is either 0 or 5. Therefore, we do not need to divide 12,476,365 by 5 to find out if it is divisible by 5. All we need to do is look at the last digit. The last digit is 5, so the number is divisible by 5. Below are listed some methods for testing whether a number is divisible by 2, 3, 4, 5, 6, 8, 9, and 10.

TESTS FOR DIVISIBILITY

A number can be divided by...

2	if the last digit can be divided by 2.
4	if the last two digits can be divided by 4.
8	if the last three digits can be divided by 8.
5	if the last digit is 0 or 5.
10	if the last digit is 0.
3	if the **sum of the digits** can be divided by 3.
6	if the number can be divided by 2 **and** by 3.
9	if the **sum of the digits** can be divided by 9.
7	**There is no simple test for divisibility by 7.**

Example 3 Which whole numbers from 1 through 10 are divisors of 9060?

Solution In the sense used in this problem, a **divisor** is the same thing as a **factor**. The number 1 is a divisor of any whole number. As we apply the tests for divisibility, we find that 9060 passes the tests for 2, 4, 5, and 10. The sum of its digits $(9 + 0 + 6 + 0)$ is 15, which can be divided by 3 but not by 9. Since 9060 is divisible by both 2 and 3, it is also divisible by 6. The only whole number from 1 to 10 we have not tried is 7, for which we have no simple test. **We have to divide 9060 by 7 to find out if 7 is a divisor.** It is not. We find that the numbers from 1 to 10 that are divisors of 9060 are **1, 2, 3, 4, 5, 6,** and **10**.

Practice List the whole numbers that are factors of each number.

a. 25 **b.** 24 **c.** 23

List the whole numbers from 1 to 10 that are factors of each number.

d. 1260

e. 73,500

f. 3600

g. List the single-digit divisors of 1356.

h. The number 7000 is divisible by which single-digit numbers?

Problem set 5

1. If the product of 10 and 20 is divided by the sum of 20 and 30, what is the quotient?

2. List the whole numbers that are factors of 30.

3. Use digits and symbols to write "Negative five is less than positive four."

4. Use digits to write four hundred seven million, six thousand, nine hundred sixty-two.

5. List the whole numbers from 1 to 10 that are divisors of 12,300.

6. Replace the circle with the proper comparison symbol. Then use words to state the same comparison.
$$-7 \bigcirc -11$$

7. The number 3456 is divisible by which single-digit numbers?

8. Show this subtraction problem on a number line: $2 - 5$

9. Use words to write 604,003,504.

Find each missing number:

10. X
       $+\ \ 4.60$
       $\overline{\$10.00}$

11. P
       $-\ 3850$
       $\overline{\ 4500}$

12.   7
       4
       8
       6
       2
       1
       6
       8
       9
       5
       4
      $+\ 5$
       $\overline{N}$

13. Z
       $\times\ \ \ \ 8$
       $\overline{\$50.00}$

14.   1426
       $-\ \ \ \ K$
       $\overline{\ \ \ 87}$

15.   45
       $\times\ P$
       $\overline{990}$

16.   32,800
       $+\ \ \ \ Z$
       $\overline{60,000}$

Add, subtract, multiply, or divide, as indicated:

17. $\dfrac{1225}{35}$

18.   800
       $\times\ 50$

19. $\$100.00$
       $-\ \ \ 48.37$

20.   46,302
       $+\ 49,998$

21. $\$45.00 \div 20$

22. $7 \cdot 11 \cdot 13$

23. $9\overline{)43,271}$

24. $3625 + 59 + 570 + 8$

25. $48¢ + \$8.49 + \$14$

26. $1000 - (430 - 58)$

27. $140(16)$

28. $25¢$
       $\times\ 24$

29. $\dfrac{\$43.50}{10}$

30. $\$0.07$
       $\times\ \ \ \ 50$

LESSON
6

<h1 style="text-align:center">Lines, Rays, and Segments</h1>

Lines   A **mathematical line** has no ends. A mathematical line goes on and on without end in both directions. To represent a mathematical line, we may draw a pencil line like this:

◄─────────────────────────────►

A line

The arrowheads indicate that the mathematical line has no ends. To name a line, we identify two points on the line. Below is line *AB*.

A   B

◄────•──────────────•────►

Line *AB* or line *BA*

This line may also be named line *BA*. The symbol ←→ above the two letters may be used instead of the word *line*, as $\overleftrightarrow{AB}$ or $\overleftrightarrow{BA}$ (read line *AB* or line *BA*).

Rays   A part of a line that has one **endpoint** is a **ray**. A ray of sunlight is a physical example of a ray. It begins at the sun and then continues on and on. To represent a ray, we may draw a figure like this:

•──────────────────────────────►

A ray

A ray is sometimes called a **half line**. To name a ray, we name the endpoint and then name one other point on the ray. The endpoint is called the **origin** of the ray and must be named first. Below is ray *AB*. It may also be named by using the symbol ──→ above the letters, as $\overrightarrow{AB}$.

A   B

•────────────────•───────────►

Ray *AB* or $\overrightarrow{AB}$

Segments   A part of a line that has two endpoints is called a **line segment** or just a **segment**.

•────────────────────────────•

A segment

We name a segment by naming its endpoints. Below is segment *AB* or segment *BA*.

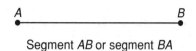

Segment *AB* or segment *BA*

An overbar with no arrowheads means segment. Segment *BA* can also be written $\overline{BA}$. A segment is part of a line. A segment may also be part of a ray or part of another segment. In this figure we can identify three segments.

The three segments are $\overline{AB}$, $\overline{BC}$, and $\overline{AC}$.

A segment has length. We can describe the length of segment *AB* by using the letters with no overbar, or we can designate the measure of (the length of) segment *AB* by writing an *m* in front, $m\overline{AB}$.

Thus both *AB* and $m\overline{AB}$ denote the length of segment *AB*. On segment *AC* above, we note that

$$AC = AB + BC \qquad \text{or} \qquad m\overline{AC} = m\overline{AB} + m\overline{BC}$$

Example 1 Describe each figure as a line, a ray, or a segment. Then use a symbol and letters to name it.

Solution (a) Figure (a) is a **line:** $\overleftrightarrow{ST}$ or $\overleftrightarrow{TS}$

(b) Figure (b) is a **segment:** $\overline{CD}$ or $\overline{DC}$

(c) Figure (c) is a **ray:** $\overrightarrow{ON}$

Example 2 Name three segments in this figure.

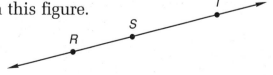

Solution The three segments are: (1) $\overline{RS}$ or $\overline{SR}$

(2) $\overline{ST}$ or $\overline{TS}$

(3) $\overline{RT}$ or $\overline{TR}$

Practice Describe each figure as a line, ray, or segment. Then use a symbol and letters to name it.

a.   **b.**   **c.**

Problem set
6

1. If the product of two one-digit whole numbers is 35, what is the sum of the same two numbers?

2. What is another name for a half line?

3. List the whole number divisors of 50.

4. Use digits and symbols to write "Two minus five equals negative three."

5. Use digits to write ninety million, five hundred thousand, thirty-five.

6. List the single-digit whole numbers that are factors of 924.

7. Arrange these numbers in order from least to greatest:
$$-10, 5, -7, 8, 0, -2$$

8. What is the name for part of a line?

9. Use words to write 10203045.

10. How many units is it from 3 to -4 on the number line?

Find each missing number:

11.   36
    ×   Z
    1224

12. $100.00
   −   K
    $17.54

13.   4
   1
   5
   6
   9
   8
   5
   7
   2
   4
   6
  +7
   N

14.   R
  + 8320
  10,000

15.   X
   ×   20
  $36.00

16.   W
  − 98
   432

Add, subtract, multiply, or divide, as indicated:

17.   36,475
  + 55,984

18.   476
  × 38

19. $\dfrac{4554}{9}$

20. $80.00 − $72.45

21. 49 + 387 + 1579 + 98

22. 4000 ÷ (200 ÷ 10)

23. (4000 ÷ 200) ÷ 10

24. (200)(400)

25. $68.00 ÷ 40

26. 8 · 7 · 5

27. $1.25
   ×   38

28. $\dfrac{770}{35}$

29.   99¢
   × 99

30. Name three segments in this figure.

LESSON
7

Fractions and Mixed Numbers

Fractions A fraction may be used to name part of a whole.

Here we use a circle to represent the number 1. A fraction may name part of 1. Here $\frac{1}{4}$ of 1 is shaded. We see that $\frac{3}{4}$ of 1 is not shaded.

A fraction may name part of a group. Here $\frac{1}{4}$ of the group is shaded and $\frac{3}{4}$ of the group is not shaded.

A fraction is written with two numbers and a division line. The division line may be a bar or a slash.

$$\text{Bar} \quad \frac{1}{4} \qquad \text{Slash} \quad 1/4$$

The slash is used when entering fractions on computers and by some people when writing fractions by hand. A slash may lead to confusion when fractions become more involved. **For this reason we will use a bar instead of a slash and we recommend that students develop a habit of using a bar instead of a slash.**

The number below the bar is the **denominator**. This number tells how many equal parts are in the whole. The number above the bar is the **numerator**. This number tells how many of the parts have been selected.

$$\begin{array}{rl} \text{Numerator} & \longrightarrow \quad 1 \\ \text{Denominator} & \longrightarrow \quad 4 \end{array}$$

To read a fraction, we say the name of the numerator, and then we say the name of the denominator. If the denominator is 2, we say "half." If the denominator is 3 or more, we say the **ordinal name** of the number. (The ordinal name of a number is its positional name, such as third, fourth, fifth, etc.) If the numerator is more than 1, the denominator is read with an "s" at the end (two thirds, three fourths, etc.).

Example 1   Draw a circle and shade $\frac{1}{3}$ of it.

Solution   We must be careful to divide the circle into 3 **equal** parts. Drawing a "spread-out Y" at the center is one way. We shade any one of the parts.

Example 2   Use words to name the fraction $\frac{3}{4}$.

Solution   We name the numerator first, then the denominator: **Three fourths**.

Mixed numbers   A whole number plus a fraction is a **mixed number**. To name the number of circles shaded below, we use the mixed number $2\frac{3}{4}$. We see that $2\frac{3}{4}$ means $2 + \frac{3}{4}$. To read a mixed number, we first say the whole number, then we say "and," and then we say the fraction.

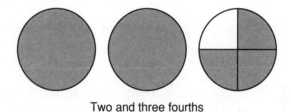

Two and three fourths

Example 3   Draw and shade circles to illustrate one and one sixth.

Solution   To illustrate one and one sixth $\left(1\frac{1}{6}\right)$, we will draw two circles and divide the second into 6 equal parts. We shade the first circle and 1 part of the second circle.

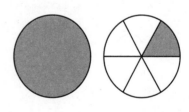

Fractions and mixed numbers on the number line   Between the points on a number line that represent whole numbers are many points that represent fractions and mixed numbers. To identify the fraction or mixed number associated with a point on a number line, it is first necessary to discover the number of segments into which each length has been divided.

Example 4   Point *A* represents what number on this number line?

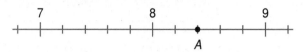

Solution   We see that point *A* represents a number greater than 8 but less than 9. It represents 8 plus a fraction. To find the fraction, we first notice that the segment from 8 to 9 has been divided into 5 smaller segments. From 8 to point *A* is 2 of the 5 segments. Thus, point *A* represents the mixed number $8\frac{2}{5}$.

> *Note*:   It is important to focus on the **number of segments** and not on the number of vertical tick marks. The four vertical tick marks divide the space between 8 and 9 into 5 segments, just as four cuts divide a candy bar into 5 pieces.

Practice   **a.** What fraction of this circle is shaded?

b. What fraction of this circle is not shaded?

Use words to name each fraction or mixed number.

c. $\dfrac{7}{12}$ **d.** $3\dfrac{1}{2}$ **e.** $6\dfrac{2}{3}$

Draw and shade circles to illustrate each fraction or mixed number.

f. $\dfrac{3}{5}$ **g.** $1\dfrac{2}{3}$ **h.** $2\dfrac{3}{4}$

Points **i** and **j** represent what mixed numbers on these number lines?

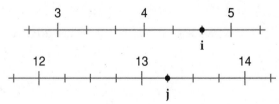

Problem set 7

1. Use digits and a comparison symbol to write "One and three fourths is greater than one and three fifths."

2. Use digits to write nine billion, forty-two.

3. What is the quotient when the product of 20 and 20 is divided by the sum of 10 and 10?

4. (a) List the factors of 39.

    (b) List the single-digit whole numbers that are divisors of 1680.

5. Point *A* represents what mixed number on this number line?

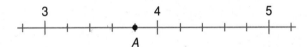

6. Replace each circle with the proper comparison symbol.

    (a) 3 + 2 ◯ 2 + 3   (b) 3 − 2 ◯ 2 − 3

7. Use words to write 32500000089.

8. (a) What fraction of the circle is shaded?

    (b) What fraction of the circle is not shaded?

9. Draw and shade circles to illustrate $2\frac{1}{3}$.

10. What is the name of the bottom number of a fraction?

Find each missing number:

11. A
     $-\ \$4.70$
     $\overline{\$2.35}$

12. B
     $+\ \$25.48$
     $\overline{\$60.00}$

13.
$$
\begin{array}{r}
C \\
\times\quad 8 \\
\hline
\$60.00
\end{array}
$$

14.
$$
\begin{array}{r}
10,000 \\
-\qquad D \\
\hline
5,420
\end{array}
$$

15.
$$
\begin{array}{r}
5376 \\
+\quad E \\
\hline
7157
\end{array}
$$

16.
$$
\begin{array}{r}
19 \\
\times\ F \\
\hline
399
\end{array}
$$

Add, subtract, multiply, or divide, as indicated:

17.
$$
\begin{array}{r}
400 \\
\times\ 500
\end{array}
$$

18.
$$
\begin{array}{r}
\$50.00 \\
-\ 48.79
\end{array}
$$

19.
$$
\begin{array}{r}
8 \\
9 \\
8 \\
8 \\
9 \\
8 \\
2 \\
8 \\
5 \\
6 \\
1 \\
+\,6
\end{array}
$$

20. 3625 + 431 + 687

21. 6000 ÷ 50

22. 20 · 10 · 5

23. $27.00 ÷ 18

24. 1000 − 11

25. 416 − (86 + 119)

26. (416 − 86) + 119

27.
$$
\begin{array}{r}
\$0.08 \\
\times\quad 75
\end{array}
$$

28. $\dfrac{3456}{6}$

29.
$$
\begin{array}{r}
79¢ \\
\times\,30
\end{array}
$$

30. Name three segments in this figure.

**LESSON
8**

Adding, Subtracting, and Multiplying Fractions

**Adding
fractions**

To add fractions with the same denominators, we add the numerators and write the sum over the same denominator.

Example 1 Add: (a) $\dfrac{4}{15} + \dfrac{7}{15}$ (b) $\dfrac{3}{7} + \dfrac{2}{7} + \dfrac{1}{7}$

Solution (a) $\dfrac{4}{15} + \dfrac{7}{15} = \mathbf{\dfrac{11}{15}}$ (b) $\dfrac{3}{7} + \dfrac{2}{7} + \dfrac{1}{7} = \mathbf{\dfrac{6}{7}}$

**Subtracting
fractions**

To subtract fractions that have the same denominators, we write the difference of the numerators over the same denominator.

Example 2 Subtract: (a) $\dfrac{5}{9} - \dfrac{1}{9}$ (b) $\dfrac{3}{5} - \dfrac{3}{5}$

Solution (a) $\dfrac{5}{9} - \dfrac{1}{9} = \mathbf{\dfrac{4}{9}}$ (b) $\dfrac{3}{5} - \dfrac{3}{5} = \mathbf{0}$

**Multiplying
fractions**

To multiply fractions, we multiply the numerators to find the numerator of the product. We multiply the denominators to find the denominator of the product.

Example 3 Multiply: (a) $\dfrac{2}{3} \times \dfrac{4}{5}$ (b) $\dfrac{1}{2} \times \dfrac{3}{4} \times \dfrac{1}{5}$

Solution (a) $\dfrac{2}{3} \times \dfrac{4}{5} = \mathbf{\dfrac{8}{15}}$ (b) $\dfrac{1}{2} \times \dfrac{3}{4} \times \dfrac{1}{5} = \mathbf{\dfrac{3}{40}}$

Practice
   a. $\dfrac{5}{9} + \dfrac{2}{9}$ **b.** $\dfrac{4}{5} - \dfrac{3}{5}$ **c.** $\dfrac{3}{5} \times \dfrac{1}{2} \times \dfrac{3}{4}$

   d. $\dfrac{5}{8} - \dfrac{5}{8}$ **e.** $\dfrac{4}{7} \times \dfrac{2}{3}$ **f.** $\dfrac{3}{10} + \dfrac{3}{10} + \dfrac{3}{10}$

Problem set 8

1. What is the quotient when the sum of 1, 2, and 3 is divided by the product of 1, 2, and 3?

2. This sign is incorrect. Show two ways to correct this sign.

Apples
0.45¢ per pound

3. Replace each circle with the proper comparison symbol. Then use words to write the same comparison.

    (a) $\frac{1}{2} \bigcirc \frac{1}{4}$

    (b) $-2 \bigcirc -4$

4. Use digits to write four hundred seventy-five billion, nine hundred forty-two thousand, ten.

5. Use words to write 406000012005.

6. (a) What fraction of the square is shaded?

    (b) What fraction of the square is not shaded?

7. Is an imaginary "line" from the earth to the moon a line, a ray, or a segment?

8. Point X represents what mixed number on this number line?

9. (a) List the factors of 18.

    (b) List the factors of 24.

    (c) Which numbers are factors of both 18 and 24?

Find each missing number:

10.   4315
$$\begin{array}{r} 4315 \\ -\quad A \\ \hline 2157 \end{array}$$

11.   85,000
$$\begin{array}{r} 85{,}000 \\ +\qquad B \\ \hline 200{,}000 \end{array}$$

12.   60
$$\begin{array}{r} 60 \\ \times\ C \\ \hline 900 \end{array}$$

13.   D
$$\begin{array}{r} D \\ +\ \$5.60 \\ \hline \$20.00 \end{array}$$

14.   E
$$\begin{array}{r} E \\ \times\quad 12 \\ \hline \$30.00 \end{array}$$

15.   F
$$\begin{array}{r} F \\ -\ \$98.03 \\ \hline \$12.47 \end{array}$$

Add, subtract, multiply, or divide, as indicated:

16. $\dfrac{11}{15} - \dfrac{3}{15}$

17. $\dfrac{3}{8} + \dfrac{4}{8}$

18.
$$\begin{array}{r} 5 \\ 7 \\ 5 \\ 7 \\ 6 \\ 8 \\ 1 \\ 2 \\ 3 \\ 4 \\ 5 \\ +6 \\ \hline \end{array}$$

19. $\dfrac{3}{4} \times \dfrac{1}{4}$

20. $5317 + 296 + 8 + 79$

21. $\$8.97 + \$110 + 53¢$

22. $(125 \div 25) \div 5$

23. $60.00
$$\begin{array}{r} \$60.00 \\ -\ 49.49 \\ \hline \end{array}$$

24.   607
$$\begin{array}{r} 607 \\ \times\ 78 \\ \hline \end{array}$$

25. $\dfrac{1802}{17}$

26. $\$6.75 \div 15$

27. $\$0.09 \times 56$

28. $50 \cdot 60 \cdot 70$

29. $\dfrac{4}{5} \times \dfrac{2}{3} \times \dfrac{1}{3}$

30. $\dfrac{1}{9} + \dfrac{2}{9} + \dfrac{4}{9}$

LESSON 9

"Some and Some More" Word Problems

In this lesson we will begin solving one-step story problems. There are hundreds of different story problems but only a few different thought patterns. One common thought pattern in story problems is that someone has some and then gets some more. We will call problems with this thought pattern **some and some more** problems. The some and some more pattern is an **addition pattern**.

We will sketch the pattern, as we show below, to help us structure our thinking. We encourage readers to sketch the patterns. Patterns occur often in mathematics, and sketching patterns can be a powerful problem-solving tool.

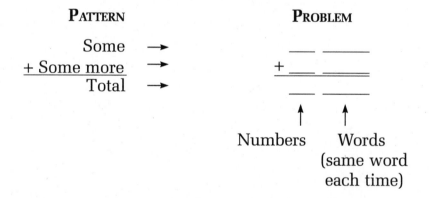

To complete an addition pattern for some and some more problems, we write the number for some on the top line, write the number for some more on the second line, and write the total on the bottom line. Thus, there are three numbers. Any one of the three numbers may be missing in the problem. To solve the problem, we need to find the missing number. Then we can write the answer. We follow these four steps:

Step 1. Read and recognize that the problem has a some and some more pattern.

Step 2. Sketch the pattern and record the given information.

Step 3. Find the missing number to complete the pattern.

Step 4. Answer the question.

We will follow these steps as we consider three examples.

Example 1 Robert had 27 dollars. For his birthday he received 18 dollars more. Now how much money does Robert have?

Solution Step 1. We **recognize** that this is a some and some more problem because Robert had some money and got some more money.

Step 2. We **sketch** the pattern and **record** the given information.

PATTERN	PROBLEM
Some	27 dollars
+ Some more	+ 18 dollars
Total	___ dollars

Step 3. We need to **find the missing number** which is the total. We find the bottom number in an addition pattern by adding.

27 dollars
+ 18 dollars
45 dollars

Step 4. We have completed the pattern. Now we answer the question. **Robert has 45 dollars**.

Example 2 At the end of the first day of camp, Marissa counted 47 mosquito bites. The next morning she counted 114 mosquito bites. How many new bites did she get?

Solution Step 1. We **recognize** this as a some and some more problem. She had some bites and then she got some more bites.

Step 2. We **sketch** the pattern and **record** the given information. She had 47 bites. She got some more bites. Then she had a total of 114 bites.

PATTERN	PROBLEM
Some	<u>47</u> bites
+ Some more	<u>+ B</u> bites
Total	114 bites

Step 3. We **find B, the missing number**. We find the first or second number of an addition pattern by subtracting. We subtract. Then we complete the pattern.

$$\begin{array}{r} 114 \\ -\ 47 \\ \hline 67 \end{array} \qquad \begin{array}{r} 47 \text{ bites} \\ +\ 67 \text{ bites} \\ \hline 114 \text{ bites} \end{array}$$

Step 4. Now we **answer the question**. Marissa got **67 new bites**.

Example 3 The first scout troop encamped in the ravine. A second troop of 137 scouts joined them, making a total of 312 scouts. How many scouts were in the first troop?

Solution Step 1. We recognize this problem has a some and some more thought pattern. There were some scouts. Then some more scouts came.

Step 2. We sketch the pattern and record the given information.

PATTERN	PROBLEM
Some	<u>S</u> scouts
+ Some more	<u>+ 137</u> scouts
Total	312 scouts

Step 3. We find S, the missing number. To find the first or second number in an addition pattern, we subtract. We subtract, then complete the pattern.

$$\begin{array}{r} 312 \\ -\ 137 \\ \hline 175 \end{array} \qquad \begin{array}{r} 175 \text{ scouts} \\ +\ 137 \text{ scouts} \\ \hline 312 \text{ scouts} \end{array}$$

Step 4. Now we answer the question. There were **175 scouts** in the first troop.

Practice   **a.** Billy stood on the scales. Billy weighed 118 pounds. Then Nathan and Billy stood on the scales. Together they weighed 230 pounds. How much did Nathan weigh?

b. Tim cranked for a number of turns. Then Dawn gave the crank 216 turns. If the total number of turns was 400, how many turns did Tim give the crank?

c. Before the game began, 87 kids were in the pool. When the whistle blew, 49 more kids jumped in. How many kids were in the pool then?

Problem set 9

1. Robin weighed 165 pounds. Little John and Robin together weighed 450 pounds. What was the weight of Little John?

2. In the morning Ricky raked 1057 leaves. That afternoon he raked 2970 leaves. In all, how many leaves did Ricky rake that day?

3. What is the difference when the sum of 2, 3, and 4 is subtracted from the product of 2, 3, and 4?

4. (a) What fraction of the rectangle is shaded?

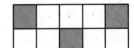

(b) What fraction of the rectangle is not shaded?

5. Use words to write the number 2,000,450,000.

6. Replace each circle with the proper comparison symbol.

(a) $2 - 2 \bigcirc 2 \div 2$ (b) $\dfrac{1}{2} + \dfrac{1}{2} \bigcirc \dfrac{1}{2} \times \dfrac{1}{2}$

7. Point *M* represents what mixed number on this number line?

8. Draw and shade circles to represent $1\frac{3}{5}$.

9. List the single-digit numbers that are divisors of 420.

Find each missing number:

10.
$$\begin{array}{r} 12{,}500 \\ + \quad X \\ \hline 36{,}275 \end{array}$$

11.
$$\begin{array}{r} 18 \\ \times \ Y \\ \hline 396 \end{array}$$

12.
$$\begin{array}{r} 8 \\ 7 \\ 5 \\ 4 \\ 5 \\ 4 \\ 7 \\ 6 \\ 4 \\ 3 \\ 2 \\ + 8 \\ \hline N \end{array}$$

13.
$$\begin{array}{r} 77{,}000 \\ - \quad Z \\ \hline 39{,}400 \end{array}$$

14.
$$\begin{array}{r} A \\ \times \quad 8 \\ \hline \$10.00 \end{array}$$

15.
$$\begin{array}{r} B \\ - \$16.25 \\ \hline \$8.75 \end{array}$$

16.
$$\begin{array}{r} C \\ + \$37.50 \\ \hline \$75.00 \end{array}$$

Add, subtract, multiply, or divide, as indicated:

17. $\dfrac{5}{7} \times \dfrac{3}{4}$

18. $\dfrac{5}{8} - \dfrac{5}{8}$

19. $\dfrac{11}{20} + \dfrac{8}{20}$

20. $2000 - (680 - 59)$

21. $436 + 2799 + 68 + 347$

22. $89¢ + 57¢ + \$15.74$

23.
$$\begin{array}{r} 800 \\ \times 300 \\ \hline \end{array}$$

24. $\dfrac{768}{16}$

25.
$$\begin{array}{r} \$40.75 \\ - \ 36.57 \\ \hline \end{array}$$

26. $8\overline{)38{,}479}$

27. $\$96.00 \div 30$

28. $\dfrac{3}{20} + \dfrac{3}{20} + \dfrac{3}{20}$ **29.** $\dfrac{2}{3} \cdot \dfrac{2}{3} \cdot \dfrac{2}{3}$

30. Describe each figure as a line, ray, or segment. Then use a symbol and letters to name each figure.

(a) M C (b) P M (c) F H

LESSON
10

"Some Went Away" Word Problems

In Lesson 9 we considered story problems that had a some and some more thought pattern. The some and some more pattern is an addition pattern.

In this lesson we will consider story problems that have a **some went away** thought pattern. The some went away pattern is a **subtraction pattern.**

PATTERN		PROBLEM
Some →		____ ____
− Some went away →		− ____ ____
What's left →		____ ____

↑   ↑
Numbers   Words
(same word
each time)

To complete the subtraction pattern for some went away problems, we write the number for some on the top line. We write the number for how many went away on the second line. The number that remains goes on the bottom line. There are three numbers in this pattern. Any one of the three numbers may be missing in the problem. To solve the problem, we find the missing number. Then we answer the question in the problem. Thus, we follow the same four steps we followed in the previous lesson:

Step 1. Read and recognize the type of thought pattern.

Step 2. Sketch the pattern and record the given information.

Step 3. Find the missing number to complete the pattern.

Step 4. Answer the question.

Example 1 Denise spent $63.45 at the grocery store. If she went to the store with $90.00, how much money did she have when she came home from the store?

Solution Step 1. We recognize that this problem has a **some went away** pattern. Denise had $90.00, then some went away.

Step 2. We sketch the pattern and record the given information.

PATTERN	PROBLEM
Some	$90.00
− Some went away	− $63.45
What's left	L

Step 3. Next we find the missing number. To find the bottom number in a subtraction pattern, we subtract.

$$\begin{array}{r} \$90.00 \\ -\ \$63.45 \\ \hline \$26.55 \end{array}$$

Step 4. We answer the question. Denise came home from the store with **$26.55**.

Example 2 Tim baked 4 dozen cookies. While they were cooling, he went to answer the phone. When he came back, only 32 cookies remained. His dog was nearby, licking her chops. How many cookies did the dog eat while Tim was answering the phone?

Solution Step 1. We recognize that this problem has a **some went away** thought pattern.

Step 2. We sketch the pattern and record the information. We know that Tim had 4 dozen, or 48, cookies.

PATTERN	PROBLEM
Some	<u>48</u> cookies
– Some went away	<u>– C</u> cookies
What's left	<u>32</u> cookies

Step 3. We find the missing number. To find the second number in a subtraction pattern, we subtract. We subtract and complete the pattern.

$$\begin{array}{r} 48 \\ -\ 32 \\ \hline 16 \end{array} \qquad \begin{array}{r} 48 \text{ cookies} \\ -\ 16 \text{ cookies} \\ \hline 32 \text{ cookies} \end{array}$$

Step 4. Now we answer the question. While Tim was answering the phone, his dog ate **16 cookies**.

Example 3 The room was full of boxes when Sharon began. Then she shipped out 56 boxes. Only 88 boxes were left. How many boxes were in the room when Sharon began?

Solution Step 1. We recognize that this problem has a some went away thought pattern.

Step 2. We sketch the pattern and record the information.

PATTERN	PROBLEM
Some	<u>B</u> boxes
– Some went away	<u>– 56</u> boxes
What's left	<u>88</u> boxes

Step 3. We find the missing number. **To find the top number in a subtraction pattern**, we add the other two numbers. Then we complete the pattern.

$$\begin{array}{r} 88 \\ +\ 56 \\ \hline 144 \end{array} \qquad \begin{array}{r} 144 \text{ boxes} \\ -\ 56 \text{ boxes} \\ \hline 88 \text{ boxes} \end{array}$$

Step 4. Now we answer the question. There were **144 boxes** in the room when Sharon began.

Practice   **a.** At dawn 254 horses were in the corral. Later that morning, Tex found the gate open and saw that only 126 horses remained. How many horses got away?

b. Cynthia had a lot of paper. After using 36 sheets for a report, only 164 sheets remained. How many sheets of paper did she have at first?

Problem set 10

1. As the day of the festival drew near, there were 200,000 people in the city. If the usual population of the city was 85,000, how many visitors had come to the city?

2. Syd returned from the store with $12.47. He had spent $98.03 on groceries. How much money did he have when he went to the store?

3. Exactly 10,000 runners began the marathon. If only 5420 runners finished the marathon, how many dropped out along the way?

4. (a) What fraction of the group is shaded?

(b) What fraction of the group is not shaded?

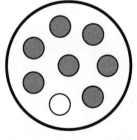

5. Arrange these numbers in order from least to greatest:

$$\frac{1}{2}, 0, -2, 1$$

6. Use words to write 407000075.

7. Use digits and symbols to write "The product of one and two is less than the sum of one and two."

8. Subtract eighty-nine million from one hundred million. Use words to write the difference.

9. (a) List the factors of 16.

   (b) List the factors of 24.

   (c) Which numbers are factors of both 16 and 24?

Find each missing number:

10.
$$\begin{array}{r} 8000 \\ -\quad K \\ \hline 5340 \end{array}$$

11.
$$\begin{array}{r} 1320 \\ +\quad M \\ \hline 1760 \end{array}$$

12.
$$\begin{array}{r} 36 \\ \times\ N \\ \hline 720 \end{array}$$

13.
$$\begin{array}{r} R \\ +\ \$126 \\ \hline \$375 \end{array}$$

14.
$$\begin{array}{r} S \\ -\ \$8.75 \\ \hline \$7.75 \end{array}$$

15.
$$\begin{array}{r} T \\ \times\quad 40 \\ \hline \$220.00 \end{array}$$

Add, subtract, multiply, or divide, as indicated:

16. $\dfrac{4}{9} + \dfrac{4}{9}$ **17.** $\dfrac{24}{25} - \dfrac{23}{25}$ **18.** $\dfrac{5}{7} \times \dfrac{2}{3}$

19. $100 - (5 \times 20)$ **20.** $(100 - 5) \times 20$

21. $29{,}214 + 6037 + 528$

22. $36{,}418 - 989$

23.
$$\begin{array}{r} 135 \\ \times\ 72 \\ \hline \end{array}$$

24. $\dfrac{1000}{40}$

25.
$$\begin{array}{r} \$100.00 \\ -\quad 81.93 \\ \hline \end{array}$$

26. $30(\$1.49)$ **27.** $\$140.70 \div 35$

28. $7\overline{)64{,}404}$

29. $\dfrac{5}{9} \cdot \dfrac{1}{3} \cdot \dfrac{1}{2}$ **30.** $\dfrac{5}{8} + \left(\dfrac{3}{8} - \dfrac{1}{8}\right)$

LESSON 11

"Larger-Smaller-Difference" Word Problems • Time Problems

Larger-smaller-difference word problems

In the previous lesson we practiced word problems that have a **some went away** thought pattern. A some went away pattern is a subtraction pattern.

The other type of problem that has a subtraction pattern is a **larger-smaller-difference** problem. In larger-smaller-difference problems we are asked to **compare** two numbers. In these problems we not only decide which number is greater and which number is less, but also **how much** greater or **how much** less. The number that describes how much greater or how much less is called the **difference**. To set up the pattern, we list the numbers in order: larger, smaller, difference.

Example 1 During the day, 1320 employees worked at the toy factory. At night, 897 employees worked there. How many more employees worked at the factory during the day than at night?

Solution Step 1. Questions such as "How many more?" or "How many fewer?" indicate that the problem has a larger-smaller-difference pattern.

Step 2. In the pattern we write the numbers in this order: the larger number, the smaller number, then the difference.

PATTERN	PROBLEM
Larger	1320 employees
– Smaller	– 897 employees
Difference	E employees

Step 3. We find the missing bottom number of a subtraction pattern by subtracting.

$$\begin{array}{r} 1320 \text{ employees} \\ -\ 897 \text{ employees} \\ \hline 423 \text{ employees} \end{array}$$

Step 4. We answer the question: **423 more employees** work at the factory during the day than work there at night.

Example 2   The number 620,000 is how much less than 1,000,000?

Solution   Step 1. The words "how much less" indicate that this problem has a larger-smaller-difference pattern.

Step 2. We sketch the pattern and record the numbers. There are no words to write this time.

PATTERN	PROBLEM
Larger	1,000,000
– Smaller	– 620,000
Difference	D

Step 3. We subtract to find the missing number.

$$\begin{array}{r} 1,000,000 \\ -\ 620,000 \\ \hline 380,000 \end{array}$$

Step 4. The difference is "how much less." We answer the question. Six hundred twenty thousand is **380,000** less than 1 million.

Time problems   Time problems are like larger-smaller-difference problems. We arrange the times in this order: later-earlier-difference. At this point we will consider time problems involving only years A.D.

Example 3   How many years were there from 1492 to 1776?

Solution   Step 1. We recognize that this problem has a later-earlier-difference thought pattern.

Step 2. We sketch the pattern and record the years.

PATTERN	PROBLEM
Later	1776
– Earlier	– 1492
Difference	Y

Step 3. We subtract to find the missing number.

$$
\begin{array}{r}
1776 \\
- \ 1492 \\
\hline
284
\end{array}
$$

Step 4. Now we answer the question. There were **284 years** from 1492 to 1776.

Example 4 Abraham Lincoln died in 1865 at the age of 56. In what year was he born?

Solution Step 1. This is a time problem. Time problems have a larger-smaller-difference pattern.

Step 2. We sketch the pattern. Age is the difference between the birth date (earlier) and the date of death (later).

PATTERN	PROBLEM
Later (death)	1865
− Earlier (birth)	− Y
Difference (age)	56

Step 3. To find the middle number in a subtraction pattern, we subtract.

$$
\begin{array}{r}
1865 \\
- \ 56 \\
\hline
1809
\end{array}
\qquad
\begin{array}{r}
1865 \\
- \ 1809 \\
\hline
56
\end{array}
$$

Step 4. Now we answer the question. Abraham Lincoln was born in **1809**.

Practice **a.** The number 1,000,000,000 is how much greater than 25,000,000?

b. How many years were there from 1215 to 1791?

c. John F. Kennedy died in 1963 at the age of 46. In what year was he born?

Problem set 11

In Problems 1–4, identify the type of problem. Then find the answer.

1. Seventy-seven thousand fans filled the stadium. As the fourth quarter began, only thirty-nine thousand, four hundred remained. How many fans left before the fourth quarter began?

2. Mary purchased 18 bananas at the store. When she got home, she discovered that she already had some bananas. If she now has 31 bananas, how many did she have before she went to the store?

3. How many years were there from 1066 to 1215?

4. The first week 77,000 fans came to the stadium. Only 49,600 came the second week. How many fewer fans came to the stadium the second week?

5. Draw and shade circles to show $2\frac{1}{4}$.

6. Use words to write 100000000042.

7. Twenty-three thousand is how much less than one million?

8. Replace each circle with the proper comparison symbol.

    (a) $2 - 3 \bigcirc -1$ (b) $\frac{1}{2} \bigcirc \frac{1}{3}$

9. Name three segments in this figure in order of length from shortest to longest.

10. (a) What fraction of the triangle is shaded?

    (b) What fraction of the triangle is not shaded?

11. The number 100 is divisible by what numbers?

Find each missing number:

12.
$$\begin{array}{r} X \\ \times\ 15 \\ \hline 630 \end{array}$$

13.
$$\begin{array}{r} Y \\ -\ 2714 \\ \hline 3601 \end{array}$$

14.
$$\begin{array}{r} 5 \\ 8 \\ 4 \\ 7 \\ 6 \\ 5 \\ 7 \\ N \\ 4 \\ +\ 6 \\ \hline 58 \end{array}$$

15.
$$\begin{array}{r} 2900 \\ -\quad P \\ \hline 64 \end{array}$$

16.
$$\begin{array}{r} \$1.53 \\ +\quad Q \\ \hline \$5.00 \end{array}$$

17.
$$\begin{array}{r} 20 \\ \times\quad R \\ \hline 1200 \end{array}$$

Add, subtract, multiply, or divide, as indicated:

18.
$$\begin{array}{r} 72{,}112 \\ -\ 64{,}309 \end{array}$$

19.
$$\begin{array}{r} 453{,}978 \\ +\ 386{,}864 \end{array}$$

20.
$$\begin{array}{r} 74 \\ \times\ 68 \end{array}$$

21. $\dfrac{5}{9} - \left(\dfrac{3}{9} + \dfrac{2}{9} \right)$

22. $\left(\dfrac{5}{9} - \dfrac{3}{9} \right) + \dfrac{2}{9}$

23. $\dfrac{1}{2} \cdot \dfrac{1}{3} \cdot \dfrac{1}{4}$

24. $\$37.20 \div 15$

25. $\dfrac{7000}{40}$

26. $9\overline{)42{,}847}$

27. $\$4.36 + \$15.96 + 76¢ + \$35$

28. $\$20.00 - \$0.89$

29. $30 \cdot 60 \cdot 900$

30. $120(\$0.15)$

LESSON 12

Equal Groups Word Problems

We have used both the addition pattern and the subtraction pattern to solve word problems. In this lesson we will use a multiplication pattern to solve word problems. Consider this problem:

> Willie packed 25 marbles in each box. If he filled 32 boxes, how many marbles did he pack in all?

This problem has a thought pattern that is different from the addition pattern or subtraction pattern. This problem has an **equal groups** thought pattern. An outline for an equal groups problem is shown below.

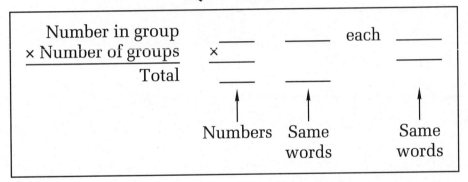

EQUAL GROUPS

Since an equal groups pattern is a multiplication pattern, we multiply the top two numbers to find the bottom number. To find one of the top two numbers, we divide. We will consider three examples.

Example 1 Willie packed 25 marbles in each box. If he filled 32 boxes, how many marbles did he pack in all?

Solution Step 1. The words "in each" are a clue to help us recognize an equal groups pattern.

Step 2. We sketch the pattern and record the information.

$$
\begin{array}{l}
\text{Number in group} \\
\times\ \text{Number of groups} \\
\hline
\hspace{2.5em}\text{Total}
\end{array}
\quad
\begin{array}{l}
\underline{25}\ \underline{\text{marbles}}\ \text{each}\ \underline{\text{box}} \\
\times 32 \hspace{6.5em} \text{boxes} \\
\hline
\underline{\ M\ }\ \underline{\text{marbles}}
\end{array}
$$

Step 3. To find the missing bottom number, we multiply.

$$
\begin{array}{r}
25 \\
\times\ 32 \\
\hline
50 \\
75 \\
\hline
800
\end{array}
$$

$$
\begin{array}{r}
\underline{25}\ \text{marbles each } \underline{\text{box}} \\
\times\ \underline{32} \qquad\qquad \underline{\text{boxes}} \\
\hline
\underline{800}\ \text{marbles}
\end{array}
$$

Step 4. We answer the question: Willie packed **800 marbles** in all.

Example 2 Movie tickets sold for $5 each. The total ticket sales were $820. How many tickets were sold?

Solution **Step 1.** The word "each" is a clue that this is an equal groups problem.

Step 2. We sketch the pattern and record the information.

$$
\begin{array}{r}
\text{Number in group} \\
\times\ \text{Number of groups} \\
\hline
\text{Total}
\end{array}
\qquad
\begin{array}{r}
\underline{5}\ \text{dollars each } \underline{\text{ticket}} \\
\times\quad \underline{T} \qquad\qquad \underline{\text{tickets}} \\
\hline
\underline{820}\ \text{dollars}
\end{array}
$$

Step 3. The second number is missing. To find a missing first or second number in a multiplication pattern, we divide.

$$
\begin{array}{r}
164 \\
5\overline{)820}
\end{array}
\qquad
\begin{array}{r}
\underline{5}\ \text{dollars each } \underline{\text{ticket}} \\
\times\ \underline{164} \qquad\qquad \underline{\text{tickets}} \\
\hline
\underline{820}\ \text{dollars}
\end{array}
$$

Step 4. We answer the question: **164 tickets** were sold.

Example 3 Every truck held the same number of cars. Six hundred new cars were delivered to the dealer by 40 identical trucks. How many cars were delivered by each truck?

Solution **Step 1.** An equal number of cars were grouped on each truck. This problem has an equal groups thought pattern.

Step 2. We sketch the pattern and record the information.

$$
\begin{array}{lll}
\text{Number in group} & \underline{C}\ \text{cars} & \text{each} \underline{\text{truck}} \\
\underline{\times \text{Number of groups}} & \underline{\times\ 40} & \underline{\text{trucks}} \\
\text{Total} & 600\ \text{cars} &
\end{array}
$$

Step 3. The first number is missing. To find a missing first or second number in a multiplication pattern, we divide.

$$
\begin{array}{ll}
\begin{array}{r} 15 \\ 40\overline{)600} \end{array} \longrightarrow &
\begin{array}{l}
\underline{15}\ \text{cars each} \underline{\text{truck}} \\
\underline{\times\ 40}\qquad \underline{\text{trucks}} \\
600\ \text{cars}
\end{array}
\end{array}
$$

Step 4. We answer the question: **15 cars** were delivered by each truck.

Practice   **a.** Beverly bought 2 dozen juice bars for 18¢ each. How much did she pay for all the juice bars?

b. Johnny planted a total of 375 trees with 25 trees in each row. How many rows of trees did he plant?

c. Every day Arnold did the same number of push-ups. If he did 1225 push-ups in one week, then how many push-ups did he do each day?

Problem set 12   In Problems 1–4, identify the type of problem. Then find the answer.

1. In 1980, the population of Ashton was 64,309. By the 1990 census, the population had increased to 72,112. The population of Ashton in 1990 was how much greater than the population in 1980?

2. Huck had 5 dozen night crawlers in his pockets. He was unhappy when all but 17 escaped through holes in his pockets. How many night crawlers had escaped?

3. President Franklin D. Roosevelt died in office in 1945 at the age of 63. In what year was he born?

4. The beach balls were packed 12 in each case. If 75 cases were delivered, how many beach balls were there in all?

5. The product of 5 and 8 is how much greater than the sum of 5 and 8?

6. Use digits to write three hundred ninety billion, five hundred seven million, forty-two.

7. How many units is it from -5 to $+5$ on the number line?

8. Use words to write 104000032.

9. Describe each figure as a line, a ray, or a segment. Then use a symbol and letters to name each figure.

(a)   (b)   (c)

10. (a) List the factors of 24.

    (b) List the factors of 36.

    (c) What whole numbers are factors of both 24 and 36?

11. What fractions or mixed numbers are represented by points A and B on this number line?

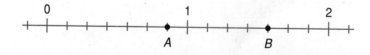

Find each missing number:

12.	3674	13.	4610	14.	36
	$-$ A		$+$ B		$\times$ C
	2159		5179		1800

15.

$$\begin{array}{r} D \\ +\ \$56.45 \\ \hline \$80.00 \end{array}$$

16.

$$\begin{array}{r} E \\ \times\ \ 30 \\ \hline 4500 \end{array}$$

17.

$$\begin{array}{r} 4 \\ 7 \\ 6 \\ 8 \\ 4 \\ 5 \\ 5 \\ 7 \\ 9 \\ 6 \\ N \\ +\ 8 \\ \hline 75 \end{array}$$

18.

$$\begin{array}{r} F \\ -\ \$1.64 \\ \hline \$3.77 \end{array}$$

Add, subtract, multiply, or divide, as indicated:

19. $\dfrac{4}{5} - \left(\dfrac{2}{5} + \dfrac{1}{5}\right)$

20. $\left(\dfrac{4}{5} - \dfrac{2}{5}\right) + \dfrac{1}{5}$

21. $\dfrac{5}{3} \cdot \dfrac{1}{2} \cdot \dfrac{1}{4}$

22. $363 + 4579 + 86 + 7$

23. $\$12.00 - \$11.37$

24. $\dfrac{600}{25}$

25.

$$\begin{array}{r} 600 \\ \times\ \ 25 \end{array}$$

26. $\$63.75 \div 5$

27. $1000 \div (100 \div 10)$

28. $(1000 \div 100) \div 10$

29. $3 \cdot 30 \cdot 300$

30. $(5 \cdot 4) \div (3 + 2)$

LESSON
13

Part-Part-Whole
Word Problems

We remember that a some and some more thought pattern is an addition pattern. Another type of pattern that has an addition pattern is the **part-part-whole** pattern. Here are two sample part-part-whole problems:

- Seventeen of the 31 students in the class were boys. How many girls were in the class?

- One third of the students earned an A on the test. What fraction of the students did not earn an A on the test?

In a part-part-whole problem, a whole group is described as being made up of two or more parts. From the information we are given, we can figure out the whole or the other part.

The part-part-whole pattern is a little different from the some and some more pattern. This is a some and some more pattern.

SOME AND SOME MORE

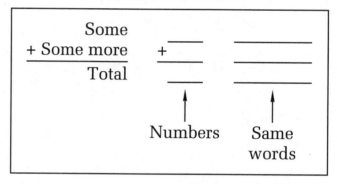

This is a part-part-whole pattern.

PART-PART-WHOLE

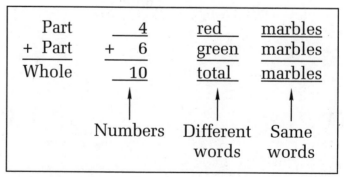

In the part-part-whole pattern there are two words to write after each number. The first word is different for each number. The second word is the same for each number. We follow the same four steps we have been practicing.

Example 1   Seventeen of the 31 students in the class were boys. How many girls were in the class?

Solution   Step 1.   We recognize that this problem has a part-part-whole pattern because the problem separates the whole class of students into two parts: boys and girls.

Step 2.   We sketch the pattern and record the given information. Notice how the words "students," "boys," and "girls" are recorded. Part of the students are boys. Part of the students are girls.

$$
\begin{array}{ll}
\text{Part} & \underline{17}\ \underline{\text{boy}}\ \text{students} \\
+\ \text{Part} & +\ \underline{}\ \underline{\text{girl}}\ \text{students} \\
\hline
\text{Whole} & \underline{31}\ \underline{\text{total}}\ \text{students}
\end{array}
$$

Step 3.   We find the missing number by subtracting. Then we complete the pattern.

$$
\begin{array}{ll}
\begin{array}{r}
31 \\
-\ 17 \\
\hline
14
\end{array}
&
\begin{array}{l}
\underline{17}\ \underline{\text{boy}}\ \text{students} \\
+\ \underline{14}\ \underline{\text{girl}}\ \text{students} \\
\hline
\underline{31}\ \underline{\text{total}}\ \text{students}
\end{array}
\end{array}
$$

Step 4.   Then we answer the question. There were **14 girls** in the class.

Example 2   One third of the students earned a B on the test. What fraction of the students did not earn a B on the test?

Solution   We are not given the number of students. We are given only the fraction of students in the whole class who earned a B on the test. A drawing may help us to visualize the problem.

All students

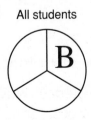

Step 1.   We recognize this as a part-part-whole problem.

Step 2. We sketch the pattern and record the information. It may seem as though we are given only one number, $\frac{1}{3}$. The drawing should remind us that the whole class of students is $\frac{3}{3}$. We will use N_B to stand for "not B" students.

$$
\begin{array}{lll}
\text{Part} & \dfrac{1}{3} & \underline{\text{B}} \quad \text{students} \\[2mm]
+\ \text{Part} & +\ N_B & \underline{\text{not B}} \ \text{students} \\[2mm]
\text{Whole} & \dfrac{3}{3} & \underline{\text{total}} \ \text{students}
\end{array}
$$

Step 3. We find the missing number, N_B, by subtracting. Then we complete the pattern

$$
\begin{array}{lll}
\dfrac{3}{3} & \dfrac{1}{3} & \underline{\text{B}} \quad \text{students} \\[3mm]
-\dfrac{1}{3} & +\dfrac{2}{3} & \underline{\text{not B}} \ \text{students} \\[3mm]
\dfrac{2}{3} & \dfrac{3}{3} & \underline{\text{total}} \ \text{students}
\end{array}
$$

Step 4. Then we answer the question. **Two thirds** $\left(\frac{2}{3}\right)$ of the students did not earn a B on the test.

Practice Use the four-step method described in this lesson to solve Problems **a** and **b**.

    a. Only 396 of the 1000 lights were on. How many of the lights were off?

    b. Two fifths of the pioneers did not survive the journey. What fraction of the pioneers did survive the journey?

Problem set 13 In Problems 1–4, identify the thought pattern used in the problem. Then find the answer.

    1. Beth fed the baby 65 grams of cereal. The baby wanted to eat 142 grams of cereal. How many additional grams of cereal did Beth need to feed the baby?

2. Seven tenths of the new recruits did not like their first haircut. What fraction of the new recruits did like their first haircut?

3. How many years were there from 1776 to 1789?

4. One hundred twenty poles were needed to construct the new pier. If each truckload contained 8 poles, how many truckloads were needed?

5. The sign shown is incorrect. Show two ways to correct this sign.

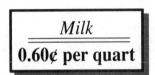

$$\underline{\underline{\textit{Milk}}}$$

0.60¢ per quart

6. Draw and shade circles to show $3\frac{1}{3}$.

7. Use digits to write four hundred seven million, forty-two thousand, six hundred three.

8. Use words to write 37,060,043.

9. (a) List the factors of 40.

(b) List the factors of 72.

(c) What is the greatest number that is a factor of both 40 and 72?

10. Name three segments in this figure in order of length from shortest to longest.

11. (a) What fraction of the group is shaded?

(b) What fraction of the group is not shaded?

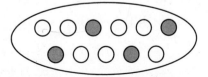

Find each missing number:

12.
$$\begin{array}{r} A \\ + \ 295 \\ \hline 1000 \end{array}$$

13.
$$\begin{array}{r} B \\ - \ 407 \\ \hline 623 \end{array}$$

14.
$$\begin{array}{r} 5 \\ 8 \\ 7 \\ 6 \\ 5 \\ 9 \\ 4 \\ 3 \\ 6 \\ 4 \\ 7 \\ 8 \\ 5 \\ N \\ + \ 6 \\ \hline 89 \end{array}$$

15.
$$\begin{array}{r} 4764 \\ + \quad D \\ \hline 9159 \end{array}$$

16.
$$\begin{array}{r} \$20.00 \\ - \qquad E \\ \hline \$3.47 \end{array}$$

17.
$$\begin{array}{r} 35 \\ \times \quad F \\ \hline 7070 \end{array}$$

Add, subtract, multiply, or divide, as indicated:

18. $\left(\dfrac{5}{7} - \dfrac{3}{7} \right) + \dfrac{2}{7}$

19. $\dfrac{5}{7} - \left(\dfrac{3}{7} + \dfrac{2}{7} \right)$

20. $\dfrac{2}{3} \cdot \dfrac{2}{3} \cdot \dfrac{2}{3}$

21. $\$3.63 + \$0.87 + 96¢$

22. $13{,}456 - 9714$

23. $\dfrac{900}{20}$

24.
$$\begin{array}{r} 145 \\ \times \ 74 \\ \hline \end{array}$$

25. $7\overline{)56{,}153}$

26. $1000 - (100 - 10)$

27. $(1000 - 100) - 10$

28. $30(65¢)$

29. $2 \cdot 3 \cdot 4 \cdot 5$

30. $(5)(5 + 5)$

LESSON
14

Fractions Equal to 1 •
Improper Fractions

Fractions   A fraction is equal to 1 if the numerator and denominator are
equal to 1   equal (and are not zero). Here we show four fractions equal
to 1.

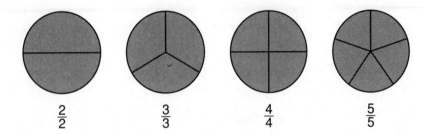

$$\frac{2}{2} \qquad \frac{3}{3} \qquad \frac{4}{4} \qquad \frac{5}{5}$$

Example 1   Which of these fractions is equal to 1?

(a) $\frac{5}{6}$ (b) $\frac{6}{6}$ (c) $\frac{7}{6}$

Solution   The fraction equal to 1 is **(b)**, or $\frac{6}{6}$. The fraction $\frac{5}{6}$ is less than
1, and the fraction $\frac{7}{6}$ is greater than 1.

Improper   A fraction that is equal to 1 or is greater than 1 is called an
fractions   **improper fraction**. Improper fractions can be rewritten either
as whole numbers or as mixed numbers. To convert an improper
fraction to a whole number or to a mixed number, we divide.

Example 2   Convert each improper fraction to either a whole number or
a mixed number.

(a) $\frac{5}{3}$ (b) $\frac{6}{3}$

Solution   We perform the division indicated by each fraction. A remainder
is written as the numerator of a fraction with the same
denominator.

(a) $\frac{5}{3} \rightarrow 3\overline{)5} \rightarrow 1\frac{2}{3}$
    $\phantom{3\overline{)5}}\underline{3}$
    $\phantom{3\overline{)5}}2$

(b) $\frac{6}{3} \rightarrow 3\overline{)6}^{\,2}$

Example 3   Draw and shade circles to show that $1\frac{3}{4} = \frac{7}{4}$.

Solution   On the left we shade a whole circle and three fourths of another circle. On the right we shade four fourths of one circle and three fourths of another circle.

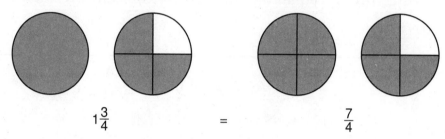

$$1\frac{3}{4} \qquad = \qquad \frac{7}{4}$$

From these circles we can see that one and three fourths equals seven fourths. If the answer to an arithmetic problem is an improper fraction, the improper fraction is usually changed to a whole number or to a mixed number.

Example 4   Simplify: (a) $\frac{4}{5} + \frac{4}{5}$ (b) $\frac{5}{2} \times \frac{3}{4}$

Solution   (a) $\frac{4}{5} + \frac{4}{5} = \frac{8}{5}$ $\frac{8}{5} = 1\frac{3}{5}$

(b) $\frac{5}{2} \times \frac{3}{4} = \frac{15}{8}$ $\frac{15}{8} = 1\frac{7}{8}$

Practice   **a.**   Write a fraction equal to 1 that has 10 as the denominator.

Convert each improper fraction to either a whole number or a mixed number.

b. $\frac{12}{5}$ **c.** $\frac{12}{6}$ **d.** $\frac{12}{7}$

e.   Draw and shade circles to illustrate that $2\frac{1}{4} = \frac{9}{4}$.

Simplify:

f. $\frac{5}{6} + \frac{1}{6}$ **g.** $\frac{7}{3} \times \frac{2}{3}$

Problem set 14   In Problems 1–4, identify the type of problem. Then find the answer.

1. Three hundred seventy answered "yes." The rest answered "no." If there were seven hundred thirty answers in all, how many answered "no"?

2. The helicopter could carry only 6 passengers on each trip. If 150 people needed to be evacuated, how many trips would it take to evacuate them all?

3. The great oak tree was nearly destroyed in the flash flood of 1970. At that time, the tree was estimated to be 550 years old. If the estimate was correct, in what year did the oak begin its life?

4. Sam spent $4.75 for the ticket. He spent $1.50 for popcorn and 85¢ for a drink. How much did he spend in all?

5. Use digits to write one hundred forty-two million, seventy-five thousand, three hundred two.

6. Replace each circle with the proper comparison symbol.

    (a) 50¢ ◯ $0.50   (b) $\frac{3}{2}$ ◯ 1

7. How many units is it from -4 to 2 on the number line?

8. List the single-digit numbers that are divisors of 27,300.

9. Draw and shade circles to show that $1\frac{2}{3}$ equals $\frac{5}{3}$.

10. (a) What fraction of the circle is shaded?

    (b) What fraction of the circle is not shaded?

11. Write a fraction that is equal to 1 and has a denominator of 12.

12. Convert each improper fraction to either a whole number or a mixed number.

(a) $\dfrac{9}{2}$ (b) $\dfrac{9}{3}$ (c) $\dfrac{9}{4}$

Find each missing number:

13.
$$\begin{array}{r} M \\ +\ 35 \\ \hline 118 \end{array}$$

14.
$$\begin{array}{r} N \\ -\ 76 \\ \hline 124 \end{array}$$

15.
$$\begin{array}{r} 5 \\ 8 \\ 6 \\ 4 \\ 5 \\ 7 \\ 3 \\ 2 \\ 9 \\ 5 \\ 6 \\ N \\ +\ 5 \\ \hline 72 \end{array}$$

16.
$$\begin{array}{r} 15 \\ \times\ Q \\ \hline 210 \end{array}$$

Add, subtract, multiply, or divide, as indicated:

17. $\dfrac{2}{9} + \dfrac{3}{9} + \dfrac{4}{9}$ 18. $\dfrac{5}{9} + \dfrac{5}{9}$

19. $\dfrac{5}{3} \cdot \dfrac{4}{3}$

20. $3617 + 98 + 249 + 77$

21. $\$100 - \$97.74$ 22. $\dfrac{900}{15}$

23.
$$\begin{array}{r} 360 \\ \times\ 50 \\ \hline \end{array}$$
24. $\$45.00 \div 20$

25. $15 \cdot 15$ 26. $(10)(10 + 10)$

27. $12(\$0.20)$ 28. $\dfrac{1}{3} + \dfrac{2}{3}$

29. $\dfrac{1}{2} + \dfrac{1}{2} + \dfrac{1}{2}$ 30. $\dfrac{3}{2} \cdot \dfrac{5}{2}$

LESSON
15

Equivalent Fractions

Different fractions that name the same number are called **equivalent fractions**. Here we show four equivalent fractions.

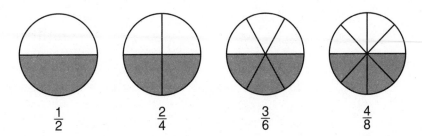

$$\frac{1}{2} \qquad \frac{2}{4} \qquad \frac{3}{6} \qquad \frac{4}{8}$$

We see in each case that half the circle is shaded. We also see that we have named the shaded portions by using four different fractions. Since each fraction has the same value, we say that the fractions are *equal fractions* or are *equivalent fractions*. **Equivalent fractions have the same value.**

$$\frac{1}{2} = \frac{2}{4} = \frac{3}{6} = \frac{4}{8}$$

We can form equivalent fractions by multiplying a number by a fraction equal to 1. We will multiply $\frac{1}{2}$ by $\frac{2}{2}$, $\frac{3}{3}$, and $\frac{4}{4}$ to show this.

$$\frac{1}{2} \times \frac{2}{2} = \frac{2}{4}$$

$$\frac{1}{2} \times \frac{3}{3} = \frac{3}{6}$$

$$\frac{1}{2} \times \frac{4}{4} = \frac{4}{8}$$

Example 1 Find three equivalent fractions for $\frac{1}{3}$ by multiplying by $\frac{3}{3}$, $\frac{4}{4}$, and $\frac{10}{10}$.

Solution   We multiply as indicated.

$$\frac{1}{3} \times \frac{3}{3} = \frac{3}{9} \qquad \text{multiplied by } \frac{3}{3}$$

$$\frac{1}{3} \times \frac{4}{4} = \frac{4}{12} \qquad \text{multiplied by } \frac{4}{4}$$

$$\frac{1}{3} \times \frac{10}{10} = \frac{10}{30} \qquad \text{multiplied by } \frac{10}{10}$$

The fractions $\frac{1}{3}$, $\frac{3}{9}$, $\frac{4}{12}$, and $\frac{10}{30}$ are equivalent fractions.

Example 2   Find an equivalent fraction for $\frac{1}{2}$ that has a denominator of 12.

Solution   We want a denominator of 12.

$$\frac{1}{2} = \frac{?}{12}$$

We have multiplied 2 by 6 to get 12. So we must also multiply 1 by 6.

$$\frac{1}{2} \times \frac{6}{6} = \frac{6}{12}$$

The fractions $\frac{1}{2}$ and $\frac{6}{12}$ are equivalent fractions.

Example 3   Find an equivalent fraction for $\frac{2}{3}$ that has a denominator of 12.

Solution   We can get the fraction we want if we multiply by $\frac{4}{4}$.

$$\frac{2}{3} \times \frac{4}{4} = \frac{8}{12}$$

The fractions $\frac{2}{3}$ and $\frac{8}{12}$ are equivalent fractions.

Practice   **a.** Form three equivalent fractions for $\frac{2}{7}$ by multiplying by $\frac{2}{2}$, $\frac{5}{5}$, and $\frac{8}{8}$.

b. Form three equivalent fractions for $\frac{3}{4}$ by multiplying by $\frac{5}{5}, \frac{7}{7}$, and $\frac{3}{3}$.

c. Find an equivalent fraction for $\frac{3}{4}$ that has a denominator of 16.

Find the number that makes these fractions equivalent fractions.

d. $\frac{4}{5} = \frac{?}{20}$ **e.** $\frac{3}{8} = \frac{9}{?}$

Problem set 15

In Problems 1–4, identify the type of problem. Then find the answer.

1. The bikers rode for several hours before stopping for lunch. Then they rode fifty-three miles after lunch. If they rode a total of one hundred twenty-one miles, how many miles did they ride before lunch?

2. The pickers filled 63 large buckets with strawberries. If each bucket held 7 pounds of strawberries, how many pounds of strawberries did they pick?

3. Twenty-three million is how much more than seven million, eight hundred thousand? Use words to write the answer.

4. Three hundred twenty-four girls and boys crowded into the auditorium. If 186 of the students were girls, how many boys were there?

5. Arrange these numbers in order from least to greatest:

$$-2, \frac{1}{2}, 0, 1, \frac{4}{3}$$

6. Draw and shade circles to show that $2\frac{1}{2}$ equals $\frac{5}{2}$.

7. Make three equivalent fractions for $\frac{2}{5}$ by multiplying by $\frac{2}{2}, \frac{3}{3}$, and $\frac{4}{4}$.

8. Write a fraction equal to 1 that has a denominator of 100.

9. (a) What fraction of the square is shaded?

(b) What fraction of the square is not shaded?

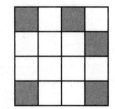

10. Complete each equivalent fraction.

(a) $\dfrac{1}{3} = \dfrac{?}{12}$ (b) $\dfrac{1}{4} = \dfrac{?}{12}$ (c) $\dfrac{5}{6} = \dfrac{?}{12}$

11. List the whole numbers that are factors of 100.

12. What mixed number names point A on this number line?

13. Convert each improper fraction to either a whole number or a mixed number.

(a) $\dfrac{9}{5}$ (b) $\dfrac{10}{5}$ (c) $\dfrac{11}{5}$

Find each missing number:

14.	15.	16.

14.
$$\begin{array}{r} A \\ \times\ \ 18 \\ \hline 4500 \end{array}$$

15.
$$\begin{array}{r} B \\ +\ 3619 \\ \hline 4087 \end{array}$$

16.
$$\begin{array}{r} 5 \\ 6 \\ 24 \\ 3 \\ 2 \\ 7 \\ 14 \\ 8 \\ 6 \\ N \\ +\ 4 \\ \hline 99 \end{array}$$

17.
$$\begin{array}{r} \$20.00 \\ -\quad\ C \\ \hline \$18.49 \end{array}$$

18.
$$\begin{array}{r} 40 \\ \times\ \ D \\ \hline 1760 \end{array}$$

Add, subtract, multiply, or divide, as indicated:

19. $\dfrac{5}{6} + \dfrac{1}{6}$ **20.** $\dfrac{3}{4} \cdot \dfrac{5}{2}$

21. $\$0.79 + \$6.48 + \$15.95$

22. 27,400 − 989

23. $\dfrac{3450}{30}$

24. $\begin{array}{r} 875 \\ \times\ 16 \\ \hline \end{array}$

25. $8\,)\overline{14{,}404}$

26. 25 · 25

27. $\dfrac{1001}{11}$

28. (6 + 7)(7)

29. $\dfrac{5}{2} + \dfrac{5}{2}$

30. $\dfrac{5}{2} \cdot \dfrac{5}{2}$

31. $\dfrac{5}{2} - \dfrac{5}{2}$

LESSON 16

Reducing Fractions, Part 1

In the preceding lesson, we formed equivalent fractions by multiplying by a fraction equal to 1.

$$\frac{1}{2} \times \mathbf{1}\,\frac{4}{4} = \frac{4}{8}$$

The fraction $\frac{1}{2}$ equals the fraction $\frac{4}{8}$ because we multiplied $\frac{1}{2}$ by $\frac{4}{4}$, which equals 1. When we multiply by 1, the value of a number is not changed. It is also true that when we **divide by 1** the value of a number is not changed. Here we divide $\frac{4}{8}$ by $\frac{4}{4}$.

$$\frac{4}{8} \div \mathbf{1}\,\frac{4}{4} = \frac{1}{2} \qquad \begin{array}{l} (4 \div 4 = 1) \\ (8 \div 4 = 2) \end{array}$$

By dividing, we have changed $\frac{4}{8}$ to $\frac{1}{2}$. When we divide, both the numerator and the denominator of $\frac{4}{8}$ become smaller. This is called **reducing**.

The numbers we use when we write a fraction are called the **terms** of the fraction. The terms of $\frac{4}{8}$ are 4 and 8. If both terms of a fraction are divisible by a number greater than 1, then the fraction can be reduced. To reduce a fraction, we divide both terms of the fraction by a number that is a divisor of both terms. In some cases the terms of a fraction are divisible by more than one number. For example, 4 and 8 are both divisible by 2 and by 4.

$$\frac{4}{8} \div \frac{2}{2} = \frac{2}{4} \qquad \frac{4}{8} \div \frac{4}{4} = \frac{1}{2}$$

Dividing $\frac{4}{8}$ by $\frac{4}{4}$ instead of by $\frac{2}{2}$ results in a fraction with lower terms, since the terms of $\frac{1}{2}$ are lower than the terms of $\frac{2}{4}$. It is customary to reduce fractions to **lowest terms**.

Example 1 Reduce $\frac{4}{6}$ to lowest terms.

Solution Both 4 and 6 are divisible by 2, so we divide by 2 over 2.

$$\frac{4}{6} \div \frac{2}{2} = \mathbf{\frac{2}{3}}$$

Example 2 Reduce $\frac{18}{24}$ to lowest terms.

Solution Both 18 and 24 are divisible by 2, so we divide by 2 over 2.

$$\frac{18}{24} \div \frac{2}{2} = \frac{9}{12}$$

This is still not in lowest terms because both 9 and 12 are divisible by 3.

$$\frac{9}{12} \div \frac{3}{3} = \mathbf{\frac{3}{4}}$$

We could have used just one step had we noticed that both 18 and 24 are divisible by 6.

$$\frac{18}{24} \div \frac{6}{6} = \mathbf{\frac{3}{4}}$$

Both methods are correct. One method took two steps, and the other took just one step.

Example 3 Reduce $3\frac{8}{12}$ to lowest terms.

Solution To reduce a mixed number, we reduce the fraction and leave the whole number unchanged.

$$\frac{8}{12} \div \frac{4}{4} = \frac{2}{3}$$

Thus
$$3\frac{8}{12} = 3\frac{2}{3}$$

Example 4 Simplify: $\frac{3}{8} + \frac{3}{8}$

Solution First we add. Then we reduce.

ADD	**REDUCE**
$\frac{3}{8} + \frac{3}{8} = \frac{6}{8}$	$\frac{6}{8} \div \frac{2}{2} = \frac{3}{4}$

Example 5 Simplify: $\frac{7}{9} - \frac{1}{9}$

Solution First we subtract. Then we reduce.

SUBTRACT	**REDUCE**
$\frac{7}{9} - \frac{1}{9} = \frac{6}{9}$	$\frac{6}{9} \div \frac{3}{3} = \frac{2}{3}$

Example 6 Reduce $4\frac{3}{10}$ to lowest terms.

Solution We cannot reduce $\frac{3}{10}$ because 3 and 10 have no common divisors. So the answer is

$$4\frac{3}{10}$$

Practice Reduce each fraction to lowest terms.

    a. $\frac{3}{6}$ **b.** $\frac{8}{10}$ **c.** $\frac{8}{12}$ **d.** $\frac{12}{16}$

e. $4\dfrac{4}{8}$ **f.** $6\dfrac{9}{12}$ **g.** $8\dfrac{16}{24}$ **h.** $12\dfrac{8}{15}$

Perform each indicated operation and reduce the result.

i. $\dfrac{5}{12} + \dfrac{5}{12}$ **j.** $3\dfrac{7}{10} - 1\dfrac{1}{10}$ **k.** $\dfrac{5}{6} \times \dfrac{2}{3}$

**Problem set
16**   In Problems 1–4, identify the type of problem. Then find the answer.

1. Great Grandpa celebrated his seventy-fifth birthday in 1989. In what year was he born?

2. Austin watched the geese fly south. He counted 27 in the first flock, 38 in the second flock, and 56 in the third flock. How many geese did Austin see in all three flocks?

3. If five twelfths of the eggs were cracked, what fraction of the eggs were not cracked?

4. The farmer harvested 9000 bushels of grain from 60 acres. The crop produced an average of how many bushels of grain for each acre?

5. Use words to write 29,705,060.

6. Use digits and symbols to write "The product of three and five is greater than the sum of three and five."

7. List the single-digit divisors of 2100.

8. Reduce each fraction or mixed number.

    (a) $\dfrac{6}{8}$ (b) $2\dfrac{6}{10}$

9. Make three equivalent fractions for $\frac{2}{3}$ by multiplying by $\frac{3}{3}$, $\frac{5}{5}$, and $\frac{6}{6}$.

10. For each fraction find an equivalent fraction that has a denominator of 20.

(a) $\dfrac{3}{5}$ (b) $\dfrac{1}{2}$ (c) $\dfrac{3}{4}$

11. Refer to this figure to

(a) name the line.

(b) name three rays.

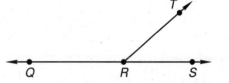

12. Convert each fraction to either a whole number or a mixed number.

(a) $\dfrac{11}{3}$ (b) $\dfrac{12}{3}$ (c) $\dfrac{13}{3}$

Find each missing number:

13.
$$\begin{array}{r} A \\ \times\quad 6 \\ \hline 3300 \end{array}$$

14.
$$\begin{array}{r} 3977 \\ +\quad B \\ \hline 5000 \end{array}$$

15.
$$\begin{array}{r} 5 \\ 21 \\ 30 \\ 6 \\ 8 \\ 4 \\ 7 \\ 6 \\ 9 \\ 5 \\ 21 \\ +\ N \\ \hline 134 \end{array}$$

16.
$$\begin{array}{r} \$3.24 \\ -\qquad C \\ \hline \$0.27 \end{array}$$

Add, subtract, multiply, or divide, as indicated:

17. $\dfrac{2}{5} + \dfrac{3}{5} + \dfrac{4}{5}$

18. $\dfrac{5}{8} - \dfrac{3}{8}$

19. $\dfrac{4}{3} \cdot \dfrac{3}{4}$

20. $35 \cdot 35$

21. $98 + 76 + 56 + 38 + 119$

22. $\$40.00 - \$19.80$

23. $\dfrac{\$26.00}{8}$

24.
$$\begin{array}{r} \$6.50 \\ \times\qquad 70 \end{array}$$

25. $12\overline{)72{,}049}$ **26.** $\dfrac{1001}{7}$ **27.** $(11)(6 + 7)$

28. $\dfrac{7}{5} + \dfrac{8}{5}$ **29.** $\dfrac{11}{12} - \dfrac{1}{12}$ **30.** $\dfrac{5}{6} \cdot \dfrac{2}{3}$

LESSON 17

Linear Measure

One of the characteristics of any civilization is the use of an agreed-upon system of measurement. The fair exchange of goods and services requires consistent units of weight, volume, and length. In a technological society the necessity for a standard system of measurement is even greater.

There are two systems of measurement currently used in the United States. The traditional system of measurement, with units such as feet, gallons, and pounds, was adopted from England. This system used to be known as the English system, but is now referred to as the **U.S. Customary System**.

The second system of measurement used in the United States is the system used in the rest of the world. It is known as the **International System** (or SI, for *Système International*) or the **metric system**. The metric system has units such as meters, liters, and kilograms.

We will consider both systems of measurement over many lessons. In this lesson we will consider units of length. The following table shows equivalent measures of length in the U.S. system. We should remember the equivalent measures, have a "feel" for the units so that we can estimate length, and be able to use rulers and read scales when measuring lengths.

UNITS OF LENGTH (U.S. SYSTEM)

12 inches (in.)	=	1 foot (ft)
3 feet	=	1 yard (yd)
1760 yards	=	1 mile (mi)
5280 feet	=	1 mile

Example 1 One yard is how many inches?

Solution One yard is 3 feet long. One foot is 12 inches long. Thus, 1 yard is 36 inches long.

$$1 \text{ yard} = 3 \times 12 \text{ inches} = \textbf{36 inches}$$

Example 2 A 10-speed bicycle is about how many feet long?

Solution We should develop a feel for various units of measure. Most 10-speed bicycles are about $5\frac{1}{2}$ feet long, so a good estimate would be **about 5 or 6 feet**.

Example 3 How long is this line segment?

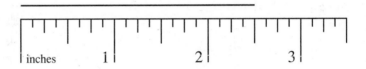

Solution Each inch on this scale has been divided into 8 equal parts. Each part is one eighth inch long. The line segment is 2 inches plus four eighths inch.

$$\text{Segment} = 2\frac{4}{8} \text{ inches long}$$

But we can reduce $\frac{4}{8}$ to $\frac{1}{2}$, so

$$\text{Segment} = \textbf{2}\frac{\textbf{1}}{\textbf{2}} \textbf{ inches}$$

The following table shows equivalent measures of commonly used units of length in the metric system.

UNITS OF LENGTH (METRIC SYSTEM)

10 millimeters (mm)	= 1 centimeter (cm)
100 centimeters	= 1 meter (m)
1000 meters	= 1 kilometer (km)

Example 4   Two meters (2 m) is how many centimeters?

Solution   One meter equals 100 centimeters (100 cm), so 2 m = **200 cm.**

Example 5   A door is about how many meters high?

Solution   The height of most doors is about **2 meters.**

Example 6   This line segment is

(a) how many centimeters long?

(b) how many millimeters long?

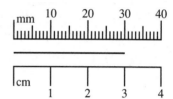

Solution   (a) **3 cm**

(b) **30 mm**

Practice   **a.** Name two systems of measure used in the United States and identify some units of measure in each system.

b. Two meters is how many millimeters?

c. Five yards is how many feet?

d. A car is about how many meters long?

e. Your shoe is about how many inches long?

f. How long is this piece of gum?

Problem set 17   In Problems 1–4, identify the type of problem. Then find the answer.

1. Thirty-five of the one hundred eighteen students who took the test earned an A. How many of the students did not earn an A on the test?

2. At Henry's egg ranch 18 eggs are packaged in each carton. How many cartons would be needed to package 4500 eggs?

3. Three hundred twenty-four ducks floated peacefully on the lake. As the first shot rang out, all but twenty-seven of the ducks flew away. How many ducks flew away?

4. The number 516,824 is how much less than 804,216?

5. Replace each circle with the proper comparison symbol.

   (a) $\dfrac{8}{10} \bigcirc \dfrac{4}{5}$ (b) $\dfrac{8}{5} \bigcirc 1\dfrac{2}{5}$

6. Find the length of the segment to the nearest eighth of an inch.

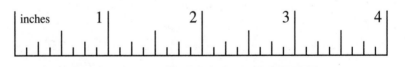

7. Reduce each fraction or mixed number.

   (a) $\dfrac{8}{12}$ (b) $\dfrac{9}{12}$ (c) $6\dfrac{10}{12}$

8. Draw and shade circles to show that $3\dfrac{1}{3}$ equals $\dfrac{10}{3}$.

9. For each fraction, find an equivalent fraction that has a denominator of 24.

   (a) $\dfrac{5}{6}$ (b) $\dfrac{3}{8}$ (c) $\dfrac{1}{4}$

10. (a) What fraction of the group is shaded?

    (b) What fraction of the group is not shaded?

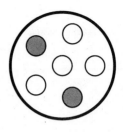

11. The number 630 is divisible by which single-digit numbers?

12. Convert each fraction to either a whole number or a mixed number.

    (a) $\dfrac{16}{7}$

    (b) $\dfrac{16}{8}$

    (c) $\dfrac{16}{9}$

Find each missing number:

13.
$$\begin{array}{r} M \\ -\ 1776 \\ \hline 87 \end{array}$$

14.
$$\begin{array}{r} K \\ +\ 2937 \\ \hline 3000 \end{array}$$

15.
$$\begin{array}{r} 43 \\ 7 \\ 86 \\ 24 \\ 7 \\ 6 \\ +\ N \\ \hline 175 \end{array}$$

16.
$$\begin{array}{r} \$16.25 \\ -\qquad B \\ \hline \$10.15 \end{array}$$

17.
$$\begin{array}{r} 42 \\ \times\quad D \\ \hline 1764 \end{array}$$

Add, subtract, multiply, or divide, as indicated:

18. $\dfrac{3}{4} - \dfrac{1}{4}$

19. $\dfrac{3}{10} + \dfrac{8}{10}$

20. $\dfrac{3}{4} \times \dfrac{1}{3}$

21. $60{,}310 - 49{,}157$

22. $\$21.56 + \$15 + 79¢$

23. $\dfrac{10{,}000}{16}$

24.
$$\begin{array}{r} 176 \\ \times\ 84 \\ \hline \end{array}$$

25. $9\overline{)70{,}000}$

26. $45 \cdot 45$

27. $\dfrac{1001}{13}$

28. $(5 + 6)(7)$

29. $\dfrac{7}{9} - \dfrac{1}{9}$

30. $\dfrac{4}{3} \cdot \dfrac{3}{2}$

LESSON
18

Pairs of Lines • Angles

Pairs of lines A desktop is a flat surface with boundaries. A desktop occupies a part of a plane. A **plane** is a flat surface that has no boundaries.

Two lines in the same plane either cross once or they do not cross at all. When two lines cross, we say that they **intersect**. They intersect at one point. Two lines that do not intersect remain the same distance apart. Lines in the same plane that are always the same distance apart are **parallel lines**.

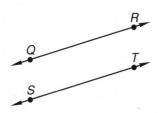

Line *AB* intersects line *CD* at point *M*.

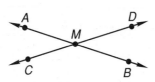

In this figure, line *QR* is parallel to line *ST*. This statement can be written with symbols, as we show here.

$$\overleftrightarrow{QR} \parallel \overleftrightarrow{ST}$$

Lines that intersect and form "square corners" are **perpendicular**. The small square in the figure below indicates a "square corner."

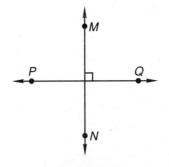

Line *MN* is perpendicular to line *PQ*.

$$\overleftrightarrow{MN} \perp \overleftrightarrow{PQ}$$

The symbol $\perp$ means **is perpendicular to**.

Angles   An angle is formed by two rays that have a common endpoint. Angle *DMB* is formed by the two rays, $\overrightarrow{MD}$ and $\overrightarrow{MB}$. The common endpoint is *M*. Ray *MD* and ray *MB* are the **sides** of the angle. Point *M* is the **vertex** of the angle.

Angles may be named in several ways:

1. Angles may be named by using three letters in this order: a point on one ray, the vertex, then a point on the other ray.

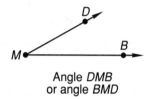
Angle *DMB*
or angle *BMD*

2. When there is no chance of confusion, an angle may be named with only one letter: the letter at the vertex.

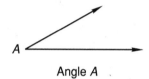
Angle *A*

3. An angle may be named by placing a small letter or number near the vertex and between the rays.

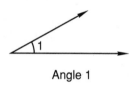
Angle 1

The symbol $\angle$ is often used instead of the word "angle." Thus, the three angles just named could be referred to as:

$\angle DMB$ read as "angle *DMB*"

$\angle A$ read as "angle *A*"

$\angle 1$ read as "angle 1"

Angles are classified by their size. An angle that is formed by perpendicular rays is a **right angle**. An angle smaller than a right angle is an **acute angle**. An angle that forms a straight line is a **straight angle**. An angle that is smaller than a straight angle but larger than a right angle is an **obtuse angle**.

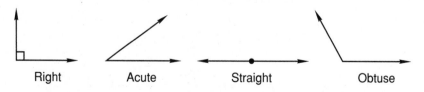

Right   Acute   Straight   Obtuse

Example 1 (a) Which line is parallel to line
AB?

(b) Which line is perpendicular
to line AB?

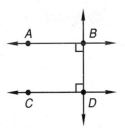

Solution (a) **Line CD** (or $\overleftrightarrow{DC}$) is parallel to line AB.

(b) **Line BD** (or $\overleftrightarrow{DB}$) is perpendicular to line AB.

Example 2 There are several angles in this figure.

(a) Name the straight angle.

(b) Name the obtuse angle.

(c) Name two right angles.

(d) Name two acute angles.

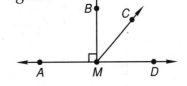

Solution (a) ∠**AMD** (or ∠**DMA**)

(b) ∠**AMC** (or ∠**CMA**)

(c) 1. ∠**AMB** (or ∠**BMA**)

2. ∠**BMD** (or ∠**DMB**)

(d) 1. ∠**BMC** (or ∠**CMB**)

2. ∠**CMD** (or ∠**DMC**)

Practice **a.** Draw two parallel lines.

b. Draw two perpendicular lines.

c. Draw two lines that intersect but are not perpendicular.

d. Draw a right angle.

e. Draw an acute angle.

f. Draw an obtuse angle.

Problem set In Problems 1–4, identify the type of problem. Then find the
18 answer.

1. Prince Caspian assembled his soldiers at the bank of the river. Two thousand, four hundred twenty had gathered by noon. An additional five thousand, ninety arrived after noon. How many soldiers arrived in all?

2. Three twentieths of the test answers were incorrect. What fraction of the answers were correct?

3. There are 210 students in the first-year physical education class. If they are equally divided into 15 squads, how many students will be in each squad?

4. How many years were there from 1492 to 1620?

5. Which of the following does not equal $1\frac{1}{3}$?

   (a) $\frac{4}{3}$ (b) $1\frac{2}{6}$ (c) $\frac{5}{3}$ (d) $1\frac{4}{12}$

6. Refer to this figure to answer (a) and (b).

   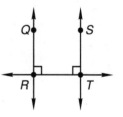

   (a) Which line is parallel to $\overleftrightarrow{ST}$?

   (b) Which line is perpendicular to $\overleftrightarrow{ST}$?

7. Reduce each fraction or mixed number.

   (a) $\frac{12}{16}$ (b) $3\frac{12}{18}$ (c) $5\frac{12}{20}$

8. Draw and shade circles to show that $2\frac{3}{4}$ equals $\frac{11}{4}$.

9. Complete each equivalent fraction.

   (a) $\frac{2}{9} = \frac{?}{18}$ (b) $\frac{1}{3} = \frac{?}{18}$ (c) $\frac{5}{6} = \frac{?}{18}$

10. Draw a triangle that has one right angle.

11. What factors of 20 are also factors of 50?

12. Draw a line and identify two points on the line as R and S. Then draw a ray ST that is perpendicular to the line.

Find each missing number:

13.
$$\begin{array}{r} W \\ \times\quad 8 \\ \hline \$30.00 \end{array}$$

14.
$$\begin{array}{r} X \\ -\ 2316 \\ \hline 1415 \end{array}$$

15.
$$\begin{array}{r} 58 \\ 4 \\ 2 \\ 62 \\ N \\ +\quad 6 \\ \hline 143 \end{array}$$

16.
$$\begin{array}{r} \$6.30 \\ +\quad Y \\ \hline \$25.00 \end{array}$$

17.
$$\begin{array}{r} 2715 \\ -\quad Z \\ \hline 1716 \end{array}$$

Add, subtract, multiply, or divide, as indicated:

18. $\dfrac{5}{6} - \dfrac{1}{6}$

19. $\dfrac{1}{2} \cdot \dfrac{2}{3}$

20. $36 + 67 + 59 + 86 + 8$

21. $\$20.25 - \$15.17$

22. $\dfrac{\$100.00}{40}$

23.
$$\begin{array}{r} 300 \\ \times\ 800 \\ \hline \end{array}$$

24. $9\overline{)31{,}805}$

25. $55 \cdot 55$

26. $\dfrac{2002}{11}$

27. $20(20 + 20)$

28. $\dfrac{3}{5} + \dfrac{3}{5} + \dfrac{3}{5}$

29. $\dfrac{14}{15} - \dfrac{4}{15}$

30. $\dfrac{1}{2} \cdot \dfrac{4}{3} \cdot \dfrac{9}{2}$

LESSON 19

Polygons

When three or more line segments are connected to enclose a portion of a plane, a **polygon** is formed. The name of a polygon tells how many sides the polygon has.

NAMES OF POLYGONS

EXAMPLE OF POLYGON	NUMBER OF SIDES	NAME OF POLYGON
	3	Triangle
	4	Quadrilateral
	5	Pentagon
	6	Hexagon
	7	Heptagon
	8	Octagon
	9	Nonagon
	10	Decagon
	11	Undecagon
	12	Dodecagon

A polygon with more than 12 sides may be referred to as an *n*-gon, with *n* being the number of sides. Thus, a polygon with 15 sides is a 15-gon.

Two sides of a polygon meet at a point that is called a **vertex** of the polygon. The plural of vertex is **vertices**. A polygon always has the same number of vertices as the number of sides.

Letters may be used to identify a particular polygon. The letters *S, T, V,* and *U* are used in the drawing below to indicate the points that are vertices of the polygon. To refer to this polygon, we give the letters of the vertices in order around the polygon. Any letter may be first. The rest of the letters can be named clockwise or counterclockwise. This polygon has eight names, which are listed here.

USTV	*TSUV*
UVTS	*TVUS*
STVU	*VTSU*
SUVT	*VUST*

If all the sides of a polygon have the same length and all the angles have the same measure, then the polygon is a **regular polygon**.

REGULAR AND IRREGULAR POLYGONS

TYPE	REGULAR	IRREGULAR
Triangle		
Quadrilateral		
Pentagon		
Hexagon		

Practice   **a.** What is the shape of a stop sign?

b. What do we usually call a regular quadrilateral?

c. What kind of angle is each angle of a regular triangle?

Problem set 19   In Problems 1–4, identify the type of problem. Then find the answer.

1. The Collins family completed the 3300-mile coast-to-coast drive in 6 days. They traveled an average of how many miles each day?

2. On their return trip, the Collins family drove four hundred fifty-six miles the first day and five hundred seventeen miles the second day. How far did they travel on the first two days of their return trip?

3. Albert ran 3977 meters of the 5000-meter race, but walked the rest of the way. How many meters of the race did Albert walk?

4. One billion is how much greater than ten million? Use words to write the answer.

5. Arrange these numbers in order from least to greatest:

$$\frac{5}{3}, -1, \frac{3}{4}, 0, 1$$

6. In a rectangle the opposite sides are parallel. In rectangle *ABCD*, which side is parallel to side *BC*?

7. (a) What is the length of segment *WX* in millimeters?

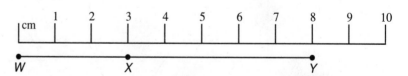

   (b) What is the length of segment *XY* in centimeters?

8. Reduce each fraction or mixed number.

   (a) $\dfrac{10}{20}$ (b) $\dfrac{12}{20}$ (c) $6\dfrac{15}{20}$

9. For each fraction find an equivalent fraction that has a denominator of 30.

(a) $\dfrac{4}{5}$ (b) $\dfrac{2}{3}$ (c) $\dfrac{1}{6}$

10. An octagon has how many more sides than a pentagon?

11. (a) Draw a triangle that has one obtuse angle.

    (b) What kind of angles are the other two angles of the triangle?

12. (a) What fraction of the circle is shaded?

    (b) What fraction of the circle is not shaded?

Find each missing number:

13.
$$\begin{array}{r} X \\ -\ 5814 \\ \hline 3286 \end{array}$$

14.
$$\begin{array}{r} Y \\ +\ 1537 \\ \hline 7351 \end{array}$$

15.
$$\begin{array}{r} 4 \\ 7 \\ 8 \\ 15 \\ 4 \\ 6 \\ 5 \\ 7 \\ 8 \\ 21 \\ +\ N \\ \hline 93 \end{array}$$

16.
$$\begin{array}{r} \$50.00 \\ -\quad M \\ \hline \$40.70 \end{array}$$

17.
$$\begin{array}{r} 300 \\ \times\quad N \\ \hline 12,000 \end{array}$$

Add, subtract, multiply, or divide, as indicated:

18. $\dfrac{7}{10} - \dfrac{3}{10}$

19. $\dfrac{3}{2} \cdot \dfrac{2}{4}$

20. $\$3.67 + \$14.39 + \$0.78$

21. $10,000 - 576$

22. $\dfrac{2025}{45}$

23.
$$\begin{array}{r} 750 \\ \times\ 80 \\ \hline \end{array}$$

24. $50\overline{)95,714}$

25. $21 \cdot 21$

26. $\dfrac{2002}{14}$

27. $30(40 + 50)$

28. $\dfrac{3}{7} + \dfrac{4}{7}$

29. $\dfrac{15}{16} - \dfrac{7}{16}$

30. $\dfrac{5}{6} \cdot \dfrac{2}{5}$

LESSON 20

Perimeter, Part 1

Perimeter   The distance around a polygon is the **perimeter** of the polygon. To find the perimeter of a polygon, we add the lengths of its sides.

Example 1   What is the perimeter of this rectangle?

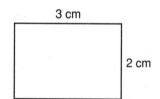

Solution   The opposite sides of a rectangle are equal in length. Tracing around the rectangle, our pencil travels 3 cm, then 2 cm, then 3 cm, then 2 cm. Thus, the perimeter is

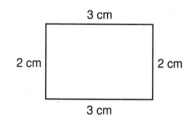

$$3 \text{ cm} + 2 \text{ cm} + 3 \text{ cm} + 2 \text{ cm} = \textbf{10 cm}$$

Example 2   What is the perimeter of this regular hexagon?

Solution   All sides of a regular polygon are equal in length. Thus the perimeter of this hexagon is

$$8 \text{ mm} + 8 \text{ mm} + 8 \text{ mm} + 8 \text{ mm} + 8 \text{ mm} + 8 \text{ mm} = \textbf{48 mm}$$

or $6 \times 8 \text{ mm} = \textbf{48 mm}$

Example 3 The perimeter of a square is 48 ft. How long is each side of the square?

Solution A square has four sides whose lengths are equal. The sum of the four lengths is 48 ft. Here are two ways to think about this problem:

1. What number added 4 times equals 48?

    _____ + _____ + _____ + _____ = 48 ft

2. What number multiplied by 4 equals 48?

    4 × _____ = 48 ft

We will use the second way and divide to find the length of each side.

$$\overset{\textstyle 12}{4\overline{)48}}$$

The length of each side of the square is **12 ft**.

Perimeter word problems

To work some story problems, we need to find a perimeter.

Example 4 Ray wants to fence some grazing land for his sheep. He made this sketch of his pasture. How many feet of wire fence does he need?

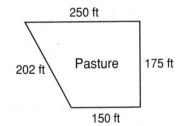

Solution If we add the lengths of the sides to find how many feet of fence Ray needs,

250 ft + 175 ft + 150 ft + 202 ft = **777 ft**

we see that Ray needs **777 ft** of wire fence.

Practice **a.** What is the perimeter of this quadrilateral?

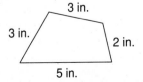

b. What is the perimeter of this regular pentagon?

5 cm

c. If each side of a regular octagon is 12 inches, what is its perimeter?

d. MacGregor has 100 feet of wire fence that he plans to use to surround a square garden. Each side of his garden will be how many feet long?

Problem set 20   In Problems 1–3, identify the type of problem. Then find the answer.

1. One eighth of the students in the class were left-handed. What fraction of the students were not left-handed?

2. The theater was full when the horror film began. Seventy-six people left before the movie ended. One hundred twenty-four people remained. How many people were in the theater when it was full?

3. The Pie King restaurant cuts each pie into 6 slices. The restaurant served 84 pies one week. How many slices of pie were served?

4. Lincoln began his speech, "Four score and seven years ago...." How many years is four score and seven? (*Hint:* A score is 20 years.)

5. Use words to write 18720501.

6. Use digits and symbols to write "Three minus seven equals negative four."

7. The ceiling in a house is about how many feet above the floor?

8. Find the perimeter of this rectangle.

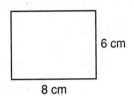

6 cm

8 cm

9. Simplify:

(a) $3\dfrac{16}{24}$

(b) $\dfrac{15}{24}$

10. Find a and b.

(a) $\dfrac{3}{4} = \dfrac{a}{36}$

(b) $\dfrac{4}{9} = \dfrac{b}{36}$

11. Draw a sketch of a regular pentagon.

12. What is the name of a polygon that has twice as many sides as a quadrilateral?

13. What kind of angle is every angle of a rectangle?

Find each missing number:

14.
$$\begin{array}{r} A \\ +\,1547 \\ \hline 8998 \end{array}$$

15.
$$\begin{array}{r} B \\ \times\quad 30 \\ \hline \$41.10 \end{array}$$

16.
$$\begin{array}{r} 3 \\ 8 \\ 7 \\ 29 \\ 4 \\ 6 \\ 8 \\ N \\ +\;5 \\ \hline 78 \end{array}$$

17.
$$\begin{array}{r} 32 \\ \times\; C \\ \hline 736 \end{array}$$

18.
$$\begin{array}{r} 2657 \\ +\quad D \\ \hline 3010 \end{array}$$

Add, subtract, multiply, or divide, as indicated:

19. $\dfrac{2}{3} + \dfrac{2}{3} + \dfrac{2}{3}$

20. $\dfrac{7}{8} - \dfrac{5}{8}$

21. $\dfrac{2}{3} \cdot \dfrac{3}{7}$

22. $\$15.00 - \$9.65$

23. $4363 + 2791 + 5814$

24. $\dfrac{3600}{18}$

25. $\begin{array}{r} \$0.79 \\ \times \quad 48 \\ \hline \end{array}$

26. $50 \cdot 50$

27. $\dfrac{100{,}100}{11}$

28. $11(12 + 13)$

29. $\dfrac{6}{7} + \dfrac{5}{7}$

30. $\dfrac{16}{20} - \dfrac{16}{20}$

LESSON 21

Solving Equations

Since Lesson 1 we have practiced finding the missing numbers in arithmetic problems such as these:

$$\begin{array}{r} X \\ + 25 \\ \hline 77 \end{array} \qquad \begin{array}{r} M \\ - 32 \\ \hline 24 \end{array} \qquad \begin{array}{r} Y \\ \times 6 \\ \hline 84 \end{array}$$

We may arrange the numbers horizontally instead of vertically. Thus, the same three problems may be written this way:

$$x + 25 = 77 \qquad m - 32 = 24 \qquad 6y = 84$$

Each of these three problems is an **equation**. We **solve** an equation by finding the number that correctly completes the equation. Note that we used lowercase letters in writing our equations. The letters in an equation can be capital letters or lowercase lettters. Also, in the equation on the right, we omitted the times sign, because $6y$ means 6 times y.

In later lessons we will learn some special rules for solving equations. For now we will solve equations in the same way we have solved missing number problems.

Example 1   Solve: $x + 25 = 77$

Solution   To find what number added to 25 equals 77, we subtract 25 from 77. Then, on the right, we check.

$$
\begin{array}{r}
77 \\
-\ 25 \\
\hline
52
\end{array}
$$

$x + 25 = 77$ equation
$(52) + 25 = 77$ replaced x with 52
$77 = 77$ check

The solution is **x = 52**.

Example 2   Solve: $m - 32 = 24$

Solution   This equation is like this missing number problem:

$$
\begin{array}{r}
m \\
-\ 32 \\
\hline
24
\end{array}
$$

To find m, we can add 24 and 32. Then we check the result.

$$
\begin{array}{r}
24 \\
+\ 32 \\
\hline
56
\end{array}
$$

$m - 32 = 24$ equation
$(56) - 32 = 24$ replaced m with 56
$24 = 24$ check

The solution is **m = 56**.

Example 3   Solve: $32 - p = 24$

Solution   This equation is like this missing number problem:

$$
\begin{array}{r}
32 \\
-\ p \\
\hline
24
\end{array}
$$

To find p, we can subtract 24 from 32. Then, on the right, we check the result.

$$
\begin{array}{r}
32 \\
-\ 24 \\
\hline
8
\end{array}
$$

$32 - p = 24$ equation
$32 - (8) = 24$ replaced p with 8
$24 = 24$ check

The solution is **p = 8**.

Example 4   Solve: $6y = 84$

Solution   The equation says that 6 times y equals 84. We can find a missing factor by dividing the product by the known factor.

Then, on the right, we check our answer.

$$\begin{array}{r} 14 \\ 6\overline{)84} \end{array} \qquad \begin{array}{ll} 6y = 84 & \text{equation} \\ 6(14) = 84 & \text{replaced } y \text{ with } 14 \\ 84 = 84 & \text{check} \end{array}$$

The solution is $y = 14$.

Practice   Solve each equation:

    a. $49 = 17 + w$ **b.** $59 - x = 18$

    c. $y - 59 = 18$ **d.** $84 = 4d$

Problem set 21   In Problems 1–4, identify the type of problem. Then find the answer.

1. There were 628 students in 4 dormitories. Each dormitory housed the same number of students. How many students were housed in each dormitory?

2. Thirty-six bright green parrots flew away while 46 parrots remained in the tree. How many parrots were in the tree before the 36 parrots flew away?

3. Two hundred twenty-five of the six hundred fish in the lake were trout. How many of the fish were not trout?

4. Twenty-one thousand, fifty swarmed in through the front door. Forty-eight thousand, nine hundred seventy-two swarmed in through the back door. Altogether, how many swarmed in through both doors?

5. This sign is written incorrectly. Show two ways to correct this sign.

   > *Pickled Peppers*
   >
   > **0.20¢ each**

6. Arrange these numbers in order from least to greatest:

$$\frac{1}{3}, \ -2, \ 1, \ -\frac{1}{2}, \ 0$$

7. Which is the best estimate of how much of this rectangle is shaded?

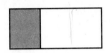

  (a) $\dfrac{1}{2}$ (b) $\dfrac{1}{3}$ (c) $\dfrac{1}{4}$ (d) $\dfrac{3}{5}$

8. Each angle of a rectangle is a right angle. Which two sides are perpendicular to side *BC*?

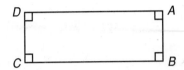

9. Simplify each fraction or mixed number.

  (a) $2\dfrac{8}{16}$ (b) $\dfrac{8}{3}$

10. For each fraction find an equivalent fraction that has a denominator of 36.

  (a) $\dfrac{2}{9}$ (b) $\dfrac{3}{4}$

11. List the factors of each number.

  (a) 10   (b) 7   (c) 1

12. The perimeter of a certain square is 2 feet. How many inches long is each side of the square?

Solve each equation:

13. $36 + a = 54$ **14.** $46 - w = 20$

15. $5x = 60$ **16.** $100 = m + 64$

17. $y - 14 = 30$ **18.** $60 = 4y$

Add, subtract, multiply, or divide, as indicated:

19. $\dfrac{9}{10} - \dfrac{3}{10}$ **20.** $\dfrac{8}{9} + \dfrac{7}{9}$ **21.** $\dfrac{5}{2} \cdot \dfrac{5}{6}$

22. $36.45 + \$15 + \$0.59$

23. $10,350 - 9764$

24.
$$41$$
$$86$$
$$49$$
$$23$$
$$51$$
$$87$$
$$93$$
$$+ 46$$

25. $\dfrac{6345}{9}$

26.
$$360$$
$$\times\ 25$$

27. $70\overline{)16{,}161}$

28. $4386 \div 21$

29. $\dfrac{3}{4} - \left(\dfrac{1}{4} + \dfrac{2}{4}\right)$

30. $\left(\dfrac{3}{4} - \dfrac{1}{4}\right) + \dfrac{2}{4}$

LESSON 22

Prime and Composite Numbers • Prime Factorization

Prime and composite numbers

We remember that the counting numbers are the numbers we use to count. They are

$$1, 2, 3, 4, 5, 6, 7, 8, 9, 10, \ldots$$

Some counting numbers are called **prime numbers** and some counting numbers are called **composite numbers.** In the following list we circle the prime numbers and leave the composite numbers uncircled. We place a box around the number 1 because it falls into neither category. The number 1 is not a composite number and is not a prime number.

$$\boxed{1}\ ②\ ③\ 4\ ⑤\ 6\ ⑦\ 8\ 9\ 10\ ⑪\ 12\ ⑬\ 14\ 15$$
$$16\ ⑰\ 18\ ⑲\ 20\ 21\ 22\ ㉓\ 24\ 25\ 26\ 27\ 28\ ㉙\ 30$$

We define a prime number as follows.

> A *prime number* is a counting number greater than 1 whose only divisors are 1 and the number itself.

A prime number has exactly two different divisors. The number 1 does not have two different divisors, so the number 1 is not a prime number.

PRIME NUMBERS	DIVISORS
2	1 and 2
3	1 and 3
5	1 and 5
7	1 and 7
11	1 and 11
13	1 and 13
17	1 and 17
19	1 and 19
23	1 and 23
29	1 and 29

All of these numbers are prime numbers because they have exactly two different divisors. Counting numbers that are greater than 1 that are not prime numbers have more than two divisors. These numbers are called composite numbers.

COMPOSITE NUMBERS	DIVISORS
4	1, 2, and 4
6	1, 2, 3, and 6
8	1, 2, 4, and 8
9	1, 3, and 9
10	1, 2, 5, and 10
12	1, 2, 3, 4, 6, and 12
14	1, 2, 7, and 14
15	1, 3, 5, and 15
16	1, 2, 4, 8, and 16
18	1, 2, 3, 6, 9, and 18

All of these numbers are composite numbers because they have more than two divisors.

Example 1   Make a list of the prime numbers that are less than 16.

Solution   First we list the counting numbers from 1 through 15.

1, 2, 3, 4, 5, 6, 7, 8, 9, 10, 11, 12, 13, 14, 15

A prime number must be greater than 1, so we cross out 1. The next number, 2, has only two divisors, so 2 is a prime number. All the even numbers greater than 2 have more than two divisors, so they are not prime. We cross these out.

$\cancel{1}$, 2, 3, $\cancel{4}$, 5, $\cancel{6}$, 7, $\cancel{8}$, 9, $\cancel{10}$, 11, $\cancel{12}$, 13, $\cancel{14}$, 15

The numbers that are left are

2, 3, 5, 7, 9, 11, 13, 15

The numbers 9 and 15 are divisible by 3 so we cross them out. Now we have

2, 3, 5, 7, $\cancel{9}$, 11, 13, $\cancel{15}$

The only divisors of **2, 3, 5, 7, 11,** and **13** are 1 and the numbers themselves. So these are the prime numbers less than 16.

Example 2   List the composite numbers between 40 and 50.

Solution   First we write the counting numbers between 40 and 50.

41, 42, 43, 44, 45, 46, 47, 48, 49

We circle the even numbers because these numbers are divisible by 2 and are composite numbers.

41, ㊷, 43, ㊹, 45, ㊻, 47, ㊽, 49

Forty-five is divisible by 5, and 49 is divisible by 7. Now we have

41, ㊷, 43, ㊹, ㊺, ㊻, 47, ㊽, ㊾

The numbers not circled are prime numbers. These numbers are divisible by only 1 and the number itself. The numbers **42, 44, 45, 46, 48,** and **49** are composite numbers.

Prime factorization   Every composite number can be formed or *composed* by multiplying two or more prime numbers. Here we show each of the first eight composite numbers written as a product of prime numbers.

$$4 = 2 \cdot 2 \qquad 6 = 2 \cdot 3 \qquad 8 = 2 \cdot 2 \cdot 2$$
$$9 = 3 \cdot 3 \qquad 10 = 2 \cdot 5 \qquad 12 = 2 \cdot 2 \cdot 3$$
$$14 = 2 \cdot 7 \qquad 15 = 3 \cdot 5$$

When we write a composite number as a product of prime numbers, we have written the **prime factorization** of the number.

Example 3   Write the prime factorization of each number.

(a) 30   (b) 81   (c) 420

Solution   We will write each number as the product of two or more prime numbers.

(a) 30 = 2 · 3 · 5   We do not use 5 · 6 or 3 · 10 because neither 6 nor 10 is prime.

(b) 81 = 3 · 3 · 3 · 3   We do not use 9 · 9 because 9 is not prime.

(c) 420 = 2 · 2 · 3 · 5 · 7   Two methods for doing this one are shown below.

Factoring composite numbers can be difficult without a method. There are two commonly used methods for factoring composite numbers. One method uses a factor tree. The other method uses repeated division. We will factor 420 using both methods.

Factor tree method   To factor a number using a factor tree, we write the number, and below the number we write any two whole numbers greater than 1 that multiply to equal the number. If these numbers are not prime, then we continue the process until there is a prime number at the end of each "branch" of the factor tree. These numbers are the prime factors of the original number.

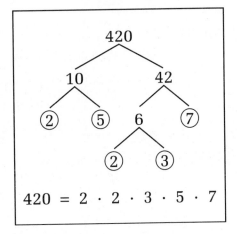

FACTOR TREE

420 = 2 · 2 · 3 · 5 · 7

Repeated division method

To factor a number using repeated division, we write the number in a division box and begin dividing by the smallest prime number that is a factor. Then we divide the answer by the smallest prime number that is a factor. We repeat this process until the answer is itself a prime number. When we use this method, we custom-arily write each division answer below the division box rather than above it. The prime factors are all the divisors and the final answer.

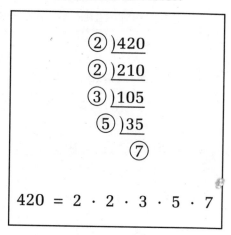

REPEATED DIVISION

②)420
②)210
③)105
⑤)35
⑦

$420 = 2 \cdot 2 \cdot 3 \cdot 5 \cdot 7$

Practice

a. List the first 10 prime numbers.

b. If a whole number greater than 1 is not prime, then what kind of number is it?

Write the prime factorization of each number.

c. 27

d. 360

Problem set 22

In Problems 1–4, identify the type of problem. Then find the answer.

1. Two thirds of the students wore green on St. Patrick's Day. What fraction of the students did not wear green on St. Patrick's Day?

2. There were 343 quills carefully placed into 7 com-partments. If each compartment held the same number of quills, how many quills were in each compartment?

3. Twenty-one million is how much less than two billion? Use words to write the answer.

4. Last year the price was $14,289. This year the price has been increased $824. What is the price this year?

5. Simplify each fraction or mixed number.

   (a) $3\frac{12}{21}$

   (b) $\frac{12}{5}$

6. List the prime numbers between 50 and 60.

7. Write the prime factorization of each number.

   (a) 50   (b) 60   (c) 300

8. Which point could represent 1610 on this number line?

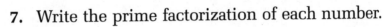

9. Complete each equivalent fraction.

   (a) $\frac{2}{3} = \frac{?}{15}$ (b) $\frac{3}{5} = \frac{?}{15}$ (c) $\frac{?}{3} = \frac{8}{12}$

10. Draw and shade circles to show that $\frac{8}{3}$ equals $2\frac{2}{3}$.

11. Draw a sketch of a regular quadrilateral.

12. This rectangle is twice as long as it is wide. What is its perimeter?

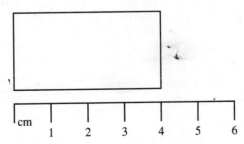

13. Find the width in millimeters of the rectangle in Problem 12.

Solve:

14. $p + 58 = 85$ **15.** $x - 46 = 20$ **16.**

17. $8y = 96$ **18.** $45 = 22 + w$

19. $51 - m = 17$ **20.** $51 = 3c$

Add, subtract, multiply, or divide, as indicated:

21. $\dfrac{2}{3} + \dfrac{2}{3} + \dfrac{2}{3}$ **22.** $\dfrac{2}{3} \cdot \dfrac{2}{3} \cdot \dfrac{2}{3}$

23. $36 + 47 + 52 + 116 + 45 + 8$

24. $\$370.47 - \$296.65$

25. $75\overline{)\$36.00}$ **26.** $960 \div 40$

27. $25(30)(40)$ **28.** $6 \cdot 5 \cdot 4 \cdot 3 \cdot 2 \cdot 1 \cdot 0$

29. $\dfrac{3}{3} - \left(\dfrac{1}{3} \cdot \dfrac{3}{1} \right)$ **30.** $\left(\dfrac{3}{3} - \dfrac{1}{3} \right) \cdot \dfrac{3}{1}$

For problem 16:

$$
\begin{array}{r}
5 \\
7 \\
8 \\
4 \\
6 \\
3 \\
7 \\
4 \\
9 \\
8 \\
N \\
+ \ 6 \\
\hline
71
\end{array}
$$

LESSON 23

Simplifying Fractions and Mixed Numbers

We can simplify some proper fractions by reducing the fractions to lowest terms.

$$\frac{9}{12} = \frac{3}{4}$$

We can simplify improper fractions by writing them as mixed numbers.

$$\frac{9}{4} = 2\frac{1}{4}$$

If an improper fraction can be reduced to lowest terms, we use both ways to simplify the fraction. It does not matter whether we reduce first or convert first. Here we simplify $\frac{12}{9}$.

<table>
<tr><td align="center">**REDUCE FIRST**</td><td align="center">**CONVERT FIRST**</td></tr>
<tr><td align="center">Reduce: $\dfrac{12}{9} = \dfrac{4}{3}$</td><td align="center">Convert: $\dfrac{12}{9} = 1\dfrac{3}{9}$</td></tr>
<tr><td align="center">Convert: $\dfrac{4}{3} = 1\dfrac{1}{3}$</td><td align="center">Reduce: $1\dfrac{3}{9} = 1\dfrac{1}{3}$</td></tr>
</table>

Example 1 Simplify: $\dfrac{24}{16}$

Solution To simplify this improper fraction, we reduce and convert. It doesn't matter which we do first. We will show both ways.

$$\text{Reduce: } \frac{24}{16} = \frac{3}{2} \qquad \text{Convert: } \frac{24}{16} = 1\frac{8}{16}$$

$$\text{Convert: } \frac{3}{2} = 1\frac{1}{2} \qquad \text{Reduce: } 1\frac{8}{16} = 1\frac{1}{2}$$

Either way we find that $\dfrac{24}{16}$ simplifies to $1\dfrac{1}{2}$.

Example 2 Simplify: $\dfrac{7}{8} + \dfrac{7}{8}$

Solution First we add. Then we simplify the result by reducing and converting to a mixed number.

$$\frac{7}{8} + \frac{7}{8} = \frac{14}{8} \qquad \text{added}$$

$$= \frac{7}{4} \qquad \text{reduced}$$

$$= 1\frac{3}{4} \qquad \text{converted}$$

Example 3 Simplify: $2\dfrac{8}{6}$

Solution The mixed number $2\frac{8}{6}$ means $2 + \frac{8}{6}$. The fraction $\frac{8}{6}$ simplifies to $1\frac{1}{3}$.

$$\frac{8}{6} = 1\frac{2}{6} = 1\frac{1}{3}$$

Thus $2\frac{8}{6}$ equals $2 + 1\frac{1}{3}$, which equals $3\frac{1}{3}$.

Practice Simplify:

    a. $\dfrac{20}{8}$ **b.** $3\dfrac{9}{6}$ **c.** $\dfrac{8}{3} \times \dfrac{5}{4}$

Problem set 23 In Problems 1–4, identify the type of problem. Then find the answer.

1. Eight hundred thirty of the one thousand, twenty villagers fled before the tsunami struck. How many villagers did not flee before the tsunami struck?

2. Tony discovered that 87 of the 402 watermelons in the field were ripe. How many of the watermelons in the field were not ripe?

3. If 112 students were equally grouped into 4 different classrooms, how many students would there be in each classroom?

4. Eight million, five hundred thousand is how much less than seventeen million, five hundred sixty thousand? Use words to write the answer.

5. Write the prime factorization of each number.
    (a) 16   (b) 525

6. Simplify each fraction or mixed number.
    (a) $\dfrac{9}{6}$ (b) $1\dfrac{16}{12}$

7. Replace each circle with the proper comparison symbol.
    (a) $5 - 3 \bigcirc 3 - 5$ (b) $\dfrac{8}{6} \bigcirc 1\dfrac{1}{3}$

8. (a) What fraction of the rectangle is shaded?

    (b) What fraction of the rectangle is not shaded?

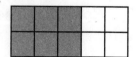

9. For each fraction, write an equivalent fraction that has a denominator of 24.

   (a) $\dfrac{7}{12}$

   (b) $\dfrac{3}{8}$

10. The perimeter of the rectangle is 20 cm. What is the width of the rectangle?

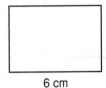

6 cm

11. Draw $\overline{AB}$ to be 1 inch long. Then draw $\overline{BC}$, also 1 inch long, perpendicular to $\overline{AB}$. Complete triangle ABC by drawing $\overline{AC}$.

12. List the prime numbers between 20 and 30.

13. Draw and shade circles to show that $\dfrac{9}{4}$ equals $2\dfrac{1}{4}$.

Solve:

14. $x + 47 = 129$

15. $y - 54 = 49$

16.
$$
\begin{array}{r}
4 \\
8 \\
3 \\
9 \\
5 \\
7 \\
N \\
6 \\
4 \\
3 \\
7 \\
5 \\
+\ 8 \\
\hline
83
\end{array}
$$

17. $108 = 18 + a$

18. $32 = 50 - d$

19. $84 = 12q$

Add, subtract, multiply, or divide, as indicated:

20. $\dfrac{5}{8} + \dfrac{5}{8}$

21. $\dfrac{5}{3} \cdot \dfrac{4}{5}$

22. $\dfrac{9}{12} - \dfrac{5}{12}$

23. $\$5.47 + \$16.79 + 85¢ + \$28 + \$0.08$

24.
$$
\begin{array}{r}
30{,}175 \\
-\ 9{,}757 \\
\hline
\end{array}
$$

25.
$$
\begin{array}{r}
678 \\
\times\ 94 \\
\hline
\end{array}
$$

26. $\dfrac{1410}{15}$

27. 400(5000)

28. 60,000 ÷ 100

29. $\dfrac{5}{6} + \left(\dfrac{1}{3} \cdot \dfrac{1}{2} \right)$ **30.** $\dfrac{5}{6} - \left(\dfrac{2}{3} \cdot \dfrac{1}{2} \right)$

LESSON 24

Writing Mixed Numbers and Whole Numbers as Fractions

Mixed numbers as fractions

We have used pictures to illustrate that mixed numbers can be rewritten as improper fractions. The illustration below shows $3\frac{1}{4}$ equals $\frac{13}{4}$.

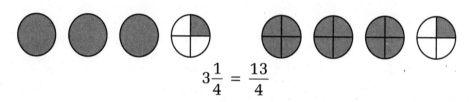

$$3\frac{1}{4} = \frac{13}{4}$$

We can use a "circular" process to convert a mixed number to an improper fraction. To convert $3\frac{1}{4}$ to a mixed number, we multiply 4 by 3 and then add 1, as we show here.

$$3\frac{1}{4} \quad \longrightarrow \quad 3 \underset{\times\,4}{\overset{+\,1}{\frown}} = \frac{4 \times 3 + 1}{4} = \frac{13}{4}$$

Example 1   Write each mixed number as an improper fraction.

     (a) $3\dfrac{1}{3}$ (b) $2\dfrac{3}{4}$ (c) $12\dfrac{1}{2}$

Solution   In each case we multiply the whole number by the denominator and then add the numerator to find the numerator of the improper fraction.

     (a) $3\dfrac{1}{3} = \dfrac{\mathbf{10}}{\mathbf{3}}$ (b) $2\dfrac{3}{4} = \dfrac{\mathbf{11}}{\mathbf{4}}$ (c) $12\dfrac{1}{2} = \dfrac{\mathbf{25}}{\mathbf{2}}$

Whole numbers as fractions The number 3 means 3 "wholes." A fraction that means 3 wholes is $\frac{3}{1}$. To write a whole number as a fraction, we simply write the whole number over a denominator of 1.

Example 2 Write each whole number as a fraction.

(a) 4   (b) 100

Solution We write each whole number over a denominator of 1.

(a) $4 = \frac{4}{1}$ (b) $100 = \frac{100}{1}$

Practice Write each number as a fraction.

a. $3\frac{5}{6}$ b. 7   c. $4\frac{3}{4}$ d. 10

Problem set 24 In Problems 1–4, identify the type of problem. Then find the answer.

1. If 600 fish were separated into 8 equal schools, how many fish would be in each school?

2. How many years were there from 789 to 1215?

3. If each ticket cost $2.75, what was the cost of 12 tickets?

4. Two fifths of the students in the class were boys. What fraction of the students in the class was made up of girls?

5. Write the prime factorization of 720.

6. What number is halfway between 20 and 50?

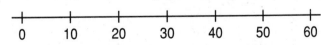

7. Simplify:

(a) $3\frac{10}{8}$ (b) $\frac{24}{10}$

8. (a) What fraction of the group is shaded?

    (b) What fraction of the group is not shaded?

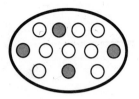

9. Complete each equivalent fraction.

    (a) $\dfrac{1}{2} = \dfrac{?}{24}$ (b) $\dfrac{5}{8} = \dfrac{?}{24}$ (c) $\dfrac{?}{4} = \dfrac{18}{24}$

10. Sketch a regular pentagon.

11. Refer to triangle ABC to answer these questions.

    (a) What kind of angle is $\angle ACB$?

    (b) What kind of angle is $\angle ABC$?

    (c) Which side of the triangle appears to be the longest side?

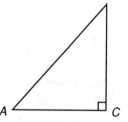

12. Write each number as an improper fraction.

    (a) $2\dfrac{1}{2}$ (b) 5   (c) $8\dfrac{2}{3}$

13. Draw and shade circles to show that $2\dfrac{5}{6}$ equals $\dfrac{17}{6}$.

Solve:

14. $m + 412 = 1000$ 15. $w - 59 = 63$ 16.

17. $300 = 30 + a$ 18. $59 = 63 - x$

19. $300 = 15t$

Add, subtract, multiply, or divide, as indicated:

20. $\dfrac{7}{9} + \dfrac{7}{9} + \dfrac{7}{9}$ 21. $\dfrac{5}{3} \cdot \dfrac{5}{4}$

16.
$$
\begin{array}{r}
8 \\
7 \\
5 \\
2 \\
6 \\
4 \\
7 \\
4 \\
N \\
+\ 5 \\
\hline
81
\end{array}
$$

22. $\dfrac{8}{10} - \dfrac{3}{10}$

23. $\dfrac{7}{8} - \left(\dfrac{3}{4} \cdot \dfrac{1}{2} \right)$

24. $476 + 380 + 767 + 1289 + 79 + 8$

25. $\begin{array}{r} \$100.00 \\ - 97.49 \\ \hline \end{array}$

26. $\begin{array}{r} 806 \\ \times\, 790 \\ \hline \end{array}$

27. $\dfrac{2718}{18}$

28. $5 \cdot 5 \cdot 5 \cdot 5$

29. $100{,}000 \div 100$

30. $\dfrac{7}{8} + \left(\dfrac{1}{4} \cdot \dfrac{1}{2} \right)$

LESSON 25

Fraction-of-a-Group Problems

One way to describe part of a group is by using a fraction. Consider this statement:

Two thirds of the students wore green on St. Patrick's Day.

We can draw a diagram of this statement. We will use a rectangle to represent all the students in the class. Next we will divide the rectangle into 3 equal parts. Then we describe the parts.

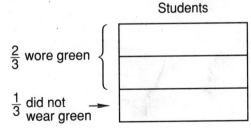

If we know how many students are in the class, we can figure out how many students are in each part.

Two thirds of the 27 students in the class wore green on St. Patrick's Day.

There are 27 students in all. If we divide 27 students into 3 equal parts, there will be 9 students in each part. We write these numbers on our diagram.

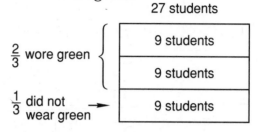

27 students

| 9 students |
| 9 students |
| 9 students |

$\frac{2}{3}$ wore green

$\frac{1}{3}$ did not wear green

Since $\frac{2}{3}$ of the students wore green, we add 2 of the parts and find that 18 students wore green. Since $\frac{1}{3}$ of the students did not wear green, we find that 9 students did not wear green.

Example 1   Draw a diagram of this statement. Then answer the questions that follow.

Two fifths of the 30 students in the class are boys.

(a) How many boys are in the class?

(b) How many girls are in the class?

Solution   We draw a rectangle to represent all 30 students. Since the statement uses fifths to describe a part of the class, we divide the 30 students into 5 equal parts with 6 in each part. Then we describe the parts.

30 students

$\frac{2}{5}$ are boys

| 6 students |
| 6 students |

$\frac{3}{5}$ are girls

| 6 students |
| 6 students |
| 6 students |

Now we answer the questions.

(a) Two of the five parts are boys. Since there are 6 students in each part, **12 students are boys.**

(b) Since 2 of the 5 parts are boys, 3 of the 5 parts must be girls. Thus, **18 students are girls**.

Another way to find the answer to (b) after finding the

answer to (a) is to subtract. Since 12 of the 30 students are boys, the rest of the students (30 − 12 = 18) are girls.

Practice   Draw a diagram of this statement. Then answer the questions that follow.

Three fourths of the 60 pumpkins were ripe.

a.   How many pumpkins were ripe?

b.   How many pumpkins were not ripe?

Problem set   In Problems 1–4, identify the type of problem. Then find the
25   answer.

1.   In Room 7 there are 28 students. In Room 9 there are 30 students. In Room 11 there are 23 students. Altogether, how many students are in all three rooms?

2.   If all the students in Problem 1 were equally divided among 3 rooms, how many students would be in each room?

3.   One hundred twenty-six thousand scurried through the colony before the edentate attacked. Afterward only seventy-nine thousand remained. How many were lost when the edentate attacked?

4.   Two thousand, seven hundred is how much less than ten thousand, three hundred thirteen?

5.   Draw a diagram of this statement. Then answer the questions that follow.

Five ninths of the 36 spectators were happy with the outcome.

(a)   How many spectators were happy with the outcome?

(b)   How many spectators were not happy with the outcome?

6. Use digits and symbols to write "Two and three fifths is less than two and three fourths."

7. (a) What fraction of the rectangle is shaded?

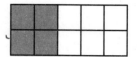

(b) What fraction of the rectangle is not shaded?

8. The top of the chalkboard in the classroom is about how many meters above the floor?

9. Simplify:

(a) $\dfrac{110}{6}$

(b) $15\dfrac{6}{4}$

10. Write each number as a fraction.

(a) $6\dfrac{2}{3}$

(b) 12

11. Draw and shade circles to show that $2\dfrac{3}{8}$ equals $\dfrac{19}{8}$.

12. (a) Draw rectangle $ABCD$ so that AB is 2 cm and BC is 1 cm.

(b) What is the perimeter of rectangle $ABCD$?

13. Write the prime factorization of each number.

(a) 32

(b) 840

14. For each fraction, write an equivalent fraction that has a denominator of 60.

(a) $\dfrac{5}{6}$

(b) $\dfrac{3}{5}$

(c) $\dfrac{7}{12}$

15. Arrange these numbers in order from least to greatest:

$$0, -\dfrac{2}{3}, 1, \dfrac{3}{2}, -2$$

Solve:

16. $475 + a = 754$ **17.** $900 - c = 90$ **18.**

19. $131 = x + 13$ **20.** $131 = y - 13$

21. $130 = 10n$

$$
\begin{array}{r}
4 \\
7 \\
8 \\
21 \\
4 \\
6 \\
7 \\
3 \\
8 \\
N \\
+\ 6 \\
\hline
81
\end{array}
$$

Add, subtract, multiply, or divide, as indicated:

22. $\dfrac{5}{6} + \dfrac{5}{6} + \dfrac{5}{6}$ **23.** $\dfrac{15}{2} \cdot \dfrac{10}{3}$

24. $\dfrac{11}{12} - \dfrac{3}{12}$

25. $\$47.58 + \$63.75 + \$114 + \$9.47 + 8¢$

26. $\begin{array}{r} 37{,}104 \\ -\ 9{,}865 \\ \hline \end{array}$ **27.** $\begin{array}{r} \$12.50 \\ \times\quad 48 \\ \hline \end{array}$ **28.** $\dfrac{\$27.00}{15}$

29. $9 \cdot 9 \cdot 9$ **30.** $1{,}000{,}000 \div 100$

LESSON 26

Adding and Subtracting Mixed Numbers

Since Lesson 8 we have practiced adding and subtracting fractions. In this lesson we will practice adding and subtracting mixed numbers.

Example 1   There are $2\frac{3}{5}$ pies on the top shelf and $1\frac{4}{5}$ pies on the bottom shelf. Altogether, how many pies are there on both shelves?

$2\frac{3}{5}$

$1\frac{4}{5}$

Solution   There are some pies on the top shelf and some more pies on the bottom shelf. To find how many pies there are in all, we add $2\frac{3}{5}$ pies and $1\frac{4}{5}$ pies. Before doing the arithmetic we will study the picture.

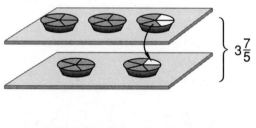

We see that there are 3 whole pies and 7 pieces, which is $3\frac{7}{5}$ pies. However, we can put 5 of the pieces together to fill another pie pan, making 4 whole pies and 2 pieces. Thus, we see that there are **$4\frac{2}{5}$ pies** on the shelves.

When adding mixed numbers, we add the fractions and add the whole numbers. Then we simplify the result. We will arrange the numbers vertically to add them.

$$\begin{array}{r} 2\dfrac{3}{5} \\ +\,1\dfrac{4}{5} \\ \hline 3\dfrac{7}{5} \end{array} \quad \begin{array}{l} \text{pies} \\[1.2em] \text{pies} \\[1.2em] \text{pies} \end{array}$$

We notice that $\frac{7}{5}$ is an improper fraction. We convert $\frac{7}{5}$ to $1\frac{2}{5}$ and add this to 3.

$$3\frac{7}{5} = 3 + 1\frac{2}{5}$$

$$= 4\frac{2}{5}$$

There are **$4\frac{2}{5}$ pies** on the shelves.

Example 2   There are $3\frac{1}{5}$ pies on the shelf. If the baker takes away $1\frac{2}{5}$

pies, how many pies will be on the shelf?

Solution   To answer this question, we subtract $1\frac{2}{5}$ from $3\frac{1}{5}$. However, before we subtract we will look at the picture again.

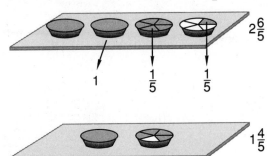

In order for the baker to remove $1\frac{2}{5}$ pies, it will be necessary to slice one of the whole pies into fifths. By cutting one pie into fifths, there are 2 whole pies and 6 fifths. Then the baker can remove $1\frac{2}{5}$ pies, which leaves $1\frac{4}{5}$ pies still on the shelf.

To perform the subtraction, we first rename $3\frac{1}{5}$ as $2\frac{6}{5}$, as we show. Then we can subtract.

$$
\begin{array}{r}
3\dfrac{1}{5} \\
-\,1\dfrac{2}{5} \\
\hline
\end{array}
\quad\longrightarrow\quad
\begin{array}{r}
2 + 1\dfrac{1}{5} \\
-\,1\dfrac{2}{5} \\
\hline
\end{array}
\quad\longrightarrow\quad
\begin{array}{r}
2\dfrac{6}{5} \\
-\,1\dfrac{2}{5} \\
\hline
1\dfrac{4}{5}
\end{array}
$$

There are **$1\frac{4}{5}$ pies** left on the shelf.

Example 3   Simplify: $3\dfrac{1}{8} + 1\dfrac{3}{8}$

Solution   We add the fractions. Then we add the whole numbers. Then we simplify.

$$
\begin{array}{r}
3\dfrac{1}{8} \\
+\,1\dfrac{3}{8} \\
\hline
4\dfrac{4}{8}
\end{array}
\;=\; 4\dfrac{1}{2}
$$

Example 4   Simplify: $6 - 1\frac{3}{4}$

Solution   First we arrange the problem vertically.

$$
\begin{array}{r}
6 \\
- 1\frac{3}{4} \\
\hline
\end{array}
$$

We need a fraction on top. Thus we rewrite 6 as $5\frac{4}{4}$.

$$
\begin{array}{r}
5\frac{4}{4} \\
- 1\frac{3}{4} \\
\hline
4\frac{1}{4}
\end{array}
$$

Practice   Add or subtract as indicated. Then simplify the answer if possible.

a. $3\frac{1}{4} + 1\frac{3}{4}$ **b.** $5\frac{1}{6} + 2\frac{1}{6}$ **c.** $3\frac{2}{3} + 6\frac{2}{3}$

d. $8\frac{5}{6} - 2\frac{1}{6}$ **e.** $7 - 2\frac{1}{3}$ **f.** $6\frac{1}{5} - 1\frac{4}{5}$

Problem set 26   In Problems 1–4, identify the type of problem. Then find the answer.

1. Willie shot eighteen rolls of film for the school annual. If there were thirty-six exposures in each roll, how many exposures were there in all?

2. Fifty million is how much greater than two hundred fifty thousand? Use words to write your answer.

3. There were 259 people who attended on opening night. On the second night 269 attended, and 307 attended on the third night. How many people attended on the first three nights?

4. The 16-pound turkey cost $14.24. What was the price for each pound?

5. Draw a diagram of this statement. Then answer the questions that follow.

> Three eighths of the 56 restaurants in town were closed on Monday.

  (a) How many of the restaurants in town were closed on Monday?

  (b) How many of the restaurants in town were open on Monday?

6. What one number can be put in both boxes to make both sides of the equation equal?

$$\square \times \square = 50 + 50$$

7. After contact was made, the spheroid sailed four thousand, one hundred forty inches. How many yards did the spheroid sail after contact was made?

8. What is the name for numbers that are less than zero?

9. What number is halfway between 2000 and 3000?

10. Replace each circle with the proper comparison symbol.

  (a) $\dfrac{2}{3} \cdot \dfrac{3}{2} \bigcirc \dfrac{5}{5}$
        (b) $\dfrac{12}{36} \bigcirc \dfrac{12}{24}$

11. Simplify each fraction or mixed number.

  (a) $3\dfrac{18}{10}$
        (b) $\dfrac{220}{12}$

12. Write each number as an improper fraction.

  (a) $2\dfrac{1}{4}$
        (b) $3\dfrac{3}{5}$

13. Complete each equivalent fraction.

(a) $\dfrac{3}{4} = \dfrac{?}{40}$ (b) $\dfrac{2}{5} = \dfrac{?}{40}$ (c) $\dfrac{?}{8} = \dfrac{15}{40}$

14. Write the prime factorization of each number.

(a) 42

(b) 600

15. Refer to this figure to answer the following questions.

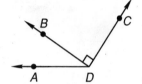

(a) What type of angle is $\angle ADB$?

(b) What type of angle is $\angle BDC$?

(c) What type of angle is $\angle ADC$?

(d) Which ray is perpendicular to $\overrightarrow{DB}$?

Solve:

16. $7w = 105$ **17.** $m - 34 = 25$

19. $36 + x = 115$

Add, subtract, multiply, or divide, as indicated:

20. $3\dfrac{1}{3} + 1\dfrac{2}{3}$ **21.** $4\dfrac{1}{8} + 1\dfrac{5}{8}$

22. $6\dfrac{3}{4} - 1\dfrac{1}{4}$ **23.** $5 - 3\dfrac{1}{3}$

24. $7\dfrac{1}{3} - 3\dfrac{2}{3}$

25. $3524 + 4617 + 2819 + 568 + 70$

18.
```
   4
   7
   8
   2
   6
   4
   9
   5
   8
   7
   4
   1
   N
 + 3
 ___
  77
```

26. $\begin{array}{r} \$50.00 \\ -\ 41.74 \\ \hline \end{array}$ **27.** $\begin{array}{r} \$0.89 \\ \times\ \ \ 76 \\ \hline \end{array}$ **28.** $\dfrac{26{,}880}{42}$

29. $\dfrac{7}{12} + \left(\dfrac{1}{4} \cdot \dfrac{1}{3} \right)$

30. $\dfrac{7}{8} - \left(\dfrac{3}{4} \cdot \dfrac{1}{2} \right)$

LESSON 27

Reciprocals

If we exchange the position of the terms in a fraction, we will write the **reciprocal** of the fraction.

The reciprocal of $\dfrac{4}{3}$ is $\dfrac{3}{4}$.

The reciprocal of $\dfrac{3}{4}$ is $\dfrac{4}{3}$.

The reciprocal of $\dfrac{1}{4}$ is $\dfrac{4}{1}$, which is 4.

The reciprocal of 4 is $\dfrac{1}{4}$.

To find the reciprocal of a mixed number, we first write the mixed number as an improper fraction.

$$3\dfrac{1}{5} = \dfrac{16}{5} \qquad \text{improper fraction}$$

Now we can write the reciprocal of $3\frac{1}{5}$, which is the reciprocal of $\frac{16}{5}$.

The reciprocal of $\dfrac{16}{5}$ is $\dfrac{5}{16}$.

We note a very important property of reciprocals. **The product of any fraction and its reciprocal is the number 1.**

$$\dfrac{16}{5} \cdot \dfrac{5}{16} = \dfrac{80}{80} = 1$$

$$\dfrac{3}{4} \cdot \dfrac{4}{3} = \dfrac{12}{12} = 1$$

$$\dfrac{1}{4} \cdot 4 = \dfrac{4}{4} = 1$$

Example 1   Find the reciprocal of each number.

(a) $\dfrac{3}{5}$ (b) 3   (c) $2\dfrac{1}{3}$

Solution   (a) The reciprocal of $\dfrac{3}{5}$ is $\dfrac{5}{3}$.

(b) The reciprocal of 3 is $\dfrac{1}{3}$.

(c) First we write the mixed number as an improper fraction.

$$2\dfrac{1}{3} = \dfrac{7}{3}$$

The reciprocal of $2\dfrac{1}{3}$ is the reciprocal of $\dfrac{7}{3}$, which is $\dfrac{3}{7}$.

Example 2   Find the missing number: $\dfrac{3}{4} \cdot N = 1$

Solution   A fraction times its reciprocal equals 1. Thus, the missing number is the reciprocal of $\frac{3}{4}$, which is $\frac{4}{3}$.

$$\dfrac{3}{4} \cdot \dfrac{4}{3} = \dfrac{12}{12} = 1$$

Practice   Write the reciprocal of each number.

a. $\dfrac{3}{5}$ b. $\dfrac{8}{7}$ c. 5   d. $2\dfrac{1}{3}$ e. $3\dfrac{3}{4}$

Find each missing number.

f. $\dfrac{5}{8} \cdot A = 1$ g. $B \cdot \dfrac{1}{6} = 1$ h. $3\dfrac{1}{3} \cdot C = 1$

Problem set 27   In Problems 1–4, identify the type of problem. Then find the answer.

1. All the lines for concert tickets were long. There were 112 people in the first line, 117 in the second line, 98 in the third line, and 105 in the fourth line. How many people were in all four lines?

2. If all the people in Problem 1 arranged themselves in 4 equal lines, how many would be in each line?

3. The Viking Bjarni Herjolfsson is believed to have seen North America in about 986. This was how many years before Columbus sighted North America in 1492?

4. Nathan left the store with $5.42 and three tapes that cost a total of $21.33. How much money did Nathan have when he entered the store?

5. Draw a diagram of this statement. Then answer the questions that follow.

> Five eighths of the 160 seats were empty during the matinee.

   (a) How many seats were empty during the matinee?

   (b) How many seats were occupied during the matinee?

6. How many thousands equal one million?

7. A *score* is 20. Three score and 10 years is how many years?

8. (a) What fraction of the square is shaded?

   (b) What fraction of the square is not shaded?

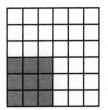

9. Simplify each fraction or mixed number.

   (a) $\dfrac{25}{6}$ (b) $4\dfrac{8}{6}$ (c) $\dfrac{75}{12}$

10. Write each number as a fraction.

   (a) $3\dfrac{3}{4}$ (b) $1\dfrac{2}{3}$ (c) 6

11. For each fraction, write an equivalent fraction that has a denominator of 45.

(a) $\dfrac{4}{5}$ (b) $\dfrac{2}{9}$

12. Write the prime factorization of 540.

13. Write the reciprocal of each number.

(a) $\dfrac{5}{8}$ (b) 4   (c) $2\dfrac{1}{3}$

14. Find each missing number.

(a) $\dfrac{2}{5} \cdot n = 1$ (b) $k \cdot 1\dfrac{1}{2} = 1$

15. This rectangle is twice as long as it is wide. What is the perimeter of the rectangle in millimeters?

15 mm

Solve:

16. $370 + d = 530$ **17.** $500 - x = 125$ **18.**
$$
\begin{array}{r}
52 \\
12 \\
4 \\
5 \\
14 \\
28 \\
7 \\
+\ N \\
\hline
126
\end{array}
$$

19. $8m = 144$

Add, subtract, multiply, or divide, as indicated:

20. $3\dfrac{3}{8} + 3\dfrac{3}{8}$ **21.** $5\dfrac{5}{6} + 1\dfrac{1}{6}$

22. $6\dfrac{7}{8} - 1\dfrac{1}{8}$ **23.** $5 - 1\dfrac{5}{6}$

24. $6\dfrac{2}{5} - 4\dfrac{4}{5}$

25. $\$96.74 + \$59.87 + \$115 + \$4.68$

26. $\dfrac{15{,}470}{14}$ **27.**
$$
\begin{array}{r}
43{,}050 \\
-\ 8{,}313 \\
\hline
\end{array}
$$
      28.
$$
\begin{array}{r}
\$6.59 \\
\times\ \ \ 78 \\
\hline
\end{array}
$$

29. $\dfrac{3}{5} + \left(\dfrac{2}{5} \cdot \dfrac{2}{1} \right)$ **30.** $\left(\dfrac{3}{5} + \dfrac{2}{5} \right) \cdot \dfrac{2}{1}$

LESSON 28

Reducing Fractions, Part 2

Reducing fractions

We have been practicing reducing fractions by dividing the numerator and the denominator by a common factor. In this lesson we will practice another method of reducing fractions. This method uses prime factorization. If we write the prime factorization of the numerator and of the denominator, we can see how to reduce a fraction easily.

Example 1 Use prime factorization to reduce $\dfrac{420}{1050}$.

Solution We rewrite the numerator and the denominator as products of prime numbers.

$$\frac{420}{1050} = \frac{2 \cdot 2 \cdot 3 \cdot 5 \cdot 7}{2 \cdot 3 \cdot 5 \cdot 5 \cdot 7}$$

Next we look for pairs of factors that equal 1. A fraction equals 1 if the numerator and denominator are equal. In this fraction there are four pairs of factors that equal 1. They are $\dfrac{2}{2}$, $\dfrac{3}{3}$, $\dfrac{5}{5}$, and $\dfrac{7}{7}$. Below we have indicated each of these pairs.

$$\frac{2 \cdot 2 \cdot 3 \cdot 5 \cdot 7}{2 \cdot 3 \cdot 5 \cdot 5 \cdot 7}$$

Thus the fraction equals $1 \cdot 1 \cdot 1 \cdot 1 \cdot \dfrac{2}{5}$, which is $\mathbf{\dfrac{2}{5}}$.

Reducing before multiplying

The terms of fractions may be reduced before they are multiplied. Reducing before multiplying is also known as **canceling**. Consider this multiplication:

$$\frac{3}{8} \cdot \frac{2}{3} = \frac{6}{24} \qquad \frac{6}{24} \text{ reduces to } \frac{1}{4}$$

We see that neither $\frac{3}{8}$ nor $\frac{2}{3}$ can be reduced. The product, $\frac{6}{24}$, can be reduced. We can avoid reducing after we multiply by reducing before we multiply. To reduce, any numerator may be paired with any denominator. Below we have paired the 3 with 3 and the 2 with 8.

Then we reduce these pairs: $\frac{3}{3}$ reduces to $\frac{1}{1}$, and $\frac{2}{8}$ reduces to $\frac{1}{4}$, as we show below. Then we multiply the reduced terms.

$$\overset{1}{\cancel{3}}_{} \cdot \underset{1}{\overset{1}{\cancel{2}}} = \frac{1}{4}$$
$$\underset{4}{\cancel{8}} \cdot \cancel{3}$$

Example 2   Simplify: $\dfrac{9}{16} \cdot \dfrac{2}{3}$

Solution   Before multiplying, we pair 9 with 3 and 2 with 16 and reduce these pairs. Then we multiply the reduced terms.

$$\overset{3}{\cancel{9}} \cdot \overset{1}{\cancel{2}} = \frac{3}{8}$$
$$\underset{8}{\cancel{16}} \cdot \underset{1}{\cancel{3}}$$

Example 3   Simplify: $\dfrac{8}{9} \cdot \dfrac{3}{10} \cdot \dfrac{5}{4}$

Solution   We mentally pair 8 with 4, 3 with 9, and 5 with 10 and reduce.

$$\overset{2}{\cancel{8}} \cdot \overset{1}{\cancel{3}} \cdot \overset{1}{\cancel{5}}$$
$$\underset{3}{\cancel{9}} \cdot \underset{2}{\cancel{10}} \cdot \underset{1}{\cancel{4}}$$

We can still reduce by pairing 2 with 2. Then we multiply.

$$\overset{\overset{1}{\cancel{2}}}{\cancel{8}} \cdot \overset{1}{\cancel{3}} \cdot \overset{1}{\cancel{5}} = \frac{1}{3}$$
$$\underset{3}{\cancel{9}} \cdot \underset{\underset{1}{\cancel{2}}}{\cancel{10}} \cdot \underset{1}{\cancel{4}}$$

Practice Use prime factorization to reduce each fraction.

a. $\dfrac{48}{144}$

b. $\dfrac{90}{324}$

Reduce before multiplying.

c. $\dfrac{5}{8} \cdot \dfrac{3}{10}$

d. $\dfrac{8}{15} \cdot \dfrac{5}{12} \cdot \dfrac{9}{10}$

e. $\dfrac{8}{3} \cdot \dfrac{6}{7} \cdot \dfrac{5}{16}$

Problem set 28 In Problems 1 and 2, identify the type of problem. Then find the answer.

1. From Hartford to Los Angeles is two thousand, eight hundred ninety-five miles. From Hartford to Portland is three thousand, twenty-six miles. The distance from Hartford to Portland is how much greater than the distance from Hartford to Los Angeles?

2. Hal ordered 15 boxes of microprocessors. If each box contained two dozen microprocessors, how many microprocessors did Hal order?

Use this information to answer questions 3 and 4.

Jason had completed reading 36 pages of a 310-page book. He read 47 more pages that morning and 139 more pages that afternoon.

3. Altogether, how many pages had Jason read?

4. How many more pages did Jason have to read to finish the book?

5. Nancy descended the 30 steps that led to the floor of the cellar. One third of the way down she paused. How many more steps were there to the cellar floor?

6. What number is halfway between 100 and 400?

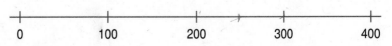

7. Write the reciprocal of each number.

   (a) 3

   (b) $2\frac{2}{3}$

8. Simplify:

   (a) $11\frac{15}{12}$

   (b) $\frac{540}{600}$

9. Write each number as a fraction.

   (a) $9\frac{1}{2}$

   (b) 8

10. Complete each equivalent fraction.

    (a) $\frac{3}{5} = \frac{?}{30}$

    (b) $\frac{7}{10} = \frac{?}{30}$

11. Write 810 as a product of prime factors.

12. Draw two parallel lines that are intersected by a third line that is perpendicular to the parallel lines.

13. The perimeter of a square is one yard. How many inches long is each side of the square?

14. (a) Estimate the length of this segment in centimeters.

    (b) Use a centimeter scale to find the length of this segment to the nearest centimeter.

Solve:

15. $514 = x + 50$

16. $w - 75 = 57$

17. $9t = 144$

Add, subtract, multiply, or divide, as indicated:

18. $5\frac{5}{8} + 3\frac{3}{8}$

19. $7\frac{2}{3} + 6\frac{2}{3}$

20. $3\frac{3}{4} + 2\frac{3}{4}$

21. $4\frac{7}{12} - 1\frac{1}{12}$

22. $6\frac{2}{3} - 4$

23. $5\frac{1}{8} - 1\frac{7}{8}$

24. $\frac{5}{6} \cdot \frac{2}{3}$

25. $\frac{3}{4} \cdot \frac{1}{2} \cdot \frac{8}{9}$

26. $13{,}513 \div 15$

27. $73{,}604 - 64{,}416$

28. $(13 \cdot 110) + 7590$

29. $13(110 + 7590)$

30. Solve mentally:

$$4 + 7 + 6 + 3 + 5 + 8 + 14 + 7 + 16 + N = 78$$

LESSON 29

Dividing Fractions

We remember that when we multiply fractions we multiply the numerators to form the new numerator. We multiply the denominators to form the new denominator.

$$\frac{4}{3} \cdot \frac{2}{5} = \frac{8}{15}$$

Sometimes when we divide fractions, we can divide the first numerator by the second numerator and divide the first denominator by the second denominator.

$$\frac{8}{9} \div \frac{2}{3} = \frac{4}{3} \qquad \begin{array}{c}(8 \div 2 = 4)\\(9 \div 3 = 3)\end{array}$$

Often the numerator and denominator of a fraction are not divisible by the numerator and denominator of the fraction by which it is divided.

$$\frac{8}{9} \div \frac{3}{4} = \frac{?}{?}$$

Thus we need to learn another method for dividing fractions. The method we will learn uses reciprocals to find the answer.

To divide two fractions, we may multiply the first fraction by the reciprocal of the second fraction. In a later lesson we will explain the reason that this procedure works.

$$\frac{8}{9} \div \frac{3}{4} \qquad \text{division}$$

$$= \frac{8}{9} \cdot \frac{4}{3} \qquad \text{multiply by reciprocal}$$

$$= \frac{32}{27} \qquad \text{result}$$

$$= 1\frac{5}{27} \qquad \text{simplified}$$

Example 1   (a)   Simplify $\frac{8}{9} \div \frac{2}{3}$ by dividing the numerators and denominators.

           (b)   Simplify $\frac{8}{9} \div \frac{2}{3}$ by multiplying $\frac{8}{9}$ by the reciprocal of the divisor, $\frac{2}{3}$.

Solution   (a) $\dfrac{8}{9} \div \dfrac{2}{3} = \dfrac{4}{3} = 1\dfrac{1}{3}$

           (b)

$$\frac{8}{9} \div \frac{2}{3} \qquad \text{division}$$

$$= \frac{8}{9} \cdot \frac{3}{2} \qquad \text{multiply by reciprocal}$$

$$= \frac{24}{18} \qquad \text{result}$$

$$= 1\frac{1}{3} \qquad \text{simplified}$$

We see that both methods produce the same result.

Example 2   Simplify: $\dfrac{2}{3} \div \dfrac{3}{4}$

Solution   To divide by $\frac{3}{4}$, we will multiply by $\frac{4}{3}$.

$$\frac{2}{3} \cdot \frac{4}{3} = \frac{8}{9}$$

Example 3 Simplify: $\dfrac{3}{4} \div \dfrac{9}{10}$

Solution To divide $\dfrac{3}{4}$ by $\dfrac{9}{10}$, we will multiply $\dfrac{3}{4}$ by $\dfrac{10}{9}$. We will simplify before we multiply.

$$\dfrac{3}{4} \div \dfrac{9}{10} \quad \longrightarrow \quad \dfrac{\overset{1}{\cancel{3}}}{\underset{2}{\cancel{4}}} \cdot \dfrac{\overset{5}{\cancel{10}}}{\underset{3}{\cancel{9}}} = \dfrac{5}{6}$$

Example 4 How many $\dfrac{2}{3}$'s are in $\dfrac{3}{4}$?

Solution This question can be restated as, "What is the answer if $\dfrac{3}{4}$ is divided by $\dfrac{2}{3}$?"

$$\dfrac{3}{4} \div \dfrac{2}{3} \qquad \text{division}$$

$$= \dfrac{3}{4} \cdot \dfrac{3}{2} \qquad \text{multiply by reciprocal}$$

$$= \dfrac{9}{8} \qquad \text{result}$$

$$= 1\dfrac{1}{8} \qquad \text{simplified}$$

Practice Simplify:

a. $\dfrac{3}{5} \div \dfrac{2}{3}$ **b.** $\dfrac{7}{8} \div \dfrac{1}{4}$ **c.** $\dfrac{5}{6} \div \dfrac{2}{3}$

d. How many $\dfrac{2}{5}$'s are in $\dfrac{3}{4}$?

Problem set 29

1. Three hundred twenty-four students were treated to ice cream for receiving A's in citizenship. If each box of ice cream contained a half dozen ice cream bars, how many boxes of ice cream were needed?

2. Martin's pockets bulged with the coins he had been paid for redeeming bottles and cans. If he had 11 quarters, 14 dimes, 15 nickels, and 17 pennies, how much money did he have in all?

Use this information to answer questions 3–5.

> The family picnic was a success, as 56 relatives attended. Half of those who attended played in the big game. However, the two teams were not equal since one team had only 7 players.

3. How many relatives played the game?

4. If one team had 7 players, how many players did the other team have?

5. If the teams were rearranged so that the number of players on each team was equal, how many players would be on each team?

6. Draw a diagram of this statement. Then answer the questions that follow.

> Jason has read $\frac{7}{10}$ of the 310 pages in the book.

(a) How many pages has Jason read?

(b) How many pages has Jason not read?

7. How many $\frac{3}{4}$'s are in $\frac{7}{8}$?

8. Which is the best estimate of how much of this rectangle is shaded?

(a) $\frac{2}{3}$ (b) $\frac{2}{4}$ (c) $\frac{2}{5}$

9. Simplify each number.

(a) $6\frac{20}{12}$ (b) $\frac{54}{8}$ (c) $\frac{84}{210}$

10. Write the reciprocal of each number.

(a) $\frac{9}{10}$ (b) 8   (c) $2\frac{3}{8}$

11. For each fraction, write an equivalent fraction that has a denominator of 20.

    (a) $\dfrac{3}{4}$ (b) $\dfrac{4}{5}$ (c) $\dfrac{7}{10}$

12. Write the prime factorization of each number.

    (a) 35   (b) 640

13. Write each number as an improper fraction.

    (a) $5\dfrac{1}{2}$ (b) 6   (c) $3\dfrac{5}{8}$

14. Points A and B represent what mixed numbers on this number line?

15. (a) Draw line AB. Then draw ray BC perpendicular to line AB.

    (b) What kind of angle is $\angle ABC$?

Solve:

16. $126 + y = 310$ 17. $35 = x - 53$

18. $175 = 7m$

Add, subtract, multiply, or divide, as indicated:

19. $4\dfrac{5}{8} + 5\dfrac{7}{8}$ 20. $6\dfrac{1}{6} + 1\dfrac{5}{6}$

21. $3 - 1\dfrac{7}{12}$ 22. $\dfrac{3}{4} \cdot \dfrac{5}{9} \cdot \dfrac{8}{15}$

23. $\dfrac{4}{5} \div \dfrac{2}{1}$ 24. $\dfrac{8}{5} \div \dfrac{6}{5}$

25. $\dfrac{3}{7} \div \dfrac{5}{6}$ 26. $\dfrac{4386}{9}$

27. $6.98
× 74

28. $120.00
− 108.89

29. $\dfrac{\$110.16}{27}$

30. Solve mentally:

$$4 + 13 + 2 + 8 + 12 + 15 + 4 + 1 + N = 65$$

LESSON 30

Multiplying and Dividing Mixed Numbers

To multiply or divide mixed numbers, we first rewrite the mixed numbers as improper fractions. Then we multiply or divide as indicated.

Example 1 Simplify: $3\dfrac{2}{3} \times 1\dfrac{1}{2}$

Solution We first rewrite $3\dfrac{2}{3}$ as $\dfrac{11}{3}$ and $1\dfrac{1}{2}$ as $\dfrac{3}{2}$. Then we multiply and simplify.

$$\dfrac{11}{\cancel{3}_1} \times \dfrac{\cancel{3}^1}{2} = \dfrac{11}{2}$$

$$= 5\dfrac{1}{2}$$

Example 2 Simplify: $3\dfrac{2}{3} \div 2$

Solution We first rewrite $3\dfrac{2}{3}$ as $\dfrac{11}{3}$ and rewrite 2 as $\dfrac{2}{1}$. Then we multiply $\dfrac{11}{3}$ by $\dfrac{1}{2}$ and simplify.

$$\dfrac{11}{3} \div \dfrac{2}{1} \qquad \text{divide}$$

$$\dfrac{11}{3} \times \dfrac{1}{2} = \dfrac{11}{6} = 1\dfrac{5}{6} \qquad \text{multiply by reciprocal}$$

Practice Simplify:

a. $6\dfrac{2}{3} \times \dfrac{3}{5}$ b. $2\dfrac{1}{3} \times 3\dfrac{1}{2}$ c. $3\dfrac{3}{4} \times 3$

d. $1\dfrac{2}{3} \div 3$ e. $3\dfrac{1}{3} \div 2\dfrac{1}{2}$ f. $5 \div \dfrac{2}{3}$

Problem set 30 In Problems 1–4, identify the type of problem. Then find the answer.

1. After the first hour of the monsoon, 23 millimeters of precipitation had fallen. After the second hour, a total of 61 millimeters of precipitation had fallen. How many millimeters of precipitation fell during the second hour?

2. Each enlargement cost 85¢ and Willie needed 26 enlargements. What was the total cost of the enlargements Willie needed?

3. The Byzantine Empire lasted from 395 to 1453. How many years did the Byzantine Empire last?

4. Dolores went to the theater with $20 and came home with $11.25. How much money did Dolores spend at the theater?

5. A gross is a dozen dozen. A gross of pencils is how many pencils?

6. Draw a diagram of this statement. Then answer the questions that follow.

    Two fifths of the 60 marbles in the bag were blue.

    (a) How many of the marbles in the bag were blue?

    (b) How many of the marbles in the bag were not blue?

7. Draw and shade circles to show that $2\dfrac{2}{5}$ equals $\dfrac{12}{5}$.

8. (a) What fraction of this square is shaded?

    (b) What fraction of this square is not shaded?

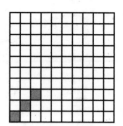

9. Simplify:

    (a) $9\dfrac{6}{4}$

    (b) $\dfrac{210}{252}$

10. Write the reciprocal of each number.

    (a) $\dfrac{5}{9}$

    (b) $5\dfrac{3}{4}$

    (c) 7

11. Complete each equivalent fraction.

    (a) $\dfrac{7}{8} = \dfrac{?}{48}$

    (b) $\dfrac{5}{16} = \dfrac{?}{48}$

    (c) $\dfrac{?}{12} = \dfrac{28}{48}$

12. List the prime numbers between 50 and 60.

13. Draw segment AB 2 cm long. Then draw $\overline{BC}$ 1 cm long perpendicular to $\overline{AB}$. Complete triangle ABC by drawing $\overline{AC}$.

14. Arrange these numbers in order from least to greatest:

$$1, -3, \dfrac{5}{6}, 0, \dfrac{4}{3}$$

Solve:

15. $250 = 700 - x$

16. $53 - y = 35$

17. $12w = 276$

Add, subtract, multiply, or divide, as indicated:

18. $8 - 7\dfrac{7}{9}$

19. $6\dfrac{5}{12} + 8\dfrac{11}{12}$

20. $9\dfrac{1}{9} - 4\dfrac{4}{9}$

21. $\dfrac{5}{8} \cdot \dfrac{3}{10} \cdot \dfrac{1}{6}$

22. $3\dfrac{1}{2} \cdot 1\dfrac{1}{3}$

23. $1\dfrac{3}{5} \div 2\dfrac{2}{3}$

24. $3\dfrac{1}{3} \div 4$

25. $5 \cdot 1\dfrac{3}{4}$

26. $4\dfrac{1}{2} \div 3$

27. $\dfrac{16{,}524}{36}$

28. $\$100 - (78¢ \times 48)$

29. $80(64)(25)$

30. Add mentally:

$14 + 4 + 11 + 12 + 5 + 6 + 8 + 42$

LESSON
31

Multiples ·
Least Common Multiple

Multiples   The **multiples** of a number are the numbers that are produced by multiplying the number by 1, by 2, by 3, by 4, and so on. Thus the multiples of 4 are

$$4, 8, 12, 16, 20, 24, 28, 32, 36, \ldots$$

The multiples of 6 are

$$6, 12, 18, 24, 30, 36, 42, 48, 54, \ldots$$

If we inspect these two lists, we see that some of the numbers in both lists are the same. A number that appears in both of these lists is a **common multiple** of 4 and 6. Here we have circled some of the common multiples of 4 and 6.

Multiples of 4: 4, 8, ⑫, 16, 20, ㉔, 28, 32, ㊱, . . .

Multiples of 6: 6, ⑫, 18, ㉔, 30, ㊱, 42, 48, 54, . . .

We see that 12, 24, and 36 are common multiples of 4 and 6. If we continued both lists, we would find many more common multiples.

Least common multiple

Of particular interest is the least (smallest) of the common multiples. The **least common multiple** of 4 and 6 is 12. It is the smallest number that is a multiple of both 4 and 6. The term "least common multiple" is often abbreviated **LCM**.

Example 1   Find the least common multiple of 6 and 8.

Solution   We will list some multiples of 6 and of 8 and circle common multiples.

Multiples of 6: 6, 12, 18, (24), 30, 36, 42, (48), . . .

Multiples of 8: 8, 16, (24), 32, 40, (48), 56, 64, . . .

The least common multiple of 6 and 8 is **24**.

Example 2   Find the LCM of 4 and 8.

Solution   The initials LCM stand for least common multiple. We list some multiples of 4 and 8 and circle the common multiples.

Multiples of 4: 4, (8), 12, (16), 20, (24), 28, . . .

Multiples of 8: (8), (16), (24), 32, 40, 48, 56, . . .

The LCM of 4 and 8 is **8**.

Example 3   Find the LCM of 3, 4, and 6.

Solution   We circle only the numbers that appear in all three lists.

Multiples of 3: 3, 6, 9, (12), 15, 18, 21, (24), . . .

Multiples of 4: 4, 8, (12), 16, 20, (24), 28, 32, . . .

Multiples of 6: 6, (12), 18, (24), 30, 36, 42, 48, . . .

The LCM of 3, 4, and 6 is **12**.

It is not necessary to list the multiples each time. Often the search for the least common multiple can be conducted mentally.

Practice   Find the least common multiple of each pair or group of numbers.

a. 8 and 10   **b.** 9 and 12   **c.** 4, 6, and 10

Problem set   In Problems 1–4, identify the type of problem. Then find the
31   answer.

1. There were three towns in the valley. The population of Brenton was 11,460. The population of Elton was 9420. The population of Jennings was 8916. What was the total population of the three towns in the valley?

2. Norman is 6 feet tall. How many inches tall is Norman?

3. If the cost of one dozen eggs was $0.96, what was the cost per egg?

4. One billion is how much greater than ten million, nine hundred thousand? Use words to write your answer.

5. Draw a diagram of this statement. Then answer the questions that follow.

   Three eighths of the 712 students bought their lunch.

   (a) How many students bought their lunch?

   (b) How many students did not buy their lunch?

6. The perimeter of this rectangle is 30 inches. What is the length of the rectangle?

   6 in.

7. (a) List the first six multiples of 6.

   (b) List the first six multiples of 10.

   (c) What is the least common multiple of 6 and 10?

8. What number is halfway between 3000 and 4000?

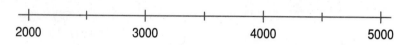

9. Simplify:

    (a) $8\dfrac{20}{6}$

    (b) $\dfrac{36}{180}$

10. Write the reciprocal of each number.

    (a) $5\dfrac{1}{8}$

    (b) 12

    (c) $\dfrac{4}{9}$

11. For each fraction, write an equivalent fraction that has a denominator of 36.

    (a) $\dfrac{5}{12}$

    (b) $\dfrac{1}{6}$

    (c) $\dfrac{7}{9}$

12. Write the prime factorization of 384.

13. Write each number as an improper fraction.

    (a) $5\dfrac{5}{6}$

    (b) 12

    (c) $12\dfrac{1}{4}$

14. Figures *ABCF* and *FCDE* are squares.

    (a) What kind of angle is ∠*ACD*?

    (b) Name two segments parallel to $\overline{FC}$.

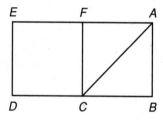

15. Refer to the figure in Problem 14. If *AB* is 3 cm, what is the perimeter of rectangle *ABDE*?

Solve:

16. $10y = 360$

17. $p + 16 = 144$

18. $56 - n = 14$

Add, subtract, multiply, or divide, as indicated:

19. $4\frac{1}{3} + 6\frac{2}{3}$

20. $5\frac{1}{8} - 1\frac{3}{8}$

21. $10 - 1\frac{3}{5}$

22. $5\frac{1}{3} \cdot 1\frac{1}{2}$

23. $3\frac{1}{3} \div \frac{5}{6}$

24. $5\frac{1}{4} \div 3$

25. $\frac{5}{6} \cdot \frac{9}{8} \cdot \frac{4}{15}$

26. $\frac{8}{9} - \left(\frac{7}{9} - \frac{5}{9}\right)$

27. $\begin{array}{r} \$16.25 \\ \times \quad\quad 8 \\ \hline \end{array}$

28. $\begin{array}{r} \$150.00 \\ - \quad 97.75 \\ \hline \end{array}$

29. $\frac{\$12.00}{16}$

30. Use mental addition to find N:

$$4 + 12 + 8 + 11 + 13 + 5 + 4 + 31 + 6 + N = 105$$

LESSON 32

Two-Step Word Problems

We have considered five types of one-step word problems thus far. Their thought patterns are:

1. Some and some more

2. Some went away

3. Larger-smaller-difference

4. Equal groups

5. Part-part-whole

Word problems often require more than one step to solve. In this lesson we will begin practicing problems that require two steps to solve. Two-step problems involve a combination of the one-step problems we have practiced.

Example   Julie went to the store with $20. If she bought 8 cans of dog food for 67 cents per can, how much money did she have left?

Solution   This is a two-step problem. First we find out how much Julie spent. This is an equal groups problem.

Number in group	$0.67 each can
Number of groups	× 8 cans
Total	$5.36

Now we can find out how much money Julie has left. This is a some went away problem.

$$\begin{array}{r} \$20.00 \\ -\;\;\; 5.36 \\ \hline \$14.64 \end{array}$$

After spending $5.36 of her $20 on dog food, Julie has **$14.64** left.

Practice   Work each problem as a two-step problem.

    a. Jody went to the store with $20 and returned home with $5.36. If all she bought was 3 bags of dog food, how much did she pay for each bag?

    b. Sam bought a pair of headphones for $15.89 and 3 tapes for $7.89 each. Altogether, how much money did he spend?

    c. Three eighths of the 32 students were girls. How many boys were in the class?

Problem set 32

    1. Quentin purchased 5 arrowheads for $1.75 each. He paid for them with a $10 bill. How much did he receive in change?

    2. In all, 379 students attended the assembly. If 198 girls were in attendance, then how many fewer boys than girls were in attendance?

3. When Gilbert stood on his toes, he was 5 feet 11 inches tall. How many inches tall was Gilbert when he stood on his toes?

4. Cynthia picked 176 apricots, but there were still 394 apricots left on the tree. How many apricots were on the tree before Cynthia picked them?

5. Draw a diagram of this statement. Then answer the questions that follow.

    Five ninths of the 270 carrot seeds sprouted.

    (a)   How many carrot seeds sprouted?

    (b)   How many carrot seeds did not sprout?

6. If the perimeter of a regular hexagon is 1 yard, how many inches long is each side?

7. (a)   List the first six multiples of 4.

    (b)   List the first six multiples of 10.

    (c)   What is the LCM of 4 and 10?

8. On this number line, 3380 is closest to

    (a)   which multiple of 100?

    (b)   which multiple of 1000?

9. Simplify:

    (a) $\dfrac{27}{6}$ (b) $5\dfrac{11}{5}$ (c) $2\dfrac{24}{8}$

10. Complete each equivalent fraction.

    (a) $\dfrac{3}{5} = \dfrac{?}{40}$ (b) $\dfrac{5}{8} = \dfrac{?}{40}$ (c) $\dfrac{?}{10} = \dfrac{12}{40}$

11. Write the prime factorization of each number.

(a) 62

(b) 312

12. Write each number as an improper fraction.

(a) $7\frac{1}{2}$

(b) 15

13. Draw rectangle *ABCD* so that *AB* is 2 cm and *BC* is 3 cm. What is the perimeter of *ABCD*?

14. In triangle *RST*,

(a) $\angle S$ is what kind of angle?

(b) $\angle T$ is what kind of angle?

(c) Which segment is perpendicular to $\overline{RS}$?

15. What mixed number is halfway between 7 and 10?

Solve:

16. $84 = 7x$

17. $210 = 21 + p$

18. $t - 56 = 14$

Add, subtract, multiply, or divide, as indicated:

19. $6\frac{4}{7} - 1\frac{6}{7}$

20. $7\frac{5}{6} + 1\frac{5}{6}$

21. $8 - \frac{5}{8}$

22. $1\frac{1}{2} \div 1\frac{3}{4}$

23. $3 \cdot 1\frac{1}{4} \cdot \frac{8}{9}$

24. $3 \div 5\frac{1}{4}$

25. $\begin{array}{r} \$375 \\ \times \quad 64 \\ \hline \end{array}$

26. $\dfrac{1,000,000}{400}$

27. $\dfrac{5}{6} + \left(\dfrac{1}{2} \cdot \dfrac{1}{3}\right)$ **28.** $\begin{array}{r} 14{,}507 \\ -\ 9{,}618 \\ \hline \end{array}$ **29.** $\dfrac{36{,}444}{12}$

30. Use mental addition to find N:

$4 + 13 + 3 + 12 + 22 + 5 + 7 + 14 + 12 + N = 99$

LESSON

33

Average, Part 1

Here we show 5 stacks of coins:

There are 15 coins in all. If we made all the stacks the same size, there would be 3 coins in each stack.

We say that the **average** number of coins in each stack is 3. Consider the following problem.

> There are 4 squads in the physical education class. Squad A has 7 players, squad B has 9 players, squad C has 6 players, and squad D has 10 players. What is the average number of players in a squad?

The average number of players in a squad is the number of players that would be on each squad if each squad had the same number of players. To find the average of a group of numbers, we begin by finding the sum of the numbers.

$$\begin{array}{r} 7 \text{ players} \\ 9 \text{ players} \\ 6 \text{ players} \\ +\ 10 \text{ players} \\ \hline 32 \text{ players} \end{array}$$

Then we divide the sum of the numbers by the number of

numbers. There are 4 squads, so we divide by 4.

$$\text{Average} = \frac{\text{sum of numbers}}{\text{number of numbers}} = \frac{32 \text{ players}}{4 \text{ squads}}$$

$$= 8 \text{ players per squad}$$

Finding an average is a two-step process. We first add the numbers to find the total. Then we divide the total to make equal groups.

Example 1 When the people were seated, there were 3 in the first row, 7 in the second row, and 20 in the third row. What was the average number of people in each of the first 3 rows?

Solution The average number of people in the first 3 rows is the number of people that would be in each row if the number in each row were equal. First we add to find the total number of people.

$$\begin{array}{r} 3 \text{ people} \\ 7 \text{ people} \\ + \ 20 \text{ people} \\ \hline 30 \text{ people} \end{array}$$

Then we divide by 3 to separate the total into 3 equal groups.

$$\frac{30 \text{ people}}{3 \text{ rows}} = 10 \text{ people per row}$$

The average was **10 people** in each of the first 3 rows.

To find the average of several numbers, we add the numbers, and then divide the total by the number of numbers.

Example 2 What is the average of 26, 42, 57, 49, and 16?

Solution There are 5 numbers. To find the average of these numbers, we first find the total. Then we divide the total by 5.

$$\begin{array}{r} 26 \\ 42 \\ 57 \\ 49 \\ + \ 16 \\ \hline 190 \end{array} \qquad \begin{array}{r} 38 \\ 5\overline{)190} \\ \underline{15} \\ 40 \\ \underline{40} \\ 0 \end{array}$$

The average of the 5 numbers is **38**.

Practice

a. In Room 1 there were 28 students, in Room 2 there were 29 students, in Room 3 there were 30 students, and in Room 4 there were 25 students. What was the average number of students in the 4 rooms?

b. What is the average of 46, 37, 34, 31, 29, and 24?

c. What is the average of 40 and 70? What number is halfway between 40 and 70?

Problem set 33

1. The 5 players on the front line weighed 242 pounds, 236 pounds, 248 pounds, 268 pounds, and 226 pounds. What was the average weight of the players on the front line?

2. Matt ran a mile in 5 minutes 14 seconds. How many seconds did it take Matt to run a mile?

3. Ginger bought a pair of pants for $24.95 and 3 blouses for $15.99 each. Altogether, how much did she spend?

4. The Italian navigator Christopher Columbus was 41 years old when he reached the Americas in 1492. In what year was he born?

5. Draw a diagram of this statement. Then answer the questions that follow.

Bill led for three fourths of the 5000-meter race.

(a) Bill led the race for how many meters?

(b) Bill did not lead the race for how many meters?

6. This rectangle is twice as long as it is wide. What is the perimeter of this rectangle?

8 cm

7. (a) List the first six multiples of 3.

(b) List the first six multiples of 4.

(c) What is the LCM of 3 and 4?

8. On this number line, 283 is closest to

(a) which multiple of 10?

(b) which multiple of 100?

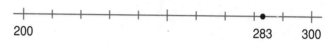

9. Simplify each fraction or mixed number.

(a) $\dfrac{56}{240}$

(b) $1\dfrac{18}{8}$

10. Write the reciprocal of each number.

(a) $\dfrac{7}{10}$

(b) $4\dfrac{3}{5}$

11. For each fraction, write an equivalent fraction that has a denominator of 24.

(a) $\dfrac{7}{8}$

(b) $\dfrac{11}{12}$

12. Write the prime factorization of each number.

(a) 27

(b) 2800

13. What is the average of 45, 36, 42, 29, 16, and 24?

14. (a) Draw square $ABCD$ so that each side is about 1 inch long.

(b) Draw segments AC and BD. Label the point at which they intersect point E.

(c) Shade triangle CDE.

15. Arrange these numbers in order from least to greatest:

$$-1, \ \dfrac{1}{10}, \ 1, \ \dfrac{11}{10}, \ 0$$

Solve:

16. $12y = 360$ **17.** $123 = m + 64$ **18.** $45 = 54 - w$

Add, subtract, multiply, or divide, as indicated:

19. $4\dfrac{5}{12} - 1\dfrac{1}{12}$

20. $8\dfrac{7}{8} + 3\dfrac{3}{8}$

21. $12 - 8\dfrac{1}{8}$

22. $6\dfrac{2}{3} \cdot 1\dfrac{1}{5}$

23. $2\dfrac{1}{4} \div 7\dfrac{1}{2}$

24. $8 \div 2\dfrac{2}{3}$

25. $\begin{array}{r} 8000 \\ \times\ 600 \\ \hline \end{array}$

26. $\dfrac{10,000}{80}$

27. $\dfrac{3}{4} - \left(\dfrac{1}{2} \div \dfrac{2}{3}\right)$

28. $\begin{array}{r} \$47.63 \\ 78.49 \\ +\ 35.24 \\ \hline \end{array}$

29. $\begin{array}{r} 37,484 \\ -\ 36,295 \\ \hline \end{array}$

30. $\begin{array}{r} \$4.56 \\ \times\qquad 9 \\ \hline \end{array}$

LESSON 34

Rounding Whole Numbers · Estimating Answers

Rounding whole numbers

The first sentence below uses an exact number to state the size of a crowd. The second sentence uses a round number.

There were 3947 fans at the game.
There were about 4000 fans at the game.

Round numbers are often used instead of exact numbers because they are easier to work with. One way to round numbers is to consider where the number is located on the number line.

Example 1 Use a number line to

(a) round 283 to the nearest hundred.

(b) round 283 to the nearest ten.

Solution   (a)   We draw a number line and mark the location of hundreds as well as the estimated location of 283.

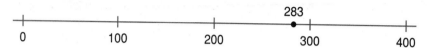

We see that 283 is between 200 and 300. Since 283 is closer to 300 than it is to 200, we say that 283 rounded to the nearest hundred is **300**.

(b)   We draw a number line and mark the location of the tens from 200 to 300 as well as the location of 283.

We see that 283 is between 280 and 290. Since 283 is closer to 280 than it is to 290, we say that 283 rounded to the nearest ten is **280**.

Sometimes we are asked to round a number to a certain place value. We can use a circle and an arrow to help us do this. We will circle the digit in the place to which we are rounding, and we will draw an arrow above the next place. Then we will follow these rules.

1.   If the arrow-marked digit is 5 or more, we increase the circled digit by 1. If the arrow-marked digit is less than 5, we leave the circled digit unchanged.

2.   We change the arrow-marked digit and all digits to the right of the arrow-marked digit to zero.

Example 2   (a)   Round 283 to the nearest hundred.

(b)   Round 283 to the nearest ten.

Solution   (a)   We circle the 2 since it is in the hundreds' place. Then we draw an arrow over the digit to its right.

②8 3

Since the arrow-marked digit is 5 or more, we increase the circled 2 to 3. Then we change the arrow-marked digit and all digits to its right to zero and get

300

(b) Since we are rounding to the nearest ten, we circle the tens' digit and mark the digit to its right with an arrow.

$$\downarrow$$

2 ⑧ 3

Since the arrow-marked digit is less than 5, we leave the 8 unchanged. Then we change the 3 to zero and get

280

Example 3   Round 5280 (a) to the nearest thousand, and (b) to the nearest hundred.

$$\downarrow$$

Solution   (a) To the nearest thousand, ⑤280 rounds to **5000**.

$$\downarrow$$

(b) To the nearest hundred, 5②80 rounds to **5300**.

Estimating answers   Rounding can help us estimate the answer to arithmetic problems. Estimating is a quick and easy way to get close to an exact answer. Sometimes a close answer is "good enough," but even when an exact answer is necessary, estimating can help us decide if our exact answer is reasonable. To estimate, we **round the numbers first**.

Example 4   Mentally estimate:

(a) 6879 + 3145   (b) 396 × 312   (c) 4160 ÷ 19

Solution   (a) We round both numbers to the same place before we add.

6879	→	7000
+ 3145	→	+ 3000
		10,000

(b) We round each number so there is one nonzero digit before we multiply.

396	→	400
× 312	→	× 300
		120,000

(c) We round each number so there is one nonzero digit before we divide.

$$\frac{4160}{19} \quad \xrightarrow{\quad} \quad \frac{4000}{20} = \textbf{200}$$

Practice

a. Sketch a number line to round 231 to the nearest hundred.

b. Round 1760 to the nearest hundred.

c. Round 186,282 to the nearest thousand.

Estimate each answer.

d. $7986 - 3074$ e. 297×31 f. $5860 \div 19$

Problem set 34

1. Larry jumped 16 feet 8 inches on his first try. How many inches did he jump on his first try?

2. If 8 pounds of bananas cost $3.68, what is the cost per pound?

3. On her first six tests Sandra's scores were 75, 70, 80, 80, 85, and 90. What was her average score on her first six tests?

4. Two hundred nineteen billion, eight hundred million is how much less than one trillion? Use words to write your answer.

5. Draw a diagram of this statement. Then answer the questions that follow.

     Two fifths of the 80 chips were blue.

(a) How many of the chips were blue?

(b) How many of the chips were not blue?

6. What is the LCM of 4, 6, and 8?

7. What is the perimeter of this square?

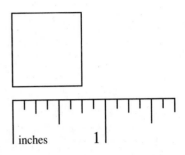

8. (a) Round 366 to the nearest hundred.

(b) Round 366 to the nearest ten.

9. Mentally estimate the sum of 6143 and 4952 by rounding each number to the nearest thousand before adding.

10. Simplify each number.

(a) $4\dfrac{60}{36}$

(b) $\dfrac{125}{500}$

11. Write each number as an improper fraction.

(a) 10

(b) $15\dfrac{3}{4}$

12. Complete each equivalent fraction.

(a) $\dfrac{2}{3} = \dfrac{?}{30}$

(b) $\dfrac{?}{6} = \dfrac{25}{30}$

13. Write the prime factorization of each number.

(a) 102

(b) 1020

14. What is the average of 374, 286, 397, and 423?

15. Draw rectangle $ABCD$ so that AB is 1 cm and BC is 2 cm. Next draw segment AC. Then shade triangle ABC.

Solve:

16. $96 = 4m$ **17.** $x + 37 = 105$ **18.** $45 = n - 54$

Add, subtract, multiply, or divide, as indicated:

19. $15 - 4\dfrac{4}{9}$

20. $3\dfrac{5}{9} + 4\dfrac{7}{9}$

21. $6\frac{1}{3} - 5\frac{2}{3}$ **22.** $\frac{3}{4} \cdot 5\frac{1}{3} \cdot 1\frac{1}{8}$

23. $6\frac{2}{3} \div 5$ **24.** $1\frac{2}{3} \div 3\frac{1}{2}$

25. $\begin{array}{r} 749 \\ \times\ 86 \end{array}$ **26.** $\dfrac{400,000}{25}$ **27.** $\frac{2}{5} + \left(\frac{4}{5} \div \frac{1}{2} \right)$

28. $\begin{array}{r} \$350.00 \\ -\ 127.48 \end{array}$ **29.** $\begin{array}{r} \$7.25 \\ \times\qquad 6 \end{array}$ **30.** $\begin{array}{r} 3741 \\ 2639 \\ 4814 \\ +\quad 68 \end{array}$

LESSON 35

Common Denominators • Adding and Subtracting Fractions with Different Denominators

Common denominators   When two fractions have the same denominator, we say they have **common denominators**.

$$\frac{3}{8} \qquad \frac{6}{8} \qquad\qquad\qquad \frac{3}{8} \qquad \frac{3}{4}$$

These two fractions have common denominators.   These two fractions do not have common denominators.

If two fractions do not have common denominators, then one or both fractions can be renamed so both fractions do have common denominators. We remember that we can rename a fraction by multiplying the fraction by a fraction equal to 1. Thus we can rename $\frac{3}{4}$ so that it has a denominator of 8 by multiplying by $\frac{2}{2}$.

$$\frac{3}{4} \cdot \frac{2}{2} = \frac{6}{8}$$

Example 1 Rename $\frac{2}{3}$ and $\frac{1}{4}$ so that they have common denominators.

Solution The denominators are 3 and 4. A common denominator for these two fractions would be any common multiple of 3 and 4. The lowest common denominator would be the least common multiple of 3 and 4, which is 12. We want to rename each fraction so that the denominator is 12.

$$\frac{2}{3} = \frac{}{12} \qquad \frac{1}{4} = \frac{}{12}$$

We multiply $\frac{2}{3}$ by $\frac{4}{4}$ and multiply $\frac{1}{4}$ by $\frac{3}{3}$.

$$\frac{2}{3} \cdot \frac{4}{4} = \frac{8}{12} \qquad \frac{1}{4} \cdot \frac{3}{3} = \frac{3}{12}$$

Thus $\frac{2}{3}$ and $\frac{1}{4}$ can be written with common denominators as

$$\frac{8}{12} \quad \text{and} \quad \frac{3}{12}$$

Fractions written with common denominators can be compared by simply comparing the numerators.

Example 2 Write these fractions with common denominators and then compare them.

$$\frac{5}{6} \bigcirc \frac{7}{9}$$

Solution The common denominator for these fractions is the LCM of 6 and 9, which is 18.

$$\frac{5}{6} \cdot \frac{3}{3} = \frac{15}{18} \qquad \frac{7}{9} \cdot \frac{2}{2} = \frac{14}{18}$$

In place of $\frac{5}{6}$ we will write $\frac{15}{18}$. In place of $\frac{7}{9}$ we will write $\frac{14}{18}$. Then we compare the fractions.

$$\frac{15}{18} \bigcirc \frac{14}{18} \qquad \text{renamed}$$

$$\frac{15}{18} > \frac{14}{18} \qquad \text{compared}$$

Adding and subtracting fractions

To add or subtract fractions that do not have common denominators, we first rename one or both fractions so that they do have common denominators. Then we can add or subtract.

Example 3 Add: $\dfrac{3}{4} + \dfrac{3}{8}$

Solution First we write the fractions with common denominators. The denominators are 4 and 8. The least common multiple of 4 and 8 is 8. We rename $\frac{3}{4}$ so that the denominator is 8 by multiplying by $\frac{2}{2}$. We do not need to rename $\frac{3}{8}$. Then we add the fractions and simplify.

$$\frac{3}{4} \cdot \frac{2}{2} = \frac{6}{8} \qquad \text{renamed } \frac{3}{4}$$
$$+ \frac{3}{8} \quad\;\; = \frac{3}{8}$$
$$\overline{\qquad\qquad \frac{9}{8}} \qquad \text{added}$$

We finish by simplifying $\frac{9}{8}$.

$$\frac{9}{8} = 1\frac{1}{8}$$

Example 4 Subtract: $\dfrac{5}{6} - \dfrac{3}{4}$

Solution First we write the fractions with common denominators. The LCM of 6 and 4 is 12. We multiply $\frac{5}{6}$ by $\frac{2}{2}$ and multiply $\frac{3}{4}$ by $\frac{3}{3}$ so that both denominators are 12. Then we subtract the renamed fractions.

$$\frac{5}{6} \cdot \frac{2}{2} = \frac{10}{12} \qquad \text{renamed } \frac{5}{6}$$
$$- \frac{3}{4} \cdot \frac{3}{3} = \frac{9}{12} \qquad \text{renamed } \frac{3}{4}$$
$$\overline{\qquad\qquad \frac{1}{12}} \qquad \text{subtracted}$$

Example 5 Subtract: $8\dfrac{2}{3} - 5\dfrac{1}{6}$

Solution We first write the fractions with common denominators. The LCM of 3 and 6 is 6. We multiply $\frac{2}{3}$ by $\frac{2}{2}$ so that the denominator is 6. Then we subtract and simplify.

$$8\frac{2}{3} = 8\frac{4}{6} \qquad \text{renamed } 8\frac{2}{3}$$

$$-5\frac{1}{6} = 5\frac{1}{6}$$

$$3\frac{3}{6} = 3\frac{1}{2} \quad \text{subtracted and simplified}$$

Example 6 Add: $\dfrac{1}{2} + \dfrac{2}{3} + \dfrac{3}{4}$

Solution The denominators are 2, 3, and 4. The LCM of 2, 3, and 4 is 12. We rename each fraction so that the denominator is 12. Then we add and simplify.

$$\frac{1}{2} \cdot \frac{6}{6} = \frac{6}{12} \qquad \text{renamed } \frac{1}{2}$$

$$\frac{2}{3} \cdot \frac{4}{4} = \frac{8}{12} \qquad \text{renamed } \frac{2}{3}$$

$$+\frac{3}{4} \cdot \frac{3}{3} = \frac{9}{12} \qquad \text{renamed } \frac{3}{4}$$

$$\frac{23}{12} = 1\frac{11}{12} \quad \text{added and simplified}$$

Practice Write the fractions with common denominators and then compare them.

a. $\dfrac{3}{5} \bigcirc \dfrac{7}{10}$ 　　　　　　　　 **b.** $\dfrac{5}{6} \bigcirc \dfrac{7}{8}$

Add or subtract.

c. $\dfrac{3}{4} + \dfrac{5}{6} + \dfrac{3}{8}$ 　　　　　　　 **d.** $7\dfrac{5}{6} - 2\dfrac{1}{2}$

e. $4\dfrac{3}{4} + 5\dfrac{5}{8}$ 　　　　　　　 **f.** $4\dfrac{5}{9} - 2\dfrac{1}{6}$

Problem set 35

1. The 5 starters on the basketball team were tall. Their heights were 76 inches, 77 inches, 77 inches, 78 inches, and 82 inches. What was the average height of the 5 starters?

2. Marie bought 6 pounds of apples for $0.87 per pound and paid for them with a $10 bill. How much should she get back in change?

3. On the first day of their 2479-mile trip, the Curtis family drove 497 miles. How many more miles do they have to drive until they complete their trip?

4. One hundred thirty-six of the two hundred sixty students in the auditorium were boys. How many girls were in the auditorium?

5. Draw a diagram of this statement. Then answer the questions that follow.

   The Daltons completed three tenths of their 2140-mile trip the first day.

   (a) How many miles did they travel the first day?

   (b) How many miles of their trip do they still have to travel?

6. If the perimeter of a square is 5 feet, how many inches long is each side of the square?

7. Rewrite $\frac{2}{3}$ and $\frac{3}{4}$ so that they have common denominators.

8. (a) Round 36,467 to the nearest thousand.

   (b) Round 36,467 to the nearest hundred.

9. Mentally estimate the quotient when 29,376 is divided by 49.

10. Simplify each number.

    (a) $15\frac{24}{12}$ (b) $\frac{90}{16}$

11. Compare: $\frac{5}{12} \bigcirc \frac{7}{15}$

12. Write the reciprocal of each number.

(a) 11

(b) $12\frac{1}{2}$

13. Write the prime factorization of each number.

(a) 51

(b) 2592

14. What is the average of 5, 7, 9, 11, 12, 13, 24, 25, 26, and 28?

15. List the single-digit divisors of 5670.

Solve:

16. $6w = 8 \cdot 9$ **17.** $356 + a = 527$ **18.** $63 - d = 35$

Add, subtract, multiply, or divide, as indicated:

19. $\frac{1}{2} + \frac{1}{3}$

20. $\frac{3}{4} - \frac{1}{3}$

21. $2\frac{5}{6} - 1\frac{1}{2}$

22. $\frac{4}{5} \cdot 1\frac{2}{3} \cdot 1\frac{1}{8}$

23. $1\frac{3}{4} \div 2\frac{2}{3}$

24. $3 \div 1\frac{7}{8}$

25. $3\frac{2}{3} + 1\frac{5}{6}$

26. $5\frac{1}{8} - 1\frac{3}{4}$

27. $\frac{1}{2} + \left(\frac{2}{3} \cdot \frac{3}{4}\right)$

28. $\begin{array}{r} \$3.87 \\ \times\ \ \ 96 \\ \hline \end{array}$

29. $\dfrac{43{,}164}{36}$

30. $\begin{array}{r} \$3.15 \\ 4.80 \\ 7.67 \\ 8.98 \\ 7.42 \\ +\ 8.65 \\ \hline \end{array}$

LESSON
36

Decimal Fractions •
Decimal Place Value

Decimal   We have used fractions to name parts of a whole. We remember
fractions   that a fraction has a numerator and a denominator. The
denominator indicates the number of equal parts in the whole.
The numerator indicates the number of parts that are being
considered.

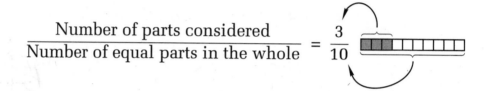

$$\frac{\text{Number of parts considered}}{\text{Number of equal parts in the whole}} = \frac{3}{10}$$

In a **common fraction** the numerator and denominator are
both written. The fraction diagramed above is three tenths.
We see that the numerator is 3. We see that the denominator
is 10.

Parts of a whole can also be named by using **decimal
fractions**. In a decimal fraction we can see the numerator,
but we cannot see the denominator. **The denominator of a
decimal fraction is indicated by place value.** Here is the decimal
fraction three tenths.

$$0.3$$

If we write one digit after a decimal point, we indicate that
the denominator is 10. We see the numerator, which is 3, but
we do not see the denominator, which is 10. The denominator
is understood to be 10 because the 3 occupies the first place
to the right of the decimal point, which is the tenths' place
$\left(\frac{1}{10}\text{ place}\right)$. The decimal 0.3 equals the fraction $\frac{3}{10}$, and both
are ways to write three tenths. It is customary to write a zero
before the decimal point.

$$0.3 = \frac{3}{10} \qquad \text{three tenths equals three tenths}$$

A decimal written with two digits after the decimal point is understood to have a denominator of 100, as we show here.

$$0.03 = \frac{3}{100} \qquad \text{three hundredths}$$

$$0.21 = \frac{21}{100} \qquad \text{twenty-one hundredths}$$

Example 1 Write seven tenths (a) as a fraction and (b) as a decimal.

Solution (a) $\dfrac{7}{10}$ (b) **0.7**

Example 2 Name the shaded part of this square

(a) as a fraction.

(b) as a decimal.

Solution (a) $\dfrac{23}{100}$ (b) **0.23**

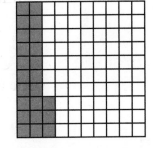

Decimal place value In our number system the place a digit occupies has a value, called **place value.** We remember that places to the left of the decimal point have values of 1, 10, 100, 1000, and so on, becoming greater and greater. Places to the right of the decimal point have values of $\frac{1}{10}$, $\frac{1}{100}$, $\frac{1}{1000}$, and so on, becoming less and less. This chart shows decimal place values from the millions' place through the millionths' place.

etc.	millions' place	hundred-thousands' place	ten-thousands' place	thousands' place	hundreds' place	tens' place	units' place	decimal point	tenths' place	hundredths' place	thousandths' place	ten-thousandths' place	hundred-thousandths' place	millionths' place	etc.
	1,000,000	100,000	10,000	1000	100	10	1	$\cdot$	$\frac{1}{10}$	$\frac{1}{100}$	$\frac{1}{1000}$	$\frac{1}{10,000}$	$\frac{1}{100,000}$	$\frac{1}{1,000,000}$	

Example 3   What is the value of the sixth decimal place to the right of the decimal point?

Solution   The sixth decimal place is six places to the right of the decimal point. The sixth decimal place has a value of one millionth, $\frac{1}{1,000,000}$.

Example 4   In the number 12.34579, which digit is in the thousandths' place?

Solution   We can name the places to the right of the decimal point by saying "tenths, hundredths, thousandths." The thousandths' place is the third place to the right of the decimal point and is occupied by the **5**.

Example 5   Name the place occupied by the 7 in 4.63471.

Solution   The 7 is in the fourth place to the right of the decimal point. This is the **ten-thousandths' place**.

Practice   **a.** Write three hundredths as a fraction. Then write three hundredths as a decimal.

b. Name the shaded part of this circle both as a fraction and as a decimal.

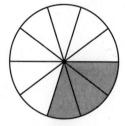

c. In the number 16.57349, which digit is in the thousandths' place?

d. Name the place occupied by the 8 in 4.634718.

e. The number 36.4375 has how many decimal places?

f. What is the value of the third decimal place?

Problem set 36

1. James and his brother are putting their money together to buy a radio that costs $89.89. James has $26.47. His brother has $32.54. How much more money do they need to buy the radio?

2. Norton read 4 books during his vacation. The first book was 326 pages, the second was 288 pages, the third was 349 pages, and the fourth was 401 pages. What was the average number of pages of the 4 books he read?

3. A one-year subscription to the monthly magazine costs $15.96. At this price, what is the cost for each issue?

4. The settlement at Jamestown began in 1607. This was how many years after Columbus reached the Americas in 1492?

5. A square and a regular hexagon share a common side. The perimeter of the square is 24 cm. What is the perimeter of the hexagon?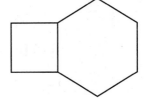

6. Draw a diagram of this statement. Then answer the questions that follow.

    Nelson correctly answered four fifths of the 20 questions on the test.

    (a) How many questions did Nelson answer correctly?

    (b) How many questions did Nelson answer incorrectly?

7. What is the least common multiple of 3, 4, and 6?

8. Round 481,462

    (a) to the nearest hundred thousand.

    (b) to the nearest thousand.

9. Mentally estimate the difference between 49,623 and 20,162.

10. Name the shaded part of this square

(a) as a fraction.

(b) as a decimal.

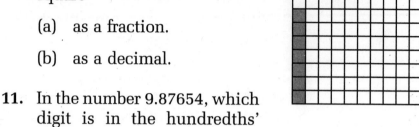

11. In the number 9.87654, which digit is in the hundredths' place?

12. Replace each circle with the proper comparison symbol.

(a) $\dfrac{3}{10}$ ◯ 0.3   (b) $\dfrac{3}{100}$ ◯ 0.3

13. Simplify each fraction or mixed number.

(a) $3\dfrac{45}{30}$ (b) $\dfrac{360}{744}$

14. Complete each equivalent fraction.

(a) $\dfrac{5}{?} = \dfrac{15}{24}$ (b) $\dfrac{7}{12} = \dfrac{?}{24}$ (c) $\dfrac{?}{6} = \dfrac{4}{24}$

15. Draw two parallel lines. Then draw two more parallel lines that are perpendicular to the first pair of lines. Label the points of intersection $A, B, C,$ and D. What kind of quadrilateral is $ABCD$?

Solve:

16. $9n = 6 \cdot 12$ **17.** $a + 57 = 220$ **18.** $60 - m = 48$

Add, subtract, multiply, or divide, as indicated:

19. $\dfrac{1}{2} + \dfrac{2}{3}$ **20.** $\dfrac{3}{4} - \dfrac{2}{3}$

21. $3\dfrac{5}{6} - \dfrac{1}{3}$ **22.** $\dfrac{5}{8} \cdot 2\dfrac{2}{5} \cdot \dfrac{4}{9}$

23. $2\dfrac{2}{3} \div 1\dfrac{3}{4}$ **24.** $1\dfrac{7}{8} \div 3$

25. $3\dfrac{1}{2} + 1\dfrac{5}{6}$ **26.** $5\dfrac{1}{4} - 1\dfrac{5}{8}$ **27.** $\dfrac{1}{2} - \left(\dfrac{3}{4} \cdot \dfrac{2}{3}\right)$

28. $\begin{array}{r} \$4.67 \\ \times \quad 87 \\ \hline \end{array}$ **29.** $\dfrac{28,308}{14}$ **30.** $\begin{array}{r} \$420.15 \\ -\ 398.75 \\ \hline \end{array}$

LESSON 37

Reading and Writing Decimal Numbers • Comparing Decimal Numbers

Reading and writing decimals

We remember that in a decimal number the digits to the left of the decimal point identify a whole number. The digits to the right of the decimal point identify a fraction. Here we show two and three tenths written as a decimal number and as a mixed number.

$$2.3 = 2\frac{3}{10}$$

To read a decimal number, we first read the whole number part, then we read the fraction part. To read the fraction part of a decimal number, we read the digits to the right of the decimal point as though we were reading a whole number. This number is the numerator of the decimal fraction. Then we say the name of the last decimal place. This number is the denominator of the decimal fraction.

Example 1 Read this decimal number: 123.123

Solution First we read the whole number part. **When we come to the decimal point, we say "and."** Then we read the fraction part, ending with the name of the last decimal place.

$$123.1\underline{23} \qquad \text{We say "and" for the decimal point.}$$

One hundred twenty-three and one hundred twenty-three thousandths.

Example 2 Use digits to write these decimal numbers.

(a) Seventy-five thousandths

(b) One hundred and eleven hundredths

Solution (a) The last word tells us the last place in the decimal number. "Thousandths" means there are three places to the right of the decimal point.

. _ _ _

We fit 75 into these places so that the 5 is in the last place. We write zero in the remaining place.

. <u>0</u> <u>7</u> <u>5</u>

Decimal numbers without a whole number part are usually written with a zero in the ones' place.

Either **.075** or **0.075**

(b) To write one hundred and eleven hundredths, we remember that the word "and" separates the whole number part of the number from the fraction part. First we write the whole number part followed by a decimal point for "and."

100.

Then we write the fraction part. We shift our attention to the last word to find out how many decimal places there are. "Hundredths" means there are two decimal places.

100. _ _

Now we fit 11 into the two decimal places.

100.11

Comparing decimals With decimal numbers it is important to consider the values of the places occupied by the digits. Each place to the left of a particular digit has a value 10 times greater than the value of the place of the particular digit. Terminal zeros to the right of the decimal point have no value.

1.3 equals 1.30 equals 1.300 equals 1.3000

When we compare decimal numbers, it is convenient to

insert terminal zeros so that both numbers will have the same number of digits to the right of the decimal point.

Example 3   Compare: 0.12 ◯ 0.012

Solution   So that each number has the same number of decimal places, we insert a terminal zero in the number on the left and get

$$0.120 \bigcirc 0.012$$

One hundred twenty thousandths is greater than twelve thousandths, so

$$0.120 > 0.012$$

Example 4   Compare: 0.4 ◯ 0.400

Solution   We may delete two terminal zeros from the number on the right and get

$$0.4 = 0.4$$

We could have added terminal zeros to the number on the left to get

$$0.400 = 0.400$$

Example 5   Compare: 1.232 ◯ 1.23185

Solution   We insert two terminal zeros in the number on the left and get

$$1.23200 \bigcirc 1.23185$$

and since 23200 is greater than 23185, we write

$$1.23200 > 1.23185$$

Practice   Use words to write each decimal number.

   a.   25.134

   b.   100.01

Use digits to write each decimal number.

   c.   One hundred two and three tenths

d. One hundred twenty-five ten-thousandths

e. Three hundred and seventy-five thousandths

Write terminal zeros as necessary so that each number has the same number of digits to the right of the decimal point. Then replace the circle with the proper comparison symbol.

f. 10.30 ◯ 10.3

g. 5.06 ◯ 5.60

h. 1.1 ◯ 1.099

Problem set 37

1. There were 3 towns on the mountain. The population of Hazelhurst was 4248. The population of Baxley was 3584. The population of Jesup was 9418. What was the average population of the 3 towns on the mountain?

2. The film was a long one, lasting 3 hours 26 minutes. How many minutes long was the film?

3. A mile is 1760 yards. Claudia ran 440 yards. How many more yards does she need to run in order to run 1 mile?

4. One sixth of the seats were empty. What fraction of the seats were not empty?

5. A square and a regular pentagon share a common side. The perimeter of the square is 20 cm. What is the perimeter of the pentagon?

6. Round 3,197,270

(a) to the nearest million.

(b) to the nearest hundred thousand.

7. Rewrite $\frac{1}{2}$, $\frac{2}{3}$, and $\frac{3}{4}$ so that they have a common denominator.

8. Mentally estimate the product of 313 and 489.

9. Draw a diagram of this statement. Then answer the questions that follow.

> Five eighths of the troubadour's 200 songs were about love and chivalry.

(a) How many of the songs were about love and chivalry?

(b) How many of the songs were not about love and chivalry?

10. (a) What fraction of the rectangle is not shaded?

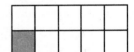

(b) What decimal part of the rectangle is not shaded?

11. Use words to write 3.025.

12. Use digits to write seventy-six and five hundredths.

13. Insert terminal zeros as necessary. Then replace each circle with the proper comparison symbol.

(a)   12.6 $\bigcirc$ 12.60

(b)   3.14159 $\bigcirc$ 3.1416

14. Simplify each fraction or mixed number.

(a) $3\dfrac{30}{18}$ 　　　　　(b) $\dfrac{81}{24}$ 　　　　　(c) $\dfrac{480}{1000}$

15. Write the prime factorization of 3008.

Solve:

16. $10 \cdot 6 = 4w$

17. $340 = m + 74$

18. $x - 28 = 82$

Add, subtract, multiply, or divide, as indicated:

19. $\dfrac{1}{4} + \dfrac{3}{8} + \dfrac{1}{2}$ **20.** $\dfrac{5}{6} - \dfrac{3}{4}$

21. $4\dfrac{5}{8} - 1\dfrac{1}{2}$ **22.** $\dfrac{8}{9} \cdot 1\dfrac{1}{5} \cdot 10$

23. $5\dfrac{2}{5} \div \dfrac{9}{10}$ **24.** $4\dfrac{5}{8} + 1\dfrac{1}{2}$

25. $7\dfrac{3}{4} + 1\dfrac{7}{8}$ **26.** $6\dfrac{1}{6} - 2\dfrac{1}{2}$

27. $\dfrac{2}{3} + \left(\dfrac{2}{3} \div \dfrac{1}{2}\right)$ **28.** $\begin{array}{r} 8600 \\ \times \quad 90 \\ \hline \end{array}$

29. $\begin{array}{r} \$52.48 \\ 78.69 \\ 47.78 \\ 86.94 \\ 42.03 \\ + \;\; 95.16 \\ \hline \end{array}$ **30.** $\dfrac{\$368.46}{18}$

LESSON 38

Rounding Decimal Numbers

To round decimal numbers, we can use the same circle and arrow procedure that we use to round whole numbers.

Example 1 Round 3.14159 to the nearest hundredth.

Solution The hundredths' place is two places to the right of the decimal point. We circle the digit in that place and mark the digit to its right with an arrow.

$$\downarrow$$
$$3.1④159$$

Since the arrow-marked digit is less than 5, we leave the circled digit unchanged. Then we change all digits to the right of the circled digit to zero.

$$3.14000$$

Terminal zeros to the right of the decimal point do not serve as placeholders as they do in whole numbers. After rounding decimal numbers, we should remove terminal zeros to the right of the decimal point.

$$3.14\cancel{000} \rightarrow \mathbf{3.14}$$

Example 2   Round 4396.4315 to the nearest hundred.

Solution   We are rounding to the nearest hundred, not to the nearest hundredth.

$$4\textcircled{3}96.4315$$

Since the arrow-marked digit is 5 or more, we increase the circled digit by 1. All of the following digits become zeros.

$$4400.0000$$

Zeros at the end of a whole number are needed as placeholders. Terminal zeros to the right of the decimal point are not needed as placeholders. We remove these zeros.

$$4400.\cancel{0000} \rightarrow \mathbf{4400}$$

Example 3   Round 38.62 to the nearest whole number.

Solution   To round a number to the nearest whole number, we round to the ones' place.

$$3\textcircled{8}.62 \rightarrow 39.00 \rightarrow \mathbf{39}$$

Practice   a.   Round 3.14159 to the nearest ten-thousandth.

   b.   Round 365.2418 to the nearest hundred.

   c.   Round 57.432 to the nearest whole number.

**Problem set
38**

1. The high jumper set a new school record when she cleared 5 feet 8 inches. How many inches is 5 feet 8 inches?

2. During the first week of November the highest daily temperatures in degrees Fahrenheit were 42°, 43°, 38°, 47°, 51°, 52°, and 49°. What was the average high daily temperature during the first week of November?

3. In 10 years the population increased from 87,196 to 120,310. By how many people did the population increase in 10 years?

4. The largest three-digit odd number is how much less than the largest four-digit even number?

5. A regular hexagon and a regular octagon share a common side. If the perimeter of the hexagon is 24 cm, what is the perimeter of the octagon?

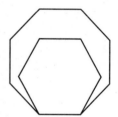

6. Draw a diagram of this statement. Then answer the questions that follow.

> Three twentieths of the 100 questions on the test were true-false.

(a) How many of the questions on the test were true-false?

(b) How many of the questions on the test were not true-false?

7. Find the LCM of 5, 6, and 10.

8. Round 15.73591

(a) to the nearest hundredth.

(b) to the nearest whole number.

9. Use words to write each of these decimal numbers.

(a) 150.035

(b) 0.0015

10. Use digits to write each of these decimal numbers.

(a) One hundred twenty-five thousandths

(b) One hundred and twenty-five thousandths

11. Insert terminal zeros as necessary. Then replace each circle with the proper comparison symbol.

(a) 0.128 $\bigcirc$ 0.14 (b) 0.03 $\bigcirc$ 0.0015

12. Find the length of this segment

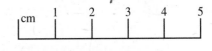

(a) in centimeters.

(b) in millimeters.

13. Simplify each fraction or mixed number.

(a) $\dfrac{810}{630}$ (b) $4\dfrac{72}{48}$

14. Write the reciprocal of each number.

(a) 7 (b) $3\dfrac{3}{8}$

15. Complete each equivalent fraction.

(a) $\dfrac{?}{9} = \dfrac{20}{36}$ (b) $\dfrac{7}{?} = \dfrac{14}{36}$

Solve:

16. $8m = 4 \cdot 18$

17. $125 + x = 500$

18. $54 = 54 - y$

Add, subtract, multiply, or divide, as indicated:

19. $\dfrac{3}{4} + \dfrac{5}{8} + \dfrac{1}{2}$ **20.** $\dfrac{3}{4} - \dfrac{1}{6}$

21. $4\dfrac{1}{2} - \dfrac{3}{8}$ **22.** $\dfrac{3}{8} \cdot 2\dfrac{2}{5} \cdot 3\dfrac{1}{3}$

23. $2\dfrac{7}{10} \div 5\dfrac{2}{5}$ **24.** $5 \div 4\dfrac{1}{6}$

25. $8\dfrac{5}{8} + 5\dfrac{3}{4}$ **26.** $6\dfrac{1}{2} - 2\dfrac{5}{6}$

27. $\dfrac{3}{4} + \left(\dfrac{1}{2} \div \dfrac{2}{3} \right)$ **28.** $\begin{array}{r} 740 \\ \times\,800 \\ \hline \end{array}$

29. $\begin{array}{r} \$500.00 \\ 47.74 \\ 865.62 \\ 478.95 \\ +\quad 30.72 \\ \hline \end{array}$ **30.** $\dfrac{48{,}080}{16}$

LESSON 39

Decimal Numbers on the Number Line

If each centimeter segment on a centimeter scale is divided into 10 equal segments, then each segment is 1 millimeter long. Each segment is also one tenth of a centimeter long.

Example 1 Find the length of this segment

(a) in millimeters.

(b) in centimeters.

Solution (a) Each centimeter is 10 mm. Thus, each small segment on the scale is 1 mm. The length of the segment is **23 mm**.

(b) Each centimeter on the scale has been divided into 10 equal parts. The length of the segment is 2 centimeters plus three tenths of a centimeter. In the metric system we use decimals rather than common fractions to indicate part of a unit. The length of the segment is **2.3 cm**.

If the distance between consecutive whole numbers on a number line is divided into 100 equal units, then numbers corresponding to the marks on the number line can be named using two decimal places.

Example 2 Find the number on the number line indicated by each arrow.

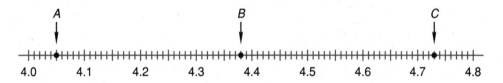

Solution We are considering a portion of the number line from 4 to 5. The distance from 4 to 5 has been divided into 100 equal segments. Tenths have been identified. The point 4.1 is one tenth of the distance from 4 to 5. However, it is also ten hundredths of the distance from 4 to 5, so 4.1 equals 4.10.

Arrow A indicates **4.05**

Arrow B indicates **4.38**

Arrow C indicates **4.73**

Practice **a.** Find the length of this segment in centimeters.

b. What point on a number line is halfway between 2.6 and 2.7?

c. What decimal number names the point marked by A on this number line?

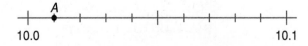

Problem set 39

1. In 3 boxes of cereal Jeff counted 188 raisins, 212 raisins, and 203 raisins. What was the average number of raisins in each box of cereal?

2. The pollen count had increased from 497 parts per million to 1032 parts per million. By how much had the pollen count increased?

3. Sylvia spent $3.95 for lunch but still had $12.55. How much money did she have before she bought lunch?

4. In 1903 the Wright brothers made the first powered airplane flight. Just 66 years later astronauts first landed on the moon. In what year did astronauts first land on the moon?

5. The perimeter of the square equals the perimeter of the regular hexagon. If each side of the hexagon is 6 inches long, how long is each side of the square?

6. Draw a diagram of this statement. Then answer the questions that follow.

   Each week Jessica saves two fifths of her $4.00 allowance.

   (a) How much allowance money does she save each week?

   (b) How much allowance money does she not save each week?

7. Estimate the product of 396 and 71.

8. Round 7.49362 to the nearest thousandth.

9. Use words to write each of these decimal numbers.

   (a) 200.02

   (b) 0.001625

10. Use digits to write each of these decimal numbers.

    (a) One hundred seventy-five millionths

    (b) Three thousand, thirty and three hundredths

11. Replace each circle with the proper comparison symbol.

    (a) 6.174 ◯ 6.17401   (b) 14.276 ◯ 1.4276

12. Find the length of this segment

    (a) in centimeters.

    (b) in millimeters.

13. What decimal number names the point marked X on this number line?

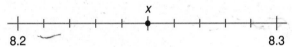

14. Simplify each fraction or mixed number.

    (a) $\dfrac{75}{300}$ (b) $3\dfrac{30}{12}$ (c) $\dfrac{288}{720}$

15. What decimal number is halfway between 7 and 8?

Solve:

16. $15 \cdot 20 = 12y$

17. $300 = 74 + c$

18. $36 = x - 24$

Add, subtract, multiply, or divide, as indicated:

19. $\dfrac{5}{6} + \dfrac{2}{3} + \dfrac{1}{2}$ **20.** $\dfrac{5}{6} - \dfrac{1}{4}$

21. $3\dfrac{11}{12} - 1\dfrac{1}{4}$ **22.** $\dfrac{1}{10} \cdot 2\dfrac{2}{3} \cdot 3\dfrac{3}{4}$

23. $5\dfrac{1}{4} \div 1\dfrac{2}{3}$ **24.** $3\dfrac{1}{5} \div 4$

25. $6\dfrac{7}{8} + 4\dfrac{1}{4}$ **26.** $5\dfrac{1}{6} - 1\dfrac{2}{3}$

27. $\dfrac{1}{8} + \left(\dfrac{5}{6} \cdot \dfrac{3}{4} \right)$ **28.** $\dfrac{90{,}900}{18}$

29.
$$\begin{array}{r} \$375.64 \\ 124.36 \\ 468.95 \\ 876.43 \\ +\ 984.86 \\ \hline \end{array}$$

30.
$$\begin{array}{r} 370 \\ \times\ 800 \\ \hline \end{array}$$

**LESSON
40**

Adding and Subtracting Decimal Numbers

When we add or subtract whole numbers, we align the ones' digits so that we add digits that have the same place value. **When we add or subtract decimal numbers, we align the decimal points vertically.** Aligning the decimal points ensures that we will be adding or subtracting digits that have the same place value. We will consider several examples.

Example 1   Add: $3.6 + 0.36 + 36$

Solution   We align the decimal points vertically. A number written without a decimal point is a whole number, so the decimal point is to the right of 36.

$$\begin{array}{r} 3.6 \\ 0.36 \\ +\ 36. \\ \hline \mathbf{39.96} \end{array}$$

Example 2   Add: $0.1 + 0.2 + 0.3 + 0.4$

Solution   We align the decimal points vertically and add. We record only one digit in each column. The sum is 1.0, not 0.10. Since 1.0 equals 1, we can simplify the answer to **1**.

$$\begin{array}{r} 0.1 \\ 0.2 \\ 0.3 \\ +\ 0.4 \\ \hline \mathbf{1.0} = \mathbf{1} \end{array}$$

Example 3   Subtract: 12.3 − 4.567

Solution   We write the first number above the second number, aligning the decimal points. We write zeros in the empty places and subtract.

$$\begin{array}{r} {\scriptstyle 1\ 12\ 9\ \ \ } \\[-2pt] {\scriptstyle 1\ 1} \\[-2pt] 12.3\cancel{0}0 \\ -\ 4.567 \\ \hline \mathbf{7.733} \end{array}$$

Example 4   Subtract: 5 − 4.32

Solution   We write the whole number 5 with a decimal point and write zeros in the two empty decimal places. Then we subtract.

$$\begin{array}{r} {\scriptstyle 4\ \ 9\ \ } \\[-2pt] {\scriptstyle 1} \\[-2pt] \cancel{5}.\cancel{0}0 \\ -\ 4.32 \\ \hline \mathbf{0.68} \end{array}$$

Practice   Add or subtract as indicated:

    a. 1.2 + 3.45 + 23.6

    b. 4.5 + 0.51 + 6 + 12.4

    c. 0.2 + 0.4 + 0.6 + 0.8

    d. 36.274 − 5.39

    e. 16.7 − 1.936

    f. 12 − 0.875

Problem set 40

1. In the first 6 months of the year the Montgomerys' monthly electricity bills were $128.45, $131.50, $112.30, $96.25, $81.70, and $71.70. What was their average monthly electricity bill during the first 6 months of the year?

2. The price was reduced from two thousand, four hundred ninety-eight dollars to one thousand, nine hundred ninety-nine dollars. By how much was the price reduced?

3. A 1-year subscription to a monthly magazine costs $15.60. The regular newsstand price is $1.75 per issue. How much is saved per issue by paying the subscription price?

4. Carlos ran one lap in 1 minute 3 seconds. Orlando ran one lap 5 seconds faster than Carlos. How many seconds did it take Orlando to run one lap?

5. The perimeter of the square equals the perimeter of the regular pentagon. Each side of the pentagon is 16 cm long. How long is each side of the square?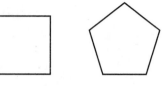

6. Draw a diagram of this statement. Then answer the questions that follow.

> Two ninths of the 54 fish in the tank were guppies.

(a) How many of the fish were guppies?

(b) How many of the fish were not guppies?

7. Find the LCM of 6, 8, and 12.

8. (a) What fraction of this square is not shaded?

(b) What decimal part of this square is not shaded?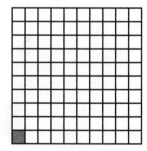

9. Round 2375.4174

(a) to the nearest hundredth.

(b) to the nearest hundred.

10. Use words to write 100.075.

11. Use digits to write twenty-five hundred-thousandths.

12. Find the length of this segment

(a) in centimeters.

(b) in millimeters.

13. What decimal number names the point marked with an arrow on this number line?

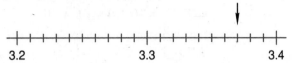

14. Simplify each of these numbers.

(a) $4\frac{44}{4}$

(b) $\frac{444}{44}$

15. What decimal number is halfway between 1.2 and 1.3?

Solve:

16. $15x = 9 \cdot 10$

17. $f + 46 = 200$

18. $512 - y = 215$

Add, subtract, multiply, or divide, as indicated:

19. $3.4 + 5.63 + 15$

20. $3.65 + 0.9 + 8 + 15.23$

21. $36.45 - 4.912$

22. $15 - 4.29$

23. $1\frac{1}{2} + 2\frac{2}{3} + 3\frac{3}{4}$

24. $1\frac{1}{2} \cdot 2\frac{2}{3} \cdot 3\frac{3}{4}$

25. $1\frac{1}{6} - \left(\frac{1}{2} + \frac{1}{3}\right)$

26. $3\frac{1}{12} - 1\frac{3}{4}$

27. $46,731 \div 30$

28. $30(40)(25)$

29. $\$36.24 + \$1.79 + 38¢ + \$15.60 + \$12$

30. $\left(3\frac{1}{2} + 1\frac{3}{4}\right) \div \left(4 - 3\frac{1}{8}\right)$

LESSON 41

Ratio

A **ratio** is a comparison of two numbers. Ratios can be written several ways. The ratio 7 to 10 can be written as follows.

With the word "to"	7 to 10
As a fraction	$\frac{7}{10}$
As a decimal number	0.7
With a colon	7:10

Usually, we will write ratios as fractions. Ratios should be written in reduced form just as we write fractions in reduced form. **A ratio should not be written as a mixed number.**

Almost every ratio problem has three ratio numbers. Usually only two of the ratio numbers are given. We must calculate the hidden ratio number. If we are told that the ratio of frogs to fish is 2 to 5, we can write

$$2 \text{ frogs}$$
$$5 \text{ fish}$$

The third ratio number is the total. Two frogs plus five fish equals a total of 7.

$$\begin{array}{l} 2 \text{ frogs} \\ \underline{5 \text{ fish}} \\ 7 \text{ total} \end{array}$$

Now we can write the ratio of frogs to fish as

$$\frac{2}{5}$$

and the ratio of frogs to the total is

$$\frac{2}{7}$$

and the ratio of fish to the total is

$$\frac{5}{7}$$

Sometimes the total is given and the hidden ratio number is one of the other numbers. If we are told that in every group of 30 dogs and cats there are 12 dogs,

$$
\begin{array}{l}
12 \text{ dogs} \\
\underline{\ ?\ \text{ cats}} \\
30 \text{ total}
\end{array}
$$

we see that the ratio number for cats must be 18.

$$
\begin{array}{l}
12 \text{ dogs} \\
\underline{18 \text{ cats}} \\
30 \text{ total}
\end{array}
$$

Whenever we work a ratio problem, it is a good idea to begin by writing all three ratio numbers.

Example 1　In a class of 28 students there are 12 boys.

(a) What is the boy-girl ratio?

(b) What is the girl-boy ratio?

Solution　We will begin by writing all three ratio numbers.

$$
\begin{array}{l}
12 \text{ boys} \\
\underline{\ ?\ \text{ girls}} \\
28 \text{ total}
\end{array}
\quad \rightarrow \quad
\begin{array}{l}
12 \text{ boys} \\
\underline{16 \text{ girls}} \\
28 \text{ total}
\end{array}
$$

Now we can write the answers.

(a) The boy-girl ratio is $\frac{12}{16}$, which reduces to $\frac{3}{4}$, a ratio of 3 to 4.

(b) The girl-boy ratio is $\frac{16}{12}$, which reduces to $\frac{4}{3}$, a ratio of 4 to 3. Note that we do not change the ratio to a mixed number.

Example 2　The team won $\frac{4}{7}$ of its games and lost the rest. What was the team's won-lost ratio?

Solution　We are not told how many games the team played. However, we are told that the team won $\frac{4}{7}$ of its games. Therefore the team lost $\frac{3}{7}$ of its games. Thus, on the average, the team won

4 out of every 7 games. So the three ratio numbers are

$$
\begin{array}{l}
4 \text{ won} \\
\underline{3 \text{ lost}} \\
7 \text{ total}
\end{array}
$$

The won-lost ratio was 4 to 3, which we write $\frac{4}{3}$.

Example 3   In the bag were red marbles and green marbles. If the ratio of red marbles to green marbles was 4 to 5, what fraction of the marbles was red?

Solution   First we write all three ratio numbers.

$$
\begin{array}{l}
4 \text{ red} \\
\underline{5 \text{ green}} \\
? \text{ total}
\end{array}
\qquad \rightarrow \qquad
\begin{array}{l}
4 \text{ red} \\
\underline{5 \text{ green}} \\
9 \text{ total}
\end{array}
$$

The question was what fraction of the marbles was red. There were 4 red marbles and a total of 9 marbles so the fraction that was red was $\frac{4}{9}$.

Practice   Begin each problem by writing all three ratio numbers.

a. In the pond were 240 little fish and 90 big fish. What was the ratio of big fish to little fish?

b. Fourteen of the 30 students in the class were girls. What was the boy-girl ratio in the class?

c. The team won $\frac{3}{8}$ of its games and lost the rest. What was the team's won-lost ratio?

d. The bag contained red marbles and blue marbles. If the ratio of red marbles to blue marbles was 5 to 3, what fraction of the marbles was blue?

Problem set 41   **1.** Fourteen of the 32 students in the class were girls. What was the ratio of boys to girls in the class?

2. During the last 3 years the annual rainfall has been 23 inches, 21 inches, and 16 inches. What has been the average annual rainfall during the last 3 years?

3. Sean reads 35 pages each night. How many pages does he read in a week?

4. Shannon swam 100 meters in 56.24 seconds. Donna swam 100 meters in 59.48 seconds. Donna took how many seconds longer to swim 100 meters than Shannon?

5. Draw a diagram of this statement. Then answer the questions that follow.

> Two fifths of the 30 players in the game had never played rugby before.

(a) How many of the players had never played rugby before?

(b) What was the ratio of those who had played rugby to those who had not played rugby?

6. *AB* is 40 mm. *AC* is 95 mm. Find *BC*.

7. The length of the rectangle is 5 cm greater than its width. What is the perimeter of the rectangle?

8. Estimate the sum of 3624, 2889, and 896 by rounding each number to the nearest hundred before adding.

9. Round 6.857142

(a) to the nearest whole number.

(b) to three decimal places.

10. Use words to write 120.0305.

11. Use digits to write each number.

(a) Twelve millionths

(b) One hundred thirty thousand and four hundredths

12. Find the length of this segment

(a) in centimeters.

(b) in millimeters.

13. What decimal number names the point marked *M* on this number line?

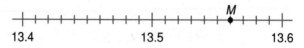

14. Simplify each of these numbers.

(a) $\dfrac{27}{18}$ (b) $8\dfrac{15}{3}$ (c) $\dfrac{144}{600}$

15. In this figure, which angle is

(a) a right angle?

(b) an acute angle?

(c) an obtuse angle?

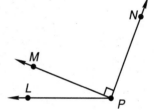

Solve:

16. $8y = 144$

17. $63 = 91 - p$

18. $213 = w + 57$

Add, subtract, multiply, or divide, as indicated:

19. $4.27 + 16.3 + 10$ **20.** $0.9 + 0.8 + 0.7 + 0.5$

21. $3\dfrac{1}{2} + 1\dfrac{1}{3} + 2\dfrac{1}{4}$ **22.** $3\dfrac{1}{2} \cdot 1\dfrac{1}{3} \cdot 2\dfrac{1}{4}$

23. $3\dfrac{5}{6} - \left(\dfrac{2}{3} - \dfrac{1}{2}\right)$ **24.** $8\dfrac{5}{12} - 3\dfrac{2}{3}$

25. $2\dfrac{3}{4} \div 4\dfrac{1}{2}$ **26.** $5 - \left(\dfrac{2}{3} \div \dfrac{1}{2}\right)$

27. (437)(86)

28. 38,015 ÷ 19

29. $6.47 + $5.24 + $11 + 53¢

30. $20.00 − (3 × $4.98)

**LESSON
42**

Perimeter, Part 2

In this lesson we will practice finding the perimeter of shapes such as the shape shown here. All the angles in the figure are right angles.

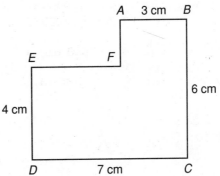

We can find the perimeter of this shape by adding the lengths of all six sides. Although we are given the lengths of only four of the sides, we can figure out the lengths of the other two sides.

To find *EF*, we observe that the length of side *EF* plus the length of side *AB* equals the length of side *DC*. Thus *EF* is 4 cm.

We find *AF* in a similar way. The length of side *AF* plus the length of side *ED* equals the length of side *BC*. Thus *AF* is 2 cm.

The perimeter of the hexagon is

3 cm + 6 cm + 7 cm + 4 cm + 4 cm + 2 cm = 26 cm

Example   Find the lengths of the two unmarked sides of this polygon. Then find the perimeter of the polygon. Dimensions are in meters. All angles are right angles.

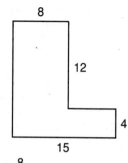

Solution   We will use the letters *a* and *b* to refer to the unmarked sides.

The length of side *a* is equal to the combined lengths of the two shorter vertical sides. Thus the length of side *a* is

$$12 \text{ m} + 4 \text{ m} = \textbf{16 m}$$

The lengths of the two shorter horizontal sides together equal 15 m. Thus the length of side *b* is

$$15 \text{ m} - 8 \text{ m} = \textbf{7 m}$$

The perimeter of the polygon is

$$8 \text{ m} + 12 \text{ m} + 7 \text{ m} + 4 \text{ m} + 15 \text{ m} + 16 \text{ m} = \textbf{62 m}$$

Practice   Find the length of each unmarked side and find the perimeter of each polygon. Dimensions are in feet. All angles are right angles.

a.
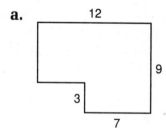

b.
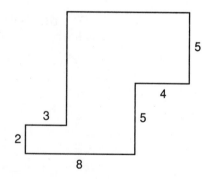

Problem set 42

1. The team won $\frac{2}{3}$ of its games and lost the rest. What was the team's won-lost ratio?

2. During the first 6 months of the year the car dealership sold 47 cars, 53 cars, 62 cars, 56 cars, 46 cars, and 48

cars. What was the average number of cars sold during the first 6 months of the year?

3. The relay team carried the baton around the track. Darren ran his part in eleven and six tenths seconds, Robert ran his part in eleven and three tenths seconds, Orlando ran his part in eleven and two tenths seconds, and Claude ran his part in ten and nine tenths seconds. What was the team's total time?

4. Jenny went to the store with $10 and returned home with 5 gallons of milk and $1.30 in change. What was the cost of each gallon of milk?

5. Draw a diagram of this statement. Then answer the questions that follow.

    Kevin shot par on two thirds of the 18 holes.

    (a) On how many holes did Kevin shoot par?

    (b) On how many holes did Kevin not shoot par?

6. Sketch this figure on your paper. Then find the length of each unmarked side. Then find the perimeter of the polygon. Dimensions are in inches. All angles are right angles.

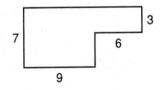

7. Complete each equivalent fraction.

    (a) $\dfrac{5}{6} = \dfrac{?}{18}$ (b) $\dfrac{?}{8} = \dfrac{9}{24}$ (c) $\dfrac{3}{4} = \dfrac{15}{?}$

8. Find the LCM of 9, 6, and 12.

9. (a) What decimal part of this square is shaded?

    (b) What decimal part of this square is not shaded?

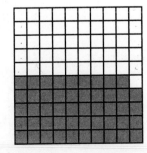

10. Round 3184.5641

    (a) to two decimal places.

    (b) to the nearest hundred.

11. Name each decimal number.

    (a) 0.001875

    (b) 600.007

12. Use digits to write each number.

    (a) Three hundred seventy-five ten-thousandths

    (b) Sixty and seven hundredths

13. Simplify each of these numbers.

    (a) $5\dfrac{15}{9}$ (b) $\dfrac{144}{30}$ (c) $\dfrac{720}{1080}$

14. Find the length of segment *AB*.

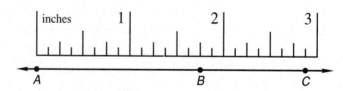

15. Draw a pair of parallel lines. Next draw another pair of parallel lines that intersect the first pair of lines but are not perpendicular to them. Then shade the region enclosed by the intersecting pairs of lines.

Solve:

16. $7 \cdot 8 = 4x$

17. $4.2 = 1.7 + y$

18. $134 = d - 27$

Add, subtract, multiply, or divide, as indicated:

19. $4.375 + 12.125 + 1.3$ 20. $0.1 + 0.2 + 0.3 + 0.4$

21. $\dfrac{3}{5} \cdot 12 \cdot 4\dfrac{1}{6}$

22. $\dfrac{5}{6} + 1\dfrac{3}{4} + 2\dfrac{1}{2}$

23. $\dfrac{5}{8} + \left(\dfrac{1}{2} + \dfrac{3}{8}\right)$

24. $3\dfrac{1}{6} - 1\dfrac{2}{3}$

25. $3\dfrac{1}{3} \div 5$

26. $4 - \left(\dfrac{3}{2} \div \dfrac{2}{3}\right)$

27. $40(60)(80)$

28. $3000 \div 24$

29. $\$12 + \$18.57 + \$7.98 + 7¢$

30. $\$10 - (4 \times 78¢)$

LESSON
43

Graphs

We use **graphs** to help us understand quantitative information. A graph may use pictures, bars, lines, or parts of circles to help the reader visualize comparisons or changes. In this lesson we will practice gathering information from graphs.

Example 1 Find the information in this pictograph to answer the following questions.

Donut Sales

Jan.	⊙ ⊙ ⊙ ⊙
Feb.	⊙ ⊙ ⊙ ⊙ ⊙ ⊙
Mar.	⊙ ⊙ ⊙ ⊙ ⊙ (

⊙ Represents 10,000 donuts

(a) About how many donuts were sold in March?

(b) About how many donuts were sold in the first 3 months of the year?

Solution   The key at the bottom of the graph shows us that each picture of a donut represents 10,000 donuts.

(a) For March we see 5 whole donuts, which represents 50,000 donuts, and half a donut, which represents 5000 donuts. Thus, the $5\frac{1}{2}$ donuts pictured mean that **about 55,000 donuts were sold in March**.

(b) We see a total of $15\frac{1}{2}$ donuts pictured for the first 3 months of the year. Fifteen times 10,000 is 150,000. Half of 10,000 is 5000. Thus, **about 155,000 donuts were sold in the first 3 months of the year**.

Example 2   Find the information in this bar graph to answer the following questions.

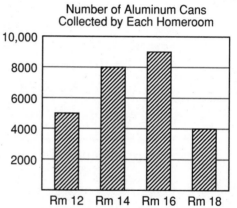

Number of Aluminum Cans
Collected by Each Homeroom

(a) About how many cans were collected by the students in Room 14?

(b) The students in Room 16 collected about as many cans as what other two homerooms combined?

Solution   We look at the scale on the left side of the graph. We see that the distance between two horizontal lines on the scale represents 2000 cans. Thus, halfway from one line to the next represents 1000 cans.

(a) The students in Room 14 collected about **8000 cans**.

(b) The students in Room 16 collected about 9000 cans. This was about as many cans as **Room 12 and Room 18 combined**.

Example 3 This line graph shows how Paul's test scores have changed during the year.

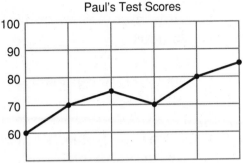

Paul's Test Scores

(a) What was Paul's score on Test 3?

(b) In general, are Paul's scores improving or getting worse?

Solution (a) We find Test 3 on the scale across the bottom of the graph and go up to the point that represents Paul's score. We see that the point is halfway between the lines that represent 70 and 80. Thus, on Test 3, Paul's score was about **75**.

(b) With only one exception, Paul scored higher on each succeeding test. So, in general, Paul's scores are **improving**.

Example 4 Find the information in this circle graph to answer the following questions.

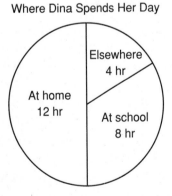

Where Dina Spends Her Day

(a) Altogether, how many hours are included in this graph?

(b) What fraction of Dina's day is spent at school?

Solution A circle graph (sometimes called a pie graph) shows the relationship between parts of a whole. This graph shows parts of a whole day.

(a) This graph includes **24 hours**, one whole day.

(b) Dina spends 8 of the 24 hours at school. We reduce $\frac{8}{24}$ to $\frac{1}{3}$.

Practice Find the information from the graphs in this lesson to answer each question.

    a. How many more donuts were sold in February than in January?

    b. How many aluminum cans were collected by all four homerooms?

    c. On which test was Paul's score lower than his score on the previous test?

    d. What fraction of Dina's day was spent somewhere other than at home or at school?

Problem set 43

    1. The ratio of soldiers to civilians at the outpost was 3 to 7. What fraction of the people at the outpost were soldiers?

    2. Denise read a 345-page book in 3 days. What was the average number of pages she read each day?

    3. Christine ran a mile in 5 minutes 52 seconds. How many seconds did it take Christine to run a mile?

Refer to the graphs in this lesson to answer questions 4 and 5.

    4. How many fewer cans were collected by the students in Room 18 than by the students in Room 16?

    5. If Paul scores 85 on Test 7, what will be his test score average for all 7 tests?

    6. Draw a diagram for this statement. Then answer the questions that follow.

        Mira read three eighths of the 384-page book before she could put it down.

    (a) How many pages did she read?

    (b) How many more pages does she need to read to be halfway through the book?

7. Sketch this figure on your paper. Then find the length of each unmarked side and find the perimeter of the polygon. Dimensions are in centimeters. All angles are right angles.

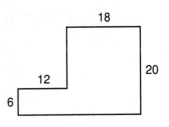

8. Complete each equivalent fraction.

(a) $\dfrac{7}{9} = \dfrac{?}{18}$ (b) $\dfrac{?}{9} = \dfrac{20}{36}$ (c) $\dfrac{4}{5} = \dfrac{24}{?}$

9. Round 2986.34157

(a) to the nearest thousand.

(b) to three decimal places.

10. Use words to write each number.

(a) 0.00325

(b) 3,280,004,000

11. Use digits to write each number.

(a) One and seventy-five thousandths

(b) Twenty and five twelfths

12. Find the length of this segment

(a) in centimeters.

(b) in millimeters.

13. What decimal number names the point marked *A* on this number line?

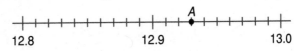

14. Simplify each of these numbers.

(a) $9\dfrac{15}{12}$ (b) $\dfrac{288}{90}$

15. In the figure shown: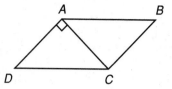

   (a) Which segment is perpendicular to $\overline{AD}$?

   (b) Which segment appears to be parallel to $\overline{AD}$?

Solve:

16. $4.3 + a = 6.7$

17. $m - 3.6 = 4.7$

18. $5m = 10 \cdot 7$

Add, subtract, multiply, or divide, as indicated:

19. $5.37 + 27.7 + 4$

20. $345.6 + 14 + 1.58$

21. $\dfrac{5}{9} \cdot 6 \cdot 2\dfrac{1}{10}$

22. $\dfrac{5}{8} + \dfrac{3}{4} + \dfrac{1}{2}$

23. $\dfrac{3}{10} + \left(\dfrac{1}{2} + \dfrac{1}{5}\right)$

24. $\dfrac{3}{10} - \left(\dfrac{1}{2} - \dfrac{1}{5}\right)$

25. $5 \div 3\dfrac{1}{3}$

26. $10 - \left(\dfrac{3}{4} \div 2\right)$

27. $470(600)$

28. $10,000 \div 16$

29. $\$0.89 + \$15 + \$5.47 + 89\cancel{c} + \$1.42$

30. $(\$20 - \$5.24) \div 6$

LESSON
44

Proportions

We remember that a ratio is a comparison of two numbers. Ratios may be written as fractions. The ratio 16 to 20 and the ratio 4 to 5 are equal ratios.

$$\frac{16}{20} = \frac{4}{5}$$

Whenever we write two ratios connected by an equals sign, we are writing a **proportion**. If we multiply the upper term of one ratio by the lower term of the other ratio, we form a **cross product**. The cross products of equal ratios are equal. We illustrate by finding the cross products of the proportion above.

$$\frac{16}{20} = \frac{4}{5}$$

$$20 \cdot 4 = 80$$

$$16 \cdot 5 = 80$$

If the cross products are equal, the ratios are equal. We will use cross products to help us find the missing terms in equal ratios.

We will follow a two-step process.

Step 1. Find the cross products.

Step 2. Divide the known product by the known factor.

Example 1 Solve the proportion: $\frac{12}{20} = \frac{n}{30}$

Solution We solve a proportion by finding the missing term.

Step 1. First we find the cross products. Since we are completing a proportion, the cross products must be equal.

$$\frac{12}{20} = \frac{n}{30}$$

$$20 \cdot n = 12 \cdot 30 \qquad \text{cross products}$$

$$20n = 360 \qquad \text{simplified}$$

Step 2. Divide the known product (360) by the known factor (20). The result is the missing term.

$$n = \frac{360}{20} \qquad \text{divide by 20}$$

$$\boldsymbol{n = 18} \qquad \text{simplified}$$

Example 2 Solve: $\dfrac{15}{x} = \dfrac{20}{32}$

Solution Step 1. $\qquad 20x = 480 \qquad$ cross products

Step 2. $\qquad \boldsymbol{x = 24} \qquad$ divided by 20

Practice Solve each proportion.

a. $\dfrac{a}{12} = \dfrac{6}{8}$ **b.** $\dfrac{30}{b} = \dfrac{20}{16}$

c. $\dfrac{14}{21} = \dfrac{c}{15}$ **d.** $\dfrac{30}{25} = \dfrac{24}{d}$

Problem set 44 Dan made a line graph to show his height on each birthday. Refer to this graph to answer questions 1 and 2.

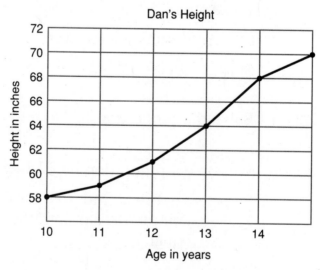

Dan's Height

1. How many inches did Dan grow between his twelfth and thirteenth birthdays?

2. Between which two birthdays did Dan grow the most?

3. There were 12 princes and 16 princesses in the palace. What was the ratio of princes to princesses in the palace?

4. On the first 4 days of their trip, the Curtis family drove 497 miles, 513 miles, 436 miles, and 410 miles. What was the average number of miles they drove on each of the first 4 days of their trip?

5. Don receives a weekly allowance of $4.50. How much allowance does he receive in a year (52 weeks)?

6. Draw a diagram for this statement. Then answer the questions that follow.

> Three sevenths of the 105 adults in the Khoikhoi clan were less than 5 feet tall.

   (a) How many of the adults were less than 5 feet tall?

   (b) How many of the adults were 5 feet tall or taller?

7. Sketch this figure on your paper. Then find the length of each unmarked side and find the perimeter of the polygon. All angles are right angles. Dimensions are in millimeters.

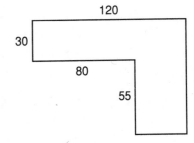

8. Find the LCM of 8, 12, and 9.

9. Name the number of shaded circles

   (a) as a decimal number.

   (b) as a mixed number.

10. Round 0.9166666

   (a) to the nearest hundredth.

   (b) to the nearest hundred-thousandth.

11. Use words to write each number.

(a) 3.0175

(b) 23,310,050,000

12. Use digits to write each number.

(a) One hundred and seventy-five thousandths

(b) Ten and eleven twelfths

13. Simplify each of these numbers.

(a) $26\dfrac{7}{3}$

(b) $\dfrac{126}{24}$

14. Compare: $\dfrac{15}{24}$ ◯ $\dfrac{10}{16}$

Solve:

15. $\dfrac{8}{12} = \dfrac{6}{x}$

16. $\dfrac{16}{y} = \dfrac{2}{3}$

17. $\dfrac{21}{14} = \dfrac{n}{4}$

18. $m + 0.36 = 0.75$

19. $1.4 - w = 0.8$

Add, subtract, multiply, or divide, as indicated:

20. $9.6 + 12 + 8.59$

21. $3.15 - (2.1 - 0.06)$

22. $\dfrac{7}{10} + \left(\dfrac{1}{2} + \dfrac{2}{5}\right)$

23. $\dfrac{7}{12} - \left(\dfrac{3}{4} \cdot \dfrac{1}{3}\right)$

24. $4\dfrac{5}{12} + 6\dfrac{5}{8}$

25. $4\dfrac{1}{4} - 1\dfrac{3}{5}$

26. $8\dfrac{1}{3} \cdot 1\dfrac{4}{5}$

27. $5\dfrac{5}{6} \div 7$

28. $970(480)$

29. $300,000 \div 24$

30. $(\$20 - \$5.24) \div 12$

LESSON 45

Multiplying Decimal Numbers

To multiply decimal numbers, we set up the problem as though we were multiplying whole numbers. (We do not need to align the decimal points.) Then we multiply. After we have multiplied, we place the decimal point in the answer so that there are the same number of decimal places in the product as there are decimal places in all the factors combined.

$$
4.2 \times 0.36 \longrightarrow
\begin{array}{r}
4.\underline{2} \quad \longleftarrow \text{ one decimal place} \\
\times\, 0.\underline{36} \quad \longleftarrow \text{ two decimal places} \\
\hline
252 \\
126 \\
\hline
1.\underline{512} \quad \longleftarrow \text{ three decimal places} \\
 \quad \text{ in the product}
\end{array}
$$

Example 1 Multiply: $(0.23)(0.4)$

Solution We set up the problem as though we were multiplying whole numbers, and then we multiply. After multiplying, we go back and count the number of decimal places in both factors. There are a total of three decimal places, so we write the product with three decimal places. We count from right to left, writing one or more zeros in front as is necessary. The product of 0.24 and 0.4 is **0.092**.

$$
\begin{array}{r}
0.23 \\
\times\ 0.4 \\
\hline
92
\end{array}
$$

$$
\begin{array}{r|l}
0.23 & 2 \text{ places} \\
\times\ 0.4 & 1 \text{ place} \\
\hline
0.092 & 3 \text{ places}
\end{array}
$$

Example 2 Multiply: 35×0.4

Solution We set up the problem as though we were multiplying whole numbers. After multiplying, we count the total number of decimal places in the factors. Then we place a decimal point in the product so that the product has the same number of decimal places as there are in the factors combined. After placing the decimal point, we simplify the result.

$$
\begin{array}{r|l}
35 & 0 \text{ places} \\
\times\ 0.4 & 1 \text{ place} \\
\hline
14.0 & 1 \text{ place}
\end{array}
$$

$$
14.0 = \mathbf{14}
$$

Example 3 Multiply: (0.2)(0.3)(0.04)

Solution Sometimes we can perform the multiplication mentally. First we multiply as though we were multiplying whole numbers: 2 · 3 · 4 = 24. Then we count decimal places. There is a total of four decimal places in the three factors. Starting from the right side of 24, we count to the left four places. We write zeros in the empty places.

$$.0024 \quad \longrightarrow \quad \textbf{0.0024}$$

Practice Multiply. Try to do Problems **d** and **e** mentally.

a. 4.2 × 0.24

b. (0.12)(0.06)

c. 5.4 × 7

d. 0.3 × 0.2 × 0.1

e. (0.04)(25)

f. 0.045 × 0.6

Problem set 45

1. The bag contained only red marbles and white marbles. If the ratio of red marbles to white marbles was 3 to 2, what fraction of the marbles was white?

2. John ran 4 laps of the track in 6 minutes 20 seconds.

   (a) How many seconds did it take John to run 4 laps?

   (b) John's average time for running each lap was how many seconds?

3. The Curtis's car traveled an average of 24 miles per gallon of gas. At that rate, how far could the car travel on a full tank of 18 gallons?

4. Normal body temperature is 98.6°F. Allan's temperature was 103.4°F. His temperature was how many degrees above normal?

5. The length of the rectangle is twice its width. What is the perimeter of the rectangle?

70 mm

6. Draw a diagram for this statement. Then answer the questions that follow.

>   Five eighths of the 200 sheep in the flock grazed in the meadow. The rest drank from the brook.

(a)   How many of the sheep grazed in the meadow?

(b)   How many of the sheep drank from the brook?

7. *AB* is 30 mm. *CD* is 45 mm. *AD* is 100 mm. Find *BC*.

8. The length of segment *CD* in Problem 7 is 45 mm. What is the length of *CD* in centimeters?

9. Round 0.083333

(a)   to the nearest thousandth.

(b)   to the nearest tenth.

10. Use words to write each number.

(a)   12.0545

(b) $10\dfrac{11}{100}$

11. Use digits to write each number.

(a)   Twenty-five millionths

(b)   Five billion, two hundred fifty thousand

12. What decimal number names the point marked *B* on this number line?

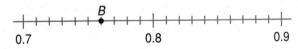

13. Write seventy and seven hundredths

(a)   as a decimal.

(b)   as a mixed number.

14. In this figure, which angle is

(a) a right angle?

(b) an obtuse angle?

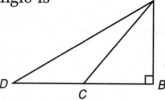

Solve:

15. $\dfrac{8}{10} = \dfrac{w}{25}$

16. $\dfrac{n}{15} = \dfrac{6}{9}$

17. $\dfrac{9}{12} = \dfrac{15}{m}$

18. $3.7 = a + 1.8$

19. $3.9 = t - 3.1$

Add, subtract, multiply, or divide, as indicated:

20. 2.7×0.18

21. $(0.15)(0.05)$

22. 15×1.5

23. $0.4 \times 0.3 \times 0.2 \times 0.1$

24. $5.6 - (4 - 1.25)$

25. $5 - (3.14 + 1.2)$

26. $6\dfrac{1}{4} \cdot 1\dfrac{3}{5}$

27. $7 \div 5\dfrac{5}{6}$

28. $4\dfrac{5}{6} + 2\dfrac{3}{8}$

29. $4\dfrac{2}{5} - 1\dfrac{3}{4}$

30. $(\$100 - \$32.50) \div 15$

**LESSON
46**

Dividing a Decimal Number by a Whole Number

Dividing a decimal number by a whole number is like dividing money. The decimal point in the answer is straight up from the decimal point in the division box.

Example 1   Divide: $3.425 \div 5$

Solution   We rewrite the problem with a division box. We place a decimal point in the answer directly above the decimal point in the division box. Then we divide as though we were dividing whole numbers. The answer is **0.685.**

$$
\begin{array}{r}
0.685 \\
5\overline{)3.425} \\
\underline{3\,0} \\
42 \\
\underline{40} \\
25 \\
\underline{25} \\
0
\end{array}
$$

Example 2   Divide: $0.0144 \div 8$

Solution   We place the decimal point in the answer directly above the decimal point inside the division box. We write a digit in every place following the decimal point until the division is completed. If we cannot perform a division, we write a zero in that place. The answer is **0.0018.**

$$
\begin{array}{r}
0.0018 \\
8\overline{)0.0144} \\
\underline{8} \\
64 \\
\underline{64} \\
0
\end{array}
$$

Example 3   Divide: $1.2 \div 5$

Solution   We do not write a decimal division answer with a remainder. Since a decimal point fixes place values, we may write a zero in the next decimal place. This zero does not change the value of the number, but it does let us continue dividing. The answer is **0.24.**

$$
\begin{array}{r}
0.2 \\
5\overline{)1.2} \\
\underline{1\,0} \\
2
\end{array}
$$

$$
\begin{array}{r}
0.24 \\
5\overline{)1.20} \\
\underline{1\,0} \\
20 \\
\underline{20} \\
0
\end{array}
$$

Example 4 Divide 126 by 8 and write the answer with a remainder. Then continue the division and write the answer as a decimal number.

$$\begin{array}{r} 15 \\ 8{\overline{\smash{\big)}\,126}} \\ \underline{8} \\ 46 \\ \underline{40} \\ 6 \end{array}$$

Solution If we divide 126 into 8 equal groups, there are 15 in each group. The remainder is 6. We may divide the remainder by placing the decimal point after 126. Then we write zeros in the following decimal places and continue to divide. The decimal point in the answer is placed directly above the decimal point we placed in the division box. The answer is **15.75**.

$$\begin{array}{r} 15.75 \\ 8{\overline{\smash{\big)}\,126.00}} \\ \underline{8} \\ 46 \\ \underline{40} \\ 60 \\ \underline{56} \\ 40 \\ \underline{40} \\ 0 \end{array}$$

Practice Find each decimal answer.

a. $13.464 \div 6$

b. $0.0288 \div 8$

c. $3.4 \div 5$

d. $145 \div 4$

e. $27.4 \div 8$

f. $371 \div 10$

Problem set 46

1. Two hundred wildebeests and 150 gazelles grazed on the savannah. What was the ratio of gazelles to wildebeests grazing on the savannah?

2. In their first 5 games the Celtics scored 105 points, 112 points, 98 points, 113 points, and 107 points. What was the average number of points the Celtics scored in their first 5 games?

3. The crowd watched with anticipation as the pole vault bar was set to 19 feet 6 inches. How many inches is 19 feet 6 inches?

4. Estimate the sum of 4387, 2914, and 796 by rounding each number to the nearest hundred before adding.

5. Draw a sketch to help with this problem. From Tad's house to John's house is 2.3 kilometers. From John's house to school is 0.8 kilometer. Tad rode from his house to John's house and then to school. Later he rode from school to John's house to his house. Altogether, how far did Tad ride?

6. About seven tenths of the earth's surface is water.

(a) About what fraction of the earth's surface is land?

(b) On the earth's surface, what is the ratio of water to land?

7. The tally ||||| ||| indicates the number 8. What number is indicated by ||||| ||||| ||||| ||||| ||| ?

8. Sketch this figure on your paper. Find the lengths of the unmarked sides and find the perimeter of the polygon. Dimensions are in feet. All angles are right angles.

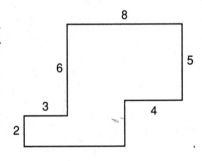

9. Name the point marked *M* on this number line

(a) as a decimal number.

(b) as a mixed number.

10. Round 5.142857142857

(a) to the nearest millionth.

(b) to the nearest hundredth.

11. What is the sum of the first four prime numbers?

12. Use words to write each number.

  (a)   1000.02

  (b)   0.102

13. Use digits to write each number.

  (a)   Sixty-seven hundred-thousandths

  (b)   One hundred and twenty-three thousandths

14. Simplify each of these numbers.

  (a) $3\dfrac{21}{12}$ (b) $\dfrac{360}{936}$

15. Write $\frac{1}{2}, \frac{3}{5},$ and $\frac{5}{7}$ with a common denominator and arrange in order from least to greatest.

Solve:

16. $\dfrac{x}{24} = \dfrac{10}{16}$ **17.** $\dfrac{18}{8} = \dfrac{m}{20}$

18. $3.45 + a = 7.6$ **19.** $2.7 - b = 1.49$

Add, subtract, multiply, or divide, as indicated:

20. $(3.4)(5.6)$ **21.** $(0.4)(0.6)(0.02)$

22. $4.315 \div 5$ **23.** $0.0144 \div 9$

24. $7.4 \div 8$ **25.** $4\dfrac{7}{12} - 3\dfrac{5}{6}$

26. $3\dfrac{1}{3} + 1\dfrac{5}{6} + \dfrac{7}{12}$ **27.** $4\dfrac{1}{6} - \left(4 - 1\dfrac{1}{4}\right)$

28. $3\dfrac{1}{5} \cdot 2\dfrac{5}{8} \cdot 1\dfrac{3}{7}$ **29.** $4\dfrac{1}{2} \div 6$

30. $6.4 + 0.27 + 12 + 9.36 + 0.144$

LESSON
47

Repeating Digits

When a decimal number is divided, the answer is sometimes a decimal number that will not end with a remainder of zero. Instead the answer will have one or more digits in a pattern that repeats indefinitely. Here we show two examples.

```
       7.1666...                    0.31818...
   6)43.0000...                 11)3.50000...
     42                            3 3
     1 0                            20
      6                            11
     40                            90
     36                           88
     40                            20
     36                           11
     40                            90
     36                           88
      4                            2
```

The repeating digits of a decimal number are called the **repetend**. In 7.1666. . ., the repetend is 6. In 0.31818. . ., the repetend is 18 (not 81). One way to indicate that a decimal number is a repeating decimal number is to write the number with a bar over the repetend where the repetend first appears to the right of the decimal point. We will write each answer above as a decimal with a bar over the repetend.

$$7.1666\ldots = 7.1\overline{6} \qquad 0.31818\ldots = 0.3\overline{18}$$

Example 1 Rewrite each of these repeating decimals with a bar over the repetend.

(a) 0.0833333. . .

(b) 5.14285714285714. . .

(c) 454.5454545. . .

Solution (a) The repeating digit is 3.

$$0.08\overline{3}$$

(b) This is a six-digit repeating pattern.

$$5.\overline{142857}$$

(c) The repetend is always to the right of the decimal point. We do not write a bar over a whole number.

$$454.\overline{54}$$

Example 2 Round each number to five decimal places.

(a) $5.31\overline{6}$　　(b) $25.\overline{405}$

Solution (a) We remove the bar and write the repeating digits to the right of the desired decimal place.

$$5.31\overline{6} = 5.316666\ldots$$

Then we round to five places.

$$5.3166\textcircled{6}6\ldots \longrightarrow \textbf{5.31667}$$

(b) We remove the bar and continue the repeating pattern beyond the fifth decimal place.

$$25.\overline{405} \longrightarrow 25.405405$$

Then we round to five places.

$$25.4054\textcircled{0}5\ldots \longrightarrow \textbf{25.40541}$$

Example 3 Divide 1.5 by 11 and write the quotient

(a) with a bar over the repetend.

(b) rounded to the nearest hundredth.

Solution (a) Since place value is fixed by the decimal point, we can write zeros in the places to the right of the decimal point. We continue dividing until the repeating pattern is apparent. The repetend is 36 (not 63). We write the quotient with a bar over 36 where it first appears.

$$0.13636\ldots = \textbf{0.1}\overline{\textbf{36}}$$

```
      0.13636...
 11)1.50000...
    1 1
    ---
      40
      33
      --
      70
      66
      --
       40
       33
       --
       70
       66
       --
        4
```

(b) The hundredths' place is the second place to the right of the decimal point.

$$\downarrow$$

$$0.1\,③\,636\ldots \;\longrightarrow\; \mathbf{0.14}$$

Practice   Write each repeating decimal with a bar over the repetend.

   a. 2.72727...   **b.** 0.816666...

Round each number to the nearest thousandth.

   c. $0.\overline{6}$ **d.** $5.3\overline{81}$

Divide 1.7 by 12 and write the quotient

   e. with a bar over the repetend.

   f. rounded to four decimal places.

Problem set 47

1. Two fifths of the children in the nursery were boys. What was the ratio of boys to girls in the nursery?

2. Four hundred thirty-two students were assigned to 16 classrooms. What was the average number of students per classroom?

3. The migrating birds flew for 7 hours at an average rate of 23 miles per hour. How far did the birds travel in 7 hours?

4. Draw a diagram for this statement. Then answer the questions that follow.

   Seven ninths of the 450 students in the assembly were enthralled by the speaker.

   (a) How many of the students were enthralled?

   (b) How many of the students were not enthralled?

5. Round each number to four decimal places.

   (a) $5.1\overline{6}$ (b) $5.\overline{27}$

Refer to this pie graph to answer questions 6 and 7.

Class Test Grades

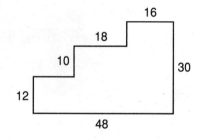

6. How many more students earned an A or B than earned a C or D?

7. What fraction of the students in the class earned an A?

8. Sketch this figure on your paper. Find the length of each unmarked side and find the perimeter of the polygon. Dimensions are in inches. All angles are right angles.

9. Divide 1.7 by 11 and write the quotient

    (a) with a bar over the repetend.

    (b) rounded to three decimal places.

10. Use digits to write the sum of twenty-seven thousandths and fifty-eight hundredths.

11. Use words to write each number.

    (a) 760.005

    (b) 3,524,000,000,000

12. Write the prime factorization of each number.

    (a) 71,000

    (b) 1296

13. Find the length of segment *BC* to the nearest eighth of an inch.

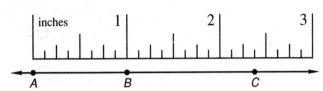

14. What is the least common multiple of 12 and 15?

Solve:

15. $\dfrac{21}{24} = \dfrac{w}{40}$

16. $\dfrac{12}{x} = \dfrac{9}{6}$

17. $\dfrac{15}{9} = \dfrac{20}{y}$

18. $m + 9.6 = 14$

19. $n - 4.2 = 1.63$

Add, subtract, multiply, or divide, as indicated:

20. 12×8.6

21. $(0.4)(0.5)(0.6)$

22. $6.165 \div 9$

23. $0.288 \div 8$

24. $7.1 \div 4$

25. $7\dfrac{5}{8} - 1\dfrac{1}{12}$

26. $6\dfrac{1}{4} + 5\dfrac{5}{12} + \dfrac{2}{3}$

27. $4 - \left(4\dfrac{1}{6} - 1\dfrac{1}{4}\right)$

28. $6\dfrac{2}{5} \cdot 2\dfrac{5}{8} \cdot 2\dfrac{6}{7}$

29. $6 \div 4\dfrac{1}{2}$

30. $8.3 + 0.72 + 15 + 3.96 + 0.108$

LESSON 48

Decimals to Fractions and Fractions to Decimals

Decimals to fractions   To write a decimal number as a fraction, we write the digits after the decimal point as the numerator of the fraction. For the denominator of the fraction we write the place value of the last digit. Then we reduce.

Example 1   Write 0.125 as a fraction.

Solution   The digits 125 form the numerator of the fraction. The denominator of the fraction is 1000 because 5 is in the thousandths' place.

$$0.125 = \frac{125}{1000}$$

Now we reduce.

$$\frac{125}{1000} = \frac{1}{8}$$

Example 2   Write 11.42 as a mixed number.

Solution   The number 11 is the whole number part. The numerator of the fraction is 42, and the denominator is 100 because 2 is in the hundredths' place.

$$11.42 = 11\frac{42}{100}$$

Now we reduce the fraction.

$$11\frac{42}{100} = 11\frac{21}{50}$$

Fractions to decimals   To change a fraction to a decimal number, we perform the division indicated by the fraction. The fraction $\frac{1}{4}$ indicates that 1 is divided by 4.

$$4\overline{)1}$$

It may appear that we cannot perform this division. However, if we fix place values with a decimal point and write zeros in the decimal places to the right of the decimal point, we can perform the division. The result is a decimal number that is equal to the fraction $\frac{1}{4}$.

$$\begin{array}{r} 0.25 \\ 4\overline{)1.00} \\ \underline{8} \\ 20 \\ \underline{20} \\ 0 \end{array} \qquad \text{Thus, } \frac{1}{4} = 0.25$$

Example 3   Write each of these numbers as a decimal number.

(a) $\dfrac{23}{100}$ (b) $\dfrac{7}{4}$ (c) $3\dfrac{4}{5}$ (d) $\dfrac{2}{3}$

Solution   (a) Fractions with denominators of 10, 100, 1000, etc., can be written directly as decimal numbers, without performing the division.

$$\frac{23}{100} = \mathbf{0.23}$$

(b) An improper fraction is equal to or greater than 1. When we change an improper fraction to a decimal number, the decimal number will be greater than or equal to 1.

$$\frac{7}{4} \longrightarrow \begin{array}{r} 1.75 \\ 4\overline{)7.00} \\ \underline{4} \\ 3\,0 \\ \underline{2\,8} \\ 20 \\ \underline{20} \\ 0 \end{array} \qquad \frac{7}{4} = \mathbf{1.75}$$

(c) To change a mixed number to a decimal number, we may change the mixed number to an improper fraction and then divide. Another way is to separate the fraction from the whole number and change the fraction to a decimal number. Then we write the whole number and the decimal number as one number. Here we show both ways.

$$3\frac{4}{5} = \frac{19}{5} \qquad \text{or} \qquad 3\frac{4}{5} = 3 + \frac{4}{5}$$

$$\begin{array}{r} 3.8 \\ 5\overline{)19.0} \\ \underline{15} \\ 40 \\ \underline{40} \\ 0 \end{array} \qquad\qquad \begin{array}{r} 0.8 \\ 5\overline{)4.0} \\ \underline{40} \\ 0 \end{array}$$

$$3\frac{4}{5} = \textbf{3.8} \qquad\qquad 3\frac{4}{5} = \textbf{3.8}$$

(d) To change $\frac{2}{3}$ to a decimal number, we divide.

$$\frac{2}{3} \quad \longrightarrow \quad \begin{array}{r} 0.666 \\ 3\overline{)2.000} \\ \underline{1\ 8} \\ 20 \\ \underline{18} \\ 20 \\ \underline{18} \\ 2 \end{array} \qquad \frac{2}{3} = \textbf{0.}\overline{\textbf{6}}$$

We will write repeating decimal numbers with a bar over the repetend unless directed otherwise.

Practice   Change each decimal number to a reduced fraction or to a mixed number.

    a. 0.24   **b.** 45.6   **c.** 2.375

Change each fraction or mixed number to a decimal number.

    d. $\frac{23}{4}$ **e.** $4\frac{3}{5}$ **f.** $\frac{5}{8}$ **g.** $\frac{5}{6}$

Problem set 48

1. The ratio of Celtic soliders to Roman soldiers was 2 to 5. What fraction of the soldiers were Celts?

2. Eric ran 8 laps in 11 minutes 44 seconds.

   (a)   How many seconds did it take Eric to run 8 laps?

   (b)   What is the average amount of time it took Eric to run each lap?

3. Some gas was still in the tank. Jan added 13.3 gallons of gas, which filled the tank. If the tank held a total of 21.0 gallons of gas, how much gas was in the tank before Jan added the gas?

4. From 1750 to 1850, the estimated population of the world increased from seven hundred twenty-five million to one billion, two hundred thousand. How many more people were living in the world in 1850 than in 1750?

5. Draw a diagram for this statement. Then answer the questions that follow.

   The Jets won two thirds of their 15 games.

   (a)   How many games did the Jets win?

   (b)   What was the Jets' won-lost ratio?

6. The tally 卌 卌 ‖ indicates the number 12. What is the tally for 16?

7. A square and a regular hexagon share a common side, as shown. The perimeter of the hexagon is 120 mm. What is the perimeter of the square?

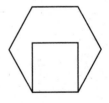

8. Write each of these numbers as a fraction or as a mixed number.

   (a)   0.375   (b)   5.55

9. Write each of these numbers as a decimal number.

   (a) $2\frac{2}{5}$ (b) $\frac{1}{8}$

10. Round each number to the nearest thousandth.

(a) $0.4\overline{5}$

(b) $3.\overline{142857}$

11. Divide 1.9 by 12 and write the quotient

(a) with a bar over the repetend.

(b) rounded to three decimal places.

12. Four and five hundredths is how much greater than one hundred sixty-seven thousandths?

13. Simplify each of these numbers.

(a) $90\dfrac{99}{10}$

(b) $\dfrac{625}{500}$

14. Draw segment AB to be 1 inch long. Draw segment AC perpendicular to $\overline{AB}$. Let segment AC be $\frac{3}{4}$ inch long. Complete triangle ABC by drawing segment $\overline{BC}$. Measure segment BC with a ruler. How long is segment BC?

Solve:

15. $\dfrac{a}{8} = \dfrac{21}{12}$

16. $\dfrac{12}{b} = \dfrac{3}{10}$

17. $\dfrac{10}{18} = \dfrac{c}{45}$

18. $1.9 = w + 0.42$

19. $7.8 = y - 6.9$

Add, subtract, multiply, or divide, as indicated:

20. 4.8×32

21. $(0.12)(0.5)(0.02)$

22. $24.156 \div 6$

23. $0.072 \div 3$

24. $6.5 \div 4$

25. $3\dfrac{3}{10} - 1\dfrac{3}{4}$

26. $5\frac{1}{2} + 6\frac{3}{10} + \frac{4}{5}$ **27.** $5 - \left(3\frac{1}{3} - 1\frac{5}{6}\right)$

28. $7\frac{1}{2} \cdot 3\frac{1}{3} \cdot \frac{4}{5}$ **29.** $10\frac{1}{2} \div 7$

30. $5.267 + 3.4 + 15 + 12.35 + 0.015$

LESSON 49

Division Answers

We have considered three ways of writing the answer to a division problem in which there is a remainder.

1. Write the answer with a remainder.

2. Write the answer as a mixed number.

3. Write the answer as a decimal number.

Example 1   Divide 54 by 4 and write the answer

(a) with a remainder.

(b) as a mixed number.

(c) as a decimal number.

Solution   (a) We divide and find the result is **13 r 2**.

$$\begin{array}{r} 13\text{ r }2 \\ 4\overline{)54} \\ \underline{4} \\ 14 \\ \underline{12} \\ 2 \end{array}$$

(b) The remainder is the numerator of a fraction and the divisor is the denominator, so this answer can be written as $13\frac{2}{4} = \mathbf{13\frac{1}{2}}$.

(c) We fix place values by placing the decimal point to the right of 54. Then we can write zeros in the the following places and continue dividing to completion. The result is **13.5**.

$$\begin{array}{r} 13.5 \\ 4\overline{)54.0} \\ \underline{4} \\ 14 \\ \underline{12} \\ 2\ 0 \\ \underline{2\ 0} \\ 0 \end{array}$$

Sometimes a division answer written as a decimal number will be a repeating decimal number or will have more decimal places than the problem requires. In this book we will show the complete division of the number unless the problem states that the answer is to be rounded.

Example 2 Divide 37.4 by 9 and round the quotient to the nearest thousandth.

Solution We continue dividing until the answer has four decimal places. Then we round to the nearest thousandth.

$$4.15\text{⑤}5 \longrightarrow \textbf{4.156}$$

```
        4.1555
     9)37.4000
        36
         1 4
           9
          50
          45
          50
          45
          50
          45
```

Practice Divide 55 by 4 and write the answer

    a. with a remainder.

    b. as a mixed number.

    c. as a decimal number.

    d. Divide 5.5 by 3 and round the answer to three decimal places.

Problem set 49

1. The rectangle was 24 inches long and 18 inches wide. What was the ratio of its length to its width?

2. Amber's test scores were 90, 95, 90, 85, 80, 85, 90, 80, 95, and 100. What was her average test score?

3. The report stated that two out of every five young people were unable to find a job. What fraction of the young people were able to find a job?

4. Rachel bought a sheet of fifty 29-cent stamps from the post office. How much did she have to pay?

5. Ninety-seven thousandths is how much less than two and ninety-eight hundredths? Write the answer in words.

6. Draw a diagram for this statement. Then answer the questions that follow.

    Five sixths of the 30 students passed the test.

    (a) How many students did not pass the test?

    (b) What was the ratio of students who passed the test to students who did not pass the test?

7. Sketch this figure on your paper. Find the length of each unmarked side and find the perimeter of the polygon. Dimensions are in meters. All angles are right angles.

    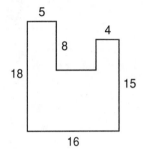

8. Write 0.75 as a fraction.

9. Write $\dfrac{5}{8}$ as a decimal number.

10. Round $123.\overline{6}$ to the nearest thousandth.

11. Divide 54 by 5 and write the answer

    (a) with a remainder.

    (b) as a mixed number.

    (c) as a decimal number.

12. Divide 5.4 by 11 and write the answer with a bar over the repetend.

13. What composite number is equal to the product of the four smallest prime numbers?

14. Arrange these numbers in order from least to greatest:

$$1.2, \ -12, \ 0.12, \ 0, \ \dfrac{1}{2}$$

15. What is the sum of the numbers marked M and N on this number line?

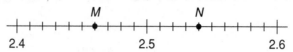

Solve:

16. $\dfrac{12}{9} = \dfrac{8}{m}$ **17.** $\dfrac{25}{15} = \dfrac{n}{12}$ **18.** $\dfrac{p}{90} = \dfrac{4}{18}$

19. $4 = 3.14 + x$ **20.** $0.1 = 1 - z$

Add, subtract, multiply, or divide, as indicated:

21. $(2.5)(2.5)$ **22.** $1.2 \times 0.4 \times 0.05$

23. $16.42 \div 8$ **24.** $0.153 \div 9$

25. $5\dfrac{3}{4} + \dfrac{5}{6} + 2\dfrac{1}{2}$ **26.** $3\dfrac{1}{3} - \left(5 - 1\dfrac{5}{6}\right)$

27. $3\dfrac{3}{4} \cdot 3\dfrac{1}{3} \cdot 8$ **28.** $7 \div 10\dfrac{1}{2}$

29. $5.46 + 2.791 + 21.4 + 10 + 0.199$

30. Compare: $\dfrac{45.0}{30} \bigcirc \dfrac{4.5}{3}$

LESSON 50

Dividing by a Decimal Number

We remember that we do not change the value of a fraction if we multiply the fraction by another fraction that has the same numerator and denominator. If we multiply $\frac{1}{2}$ by 10 over 10,

$$\frac{1}{2} \times \frac{10}{10} = \frac{10}{20}$$

we get 10 over 20, which is another name for one half. We use this fact to change division by a decimal number into a division by a whole number. If we want to divide 1.36 by 0.4, we have

$$\frac{1.36}{0.4}$$

We can change the divisor to the whole number 4 by multiplying by 10 over 10.

$$\frac{1.36}{0.4} \times \frac{10}{10} = \frac{13.6}{4}$$

The value of 13.6 divided by 4 is the same as the value of 1.36 divided by 0.4. This means that both of these divisions have the same answer.

$$0.4\overline{)1.36} \quad \text{equals} \quad 4\overline{)13.6}$$

To divide by a decimal number, we move the decimal point in the divisor to the right to make the divisor a whole number. Then we move the decimal point in the dividend the same number of places to the right.

Example 1   Divide: 3.36 ÷ 0.06

Solution   We use a division box and write

$$0.06\overline{)3.36}$$

First we move the decimal point in 0.06 two places to the right to make it 6.

$$0\underset{\smile}{0}6.\overline{)3.36}$$

Then we move the decimal point in 3.36 the same number of places to the right.

$$0\underset{\smile}{0}6.\overline{)3\underset{\smile}{3}6.}$$

The decimal point in the answer is just above the new location of the decimal point.

$$6\overline{)336.}\,\overset{\cdot}{}$$

Now we divide.

$$\begin{array}{r} 56. \\ 6\overline{)336.} \\ \underline{30} \\ 36 \\ \underline{36} \\ 0 \end{array}$$

Thus, $3.36 \div 0.06 = \textbf{56}$.

Example 2   Divide: $0.144 \div 0.8$

Solution   We need to move the decimal point in 0.8 one place to the right to get the whole number 8. Then we move the decimal point one place to the right in 0.144 to get 1.44. Then we divide.

$$\begin{array}{r} 0.18 \\ 08.\overline{)1.44} \\ \underline{8} \\ 64 \\ \underline{64} \\ 0 \end{array}$$

Example 3   Divide: $15.4 \div 0.07$

Solution   We move both decimal points two places. This makes an empty place in the division box, which we fill with a zero. We keep dividing until we reach the decimal point. We find **220** as the answer.

$$\begin{array}{r} 220. \\ 007.\overline{)1540.} \\ \underline{14} \\ 14 \\ \underline{14} \\ 0 \end{array}$$

Example 4   Divide: $21 \div 0.5$

Solution   We move the decimal point in 0.5 one place to the right. Then we move the other decimal point one place to the right. We find **42** as the answer.

$$\begin{array}{r} 42. \\ 05.\overline{)210.} \\ \underline{20} \\ 10 \\ \underline{10} \\ 0 \end{array}$$

Example 5   Divide: $1.54 \div 0.8$

Solution   We do not write a remainder. We write zeros in the places to the right of 4 and continue dividing until the remainder is zero.

$$\begin{array}{r} 1.925 \\ 08.\overline{)15.400} \\ \underline{8} \\ 7\ 4 \\ \underline{7\ 2} \\ 20 \\ \underline{16} \\ 40 \\ \underline{40} \\ 0 \end{array}$$

Practice   Divide:

     a. $5.16 \div 0.6$ **b.** $0.144 \div 0.09$

     c. $23.8 \div 0.07$ **d.** $24 \div 0.08$

Problem set 50

1. Raisins and nuts were mixed in a bowl. If five eighths of the mixture was made up of nuts, what was the ratio of raisins to nuts?

2. The new coupe traveled 702 kilometers down the autobahn in 6 hours. The new coupe averaged how many kilometers per hour?

3. Fifty-four and five hundredths is how much greater than fifty and forty thousandths?

Refer to this election tally sheet to answer questions 4 and 5.

VOTE TOTALS

Judy	ЖІ ЖІ ЖІ І
Carlos	ЖІ ЖІ ІІІІ
Yolanda	ЖІ ЖІ ЖІ ЖІ ІІ
Khanh	ЖІ ЖІ ЖІ ІІІ

4. The winner of the election received how many more votes than the runner-up?

5. What fraction of the votes did Carlos receive?

6. Draw a diagram for this statement. Then answer the questions that follow.

     Four sevenths of those who rode the Giant Gyro at the fair were euphoric. All the rest were vertiginous.

     (a)   What fraction of those who rode the ride were vertiginous?

     (b)   What was the ratio of euphoric to vertiginous riders?

7. What is the least common multiple of 10 and 16?

8. The perimeter of this rectangle is 56 cm. What is the length of the rectangle?

10 cm

9. Write 62.5 as a mixed number.

10. Write $\dfrac{9}{100}$ as a decimal number.

11. Round each number to five decimal places.

(a) $23.\overline{54}$ (b) $0.91\overline{6}$

12. Divide 51 by 6 and write the answer

(a) with a remainder.

(b) as a mixed number.

(c) as a decimal number.

13. Divide 5.1 by 9 and write the quotient rounded to the nearest thousandth.

14. Use digits to write the product of twelve hundredths and eight tenths. Then write the answer in words.

15. Draw segment XY to be 2 cm long. Draw segment YZ perpendicular to $\overline{XY}$ and 1.5 cm long. Complete triangle XYZ by drawing segment XZ. How long is segment XZ?

Solve:

16. $\dfrac{30}{25} = \dfrac{t}{10}$ 17. $\dfrac{3}{w} = \dfrac{7}{28}$ 18. $\dfrac{12}{44} = \dfrac{3}{a}$

19. $m + 0.23 = 1.2$ 20. $r - 1.97 = 0.65$

Add, subtract, multiply, or divide, as indicated:

21. $(0.15)(0.15)$ 22. $1.2 \times 2.5 \times 4$

23. $14.14 \div 5$ **24.** $0.096 \div 0.12$

25. $\dfrac{5}{8} + \dfrac{5}{6} + \dfrac{5}{12}$ **26.** $4\dfrac{1}{2} - \left(2\dfrac{1}{3} - 1\dfrac{1}{4}\right)$

27. $\dfrac{7}{15} \cdot 10 \cdot 2\dfrac{1}{7}$ **28.** $6\dfrac{3}{5} \div 1\dfrac{1}{10}$

29. Add mentally:

$$4 + 6 + 9 + 8 + 7 + 5 + 3 + 4 + 1 + 7 + 4 + 3$$

30. Solve mentally:

$$4 + 14 + 8 + 9 + 12 + 14 + 5 + 3 + 7 + N = 84$$

LESSON
51

Unit Price

As an aid to grocery store customers, the unit price for various products is often posted. The unit price is the cost for a single unit measurement of the product. The unit price can be found by dividing.

Example 1   What is the unit price of a 24-ounce box of cereal that costs $3.60?

Solution   The cereal is measured in ounces. The unit price is the cost of 1 ounce. We divide the price by 24 ounces.

$$\frac{\$3.60}{24 \text{ ounces}} = \frac{\$0.15}{1 \text{ ounce}}$$

The unit price is $0.15 per ounce, which is **15¢ per ounce.**

Example 2   What is the unit price of a 36-ounce box of cereal that costs $4.50?

Solution　The unit price for the cereal is the price per ounce. We divide the price by 36 ounces.

$$\frac{\$4.50}{36 \text{ ounces}} = \frac{\$0.125}{1 \text{ ounce}}$$

The unit price is **12.5¢ per ounce.**

　　　　Unit pricing helps customers determine which brand or which size package provides the better buy. From the two examples in this lesson, we see that the larger box of cereal was the better buy because it cost less per ounce.

Practice　**a.** What is the unit price of a 28-ounce box of cereal that costs $1.12?

b. What is the unit price of an 11-ounce can of soup that costs 55¢?

c. Which is the better buy: an 18-ounce jar of jelly that costs $1.98, or a 24-ounce jar of jelly that costs $2.28?

**Problem set
51**

1. Brand X costs $2.40 for 16 ounces. Brand Y costs $1.92 for 12 ounces. Find the unit price for each brand. Which brand is the better buy?

2. The taxi ride cost 1 dollar plus 40¢ more for each quarter mile traveled. What was the total cost for a 2-mile trip?

3. Forty-eight sheep were on the farm. Thirty-six cows were also on the farm. What was the ratio of sheep to cows?

4. At 4 different stores the price of 1 gallon of milk was $1.86, $1.83, $1.98, and $2.09. Find the average price per gallon rounded to the nearest cent.

5. Two and three hundredths is how much less than three and two tenths? Write the answer in words.

6. Draw a diagram for this statement. Then answer the questions that follow.

    Three eighths of the 48 roses were red.

    (a)   How many roses were red?

    (b)   How many roses were not red?

    (c)   What fraction of the roses were not red?

7. Replace each circle with the proper comparison symbol.

    (a)   3.0303 $\bigcirc$ 3.303   (b)   0.6 $\bigcirc$ 0.600

8. From goal line to goal line, a football field is 100 yards long. How many feet long is a football field?

9. Write 0.080 as a fraction.

10. Divide 48 by 5 and write the answer

    (a)   with a remainder.

    (b)   as a mixed number.

11. Round $14.\overline{285714}$ to three decimal places.

12. Estimate the sum of 37,142 and 28,519 and 43,456 by rounding each number to the nearest thousand before adding.

13. Write $\dfrac{1}{11}$ as a decimal number.

14. What is the average of the first five prime numbers?

Solve:

15. $\dfrac{10}{12} = \dfrac{25}{a}$

16. $\dfrac{6}{8} = \dfrac{b}{100}$

17. $4.7 - w = 1.2$

18. $43 = 821 - m$

19. Sketch this figure on your paper. Find the length of each unmarked side. Then find the perimeter of the polygon. Dimensions are in inches. All angles are right angles.

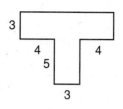

Add, subtract, multiply, or divide, as indicated:

20. $(5 \cdot 5 \cdot 5 \cdot 5) - (5 \cdot 5)$

21. $10 \cdot 10 \cdot 10 \cdot 10 \cdot 10$

22. $3\frac{3}{8} + 4\frac{3}{4} + 1\frac{1}{2}$ **23.** $5\frac{5}{6} - \left(3 - 1\frac{1}{3}\right)$

24. $\frac{2}{3} \times 4 \times 1\frac{1}{8}$ **25.** $6\frac{2}{3} \div 4$

26. $3.45 + 6 + (5.2 - 0.57)$ **27.** $2.4 \div 0.016$

28. 0.35×2.4 **29.** $4.26 \div 40$

30. Add mentally:

$4 + 6 + 5 + 8 + 12 + 14 + 3 + 6 + 8 + 9 + 4$

LESSON 52

Exponents

We remember that we can show repeated addition by using multiplication.

$5 + 5 + 5 + 5$ has the same value as 4×5

There is also a way to show repeated multiplication. We can show repeated multiplication by using an **exponent**.

$$5 \cdot 5 \cdot 5 \cdot 5 = 5^4$$

In the expression 5^4, the 4 is the exponent and the 5 is the base. The exponent shows how many times the base is to be used as a factor.

$$\text{base} \longrightarrow 5^4 \longleftarrow \text{exponent}$$

The following examples show how we read expressions with exponents, which we call **exponential expressions.**

4^2 "four squared" or "four to the second power"

2^3 "two cubed" or "two to the third power"

5^4 "five to the fourth power"

10^5 "ten to the fifth power"

To find the value of an expression with an exponent, we write the base the number of times shown by the exponent. Then we **multiply**.

$$5^4 = 5 \cdot 5 \cdot 5 \cdot 5 = 625$$

Example 1   Simplify: (a) 4^2 (b) 2^3 (c) 10^5

Solution   (a) $4^2 = 4 \cdot 4 = \mathbf{16}$

(b) $2^3 = 2 \cdot 2 \cdot 2 = \mathbf{8}$

(c) $10^5 = 10 \cdot 10 \cdot 10 \cdot 10 \cdot 10 = \mathbf{100,000}$

Example 2   Simplify: $4^2 - 2^3$

Solution   We first find the value of each expression. Then we subtract.

$$4^2 - 2^3$$

$$16 - 8 = 8$$

Practice   Use words to show how each exponential expression is read.

a. 4^3

b. 5^2

c. 10^6

d. In the expression 10^3, what number is the base and what number is the exponent?

Simplify:

e. 5^3 **f.** 10^4 **g.** $3^2 - 2^3$ **h.** $\dfrac{6^3}{3^2}$

Problem set 52

1. In 1803, the United States purchased the Louisiana territory from France for $15 million. In 1867, the United States purchased Alaska from Russia for $7.2 million. The purchase of Alaska occurred how many years after the purchase of the Louisiana territory?

2. Red and blue marbles were in the bag. Five twelfths of the marbles were red.

   (a) What fraction of the marbles were blue?

   (b) What was the ratio of red marbles to blue marbles?

3. A 6-ounce can of peaches sells for 90¢. A 9-ounce can of peaches sells for $1.26. Find the unit price for each size. Which size is the better buy?

4. The average of two numbers is the number halfway between the two numbers. What number is halfway between two thousand, five hundred fifty and two thousand, nine hundred?

5. Five hundred thirty-three thousandths is how much more than forty-five hundredths? Use words to write the answer.

6. Draw a diagram for this statement. Then answer the questions that follow.

   Five sixths of the 30 students smiled when their teacher told the joke.

   (a) How many students smiled?

   (b) How many students did not smile?

7. Use digits and symbols to write "Twenty-five thousandths is less than three hundredths."

8. (a) Estimate the length of segment *AB* in centimeters.

   (b) Use a centimeter scale to find the length of segment *AB* to the nearest centimeter.

9. Write each of these numbers as a decimal number.

   (a) $3\dfrac{1}{3}$ (b) $\dfrac{5}{8}$

10. Divide 2.5 by 22 and write the answer

    (a) with a bar over the repetend.

    (b) rounded to the nearest thousandth.

11. Estimate the product of 596 and 306.

12. In the expression 5^3, what number is the exponent and what number is the base?

13. If the perimeter of a regular hexagon is 1 foot, each side is how many inches long?

14. (a) What fraction of this rectangle is shaded?

    (b) What fraction of this rectangle is not shaded?

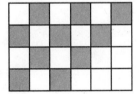

15. Refer to quadrilateral *ABCD* to answer the following questions.

    (a) Which angle is a right angle?

    (b) Which angle appears to be obtuse?

    (c) What kind of angle is $\angle C$?

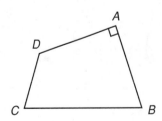

Solve:

16. $\dfrac{8}{m} = \dfrac{28}{49}$ **17.** $\dfrac{50}{100} = \dfrac{n}{12}$

18. $72 + m = 340$ **19.** $7.2 = n - 0.27$

Add, subtract, multiply, or divide, as indicated:

20. $3^2 + 2^3$ **21.** $10^4 - 10^3$

22. $\dfrac{3}{5} + \dfrac{3}{4} + \dfrac{3}{3}$ **23.** $3\dfrac{1}{3} - \left(2 - 1\dfrac{1}{4}\right)$

24. $3\dfrac{3}{4} \times 1\dfrac{1}{9} \times 6$ **25.** $4 \div 6\dfrac{2}{3}$

26. $24 - 15.8 + (12 - 3.64)$

27. 0.12×0.15 **28.** 100×0.0125

29. $0.1 \div 4$ **30.** $10 \div 0.25$

LESSON 53

Powers of 10

The positive powers of 10 are easy to write. The exponent matches the number of zeros in the product.

$10^2 = 10 \cdot 10 = 100$ (two zeros)

$10^3 = 10 \cdot 10 \cdot 10 = 1000$ (three zeros)

$10^4 = 10 \cdot 10 \cdot 10 \cdot 10 = 10,000$ (four zeros)

Place value   We can use powers of 10 to show place value, as we see in the chart below. Notice that 10^0 equals 1.

Etc.	Trillions			Billions			Millions			Thousands			Units (ones)			Decimal point
	Hundreds	Tens	Ones	Hundreds	Tens	Ones	Hundreds	Tens	Ones	Hundreds	Tens	Ones	Hundreds	Tens	Ones	
	10^{14}	10^{13}	10^{12}	10^{11}	10^{10}	10^9	10^8	10^7	10^6	10^5	10^4	10^3	10^2	10^1	10^0	.

Powers of 10 are sometimes used to write numbers in expanded notation. In expanded notation we write a number as the sum of each nonzero digit times its place value.

Example 1 Write 5206 in expanded notation using powers of 10.

Solution The number 5206 means 5000 + 200 + 6. We will write each number as a digit times its place value.

$$5000 \quad + \quad 200 \quad + \quad 6$$

$$(5 \times 10^3) + (2 \times 10^2) + (6 \times 10^0)$$

Multiplying by powers of 10 When we multiply a decimal number by a power of 10, the answer has the same digits in the same order. Only their place values are changed.

Example 2 Multiply: 46.235×10^2

Solution This time we will write 10^2 as 100 and multiply.

$$
\begin{array}{r}
46.235 \\
\times \quad 100 \\
\hline
\mathbf{4623.500}
\end{array}
$$

We see that the same digits occur in the same order. Only the place values have changed as the decimal point has been shifted two places to the right. **Thus, to multiply by a power of 10, we need only to shift the decimal point to the right the number of places indicated by the exponent.**

Example 3 Multiply: 3.14×10^4

Solution The power of 10 shows us the number of places to move the decimal point to the right. We move the decimal point four places to the right.

$$3.14 \times 10^4 = \mathbf{31,400}$$

Practice Write each number in expanded notation by using powers of 10.

    a. 456

b. 1760

c. 186,000

Multiply:

d. 24.25×10^3

e. 25×10^6

**Problem set
53**

Refer to the graph to answer questions 1–3.

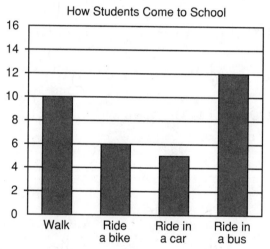

How Students Come to School

1. Answer true or false.

(a) Twice as many students walk to school as ride to school in a car.

(b) The majority of the students ride to school in a bus or car.

2. What is the ratio of those who walk to school to those who ride the bus?

3. What fraction of the students ride the bus?

4. What is the average of these numbers?

1.2, 1.4, 1.5, 1.7, 2

5. What is the product of twelve thousandths and one and two tenths? Write the answer in words.

6. Draw a diagram of this statement. Then answer the questions that follow.

> Only one eighth of the 40 students correctly answered question 5.

   (a) How many students correctly answered question 5?

   (b) How many students did not correctly answer question 5?

7. Replace each circle with the proper comparison symbol.

   (a) 4.0102 $\bigcirc$ 4.0120   (b) 5.014 $\bigcirc$ 50.140

8. A cubit is the distance from the elbow to the fingertips.

   (a) Estimate the number of inches from your elbow to your fingertips.

   (b) Measure the distance from your elbow to your fingertips to the nearest inch.

9. Write 0.375 as a fraction.

10. Divide 59 by 4 and write the answer as a decimal number.

11. Round 53714.$\overline{54}$ to the nearest

   (a) thousandth.

   (b) thousand.

12. Write 5280 in expanded notation using powers of 10.

13. The point marked by the arrow represents what decimal number?

Solve:

14. $\dfrac{6}{10} = \dfrac{w}{100}$ **15.** $\dfrac{36}{x} = \dfrac{16}{24}$

16. $9.8 = x + 8.9$ **17.** $400 - y = 263$

18. In figure *ABCDEF*, all angles are right angles and $AF = AB = BC$. Segment *BC* is twice the length of segment *CD*. If *CD* is 3 cm, what is the perimeter of the figure?

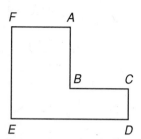

19. Sketch a circle. Within the circle sketch a regular hexagon so that each vertex of the hexagon "touches" the circle.

Add, subtract, multiply, or divide, as indicated:

20. $5^3 - 9^2$ **21.** 3.6×10^3

22. $4\frac{1}{5} + 5\frac{1}{3} + \frac{1}{2}$ **23.** $6\frac{1}{8} - \left(5 - 1\frac{2}{3}\right)$

24. $8\frac{1}{3} \times 3\frac{3}{5} \times \frac{1}{3}$ **25.** $3\frac{1}{8} \div 6\frac{1}{4}$

26. $26.7 + 3.45 + 0.036 + 12 + 8.7$

27. $5 - (0.4 - 0.032)$ **28.** $0.06 \times \$12.50$

29. $3.625 \div 100$ **30.** $3.8 \div 0.16$

LESSON 54

Rectangular Area, Part 1

The diagram below represents the floor of a hallway that has been covered with square floor tiles that are 1 foot on each side. How many 1-ft square tiles does it take to cover the floor of the hallway?

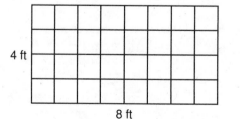

We see that there are 8 floor tiles in each row and 4 rows. So there are 32 1-ft square tiles.

The floor tiles cover the **area** of the hallway. Area is an amount of surface. Floors, ceilings, walls, sheets of paper, and polygons all have an area. If a square is 1 foot on each side, it is a **square foot**. Thus the area of the hallway is 32 square feet.

Other standard square units in the U.S. system include square inches, square yards, and square miles. Units of area in the metric system include square centimeters, square meters, and square kilometers. It is important to distinguish between a unit of length and a unit of area. Units of length, such as an inch or a centimeter, are used for measuring distances, not for measuring areas. To measure area, we use units that take up area. **Square centimeters** and **square inches** take up area and are used to measure area. We include the word "square" or the exponent 2 when we designate units of area.

UNITS OF LENGTH	UNITS OF AREA

1 cm   1 in.

1 cm^2 =
1 square
centimeter

1 in.^{2} =
1 square
inch

Example 1 How many square floor tiles 1 foot on each side would be needed to cover the floor of a rectangular room 12 feet long and 10 feet wide?

12 ft

10 ft

Solution We use parallel lines to draw squares. Twelve tiles will fit in each row. There are 10 rows. Ten rows with 12 tiles in each row equals **120 tiles**. Since each tile is 1 square foot, the area of the room is 120 square feet (120 ft^2).

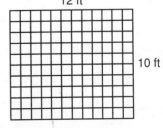

12 ft

10 ft

Notice that the area of the rectangular room equals the length of the room times the width.

Area of rectangle = length × width

Example 2   What is the area of this rectangle?

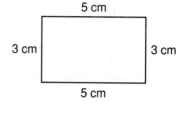

Solution   The area of the rectangle is the number of square centimeters it takes to cover the rectangle. We can find this number by multiplying the length (5 cm) times the width (3 cm).

Area of rectangle = 5 cm · 3 cm
= **15 cm²**

Example 3   The perimeter of a certain square is 12 inches. What is the area of the square?

Solution   To find the area of the square, we first need to know the length of the sides. A square has 4 equal sides, so we divide 12 inches by 4 and find that each side is 3 inches. Then we multiply the length (3 in.) by the width (3 in.) to find the area.

Area = 3 in. × 3 in.
= **9 in.²**

Practice   Find the area of each rectangle.

a.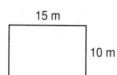

b.

c.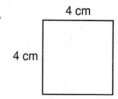

d.   If the perimeter of a square is 20 cm, what is its area?

**Problem set
54**

1. During January the precipitation was 4.5 inches. During February the precipitation was 5.7 inches and during March, 4.2 inches. What was the average amount of precipitation per month for the 3-month period?

2. If 6 ounces of sushi costs $1.86, what is the price per ounce?

3. The parking lot charges $2 for the first hour plus 50¢ for each additional half hour. What is the total charge for parking a car in the lot for 4 hours?

4. Sixty girls and 75 boys were seated on the bleachers. What was the ratio of boys to girls seated on the bleachers?

5. Five billion, three hundred ten million is how much more than two billion ninety-seven million?

6. The Smiths are covering their kitchen floor with tiles 1 foot square. If the kitchen is 12 feet long and 8 feet wide, how many tiles will they need?

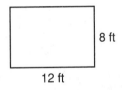

8 ft

12 ft

7. Draw a diagram of this statement. Then answer the questions that follow.

   Three fifths of the 120 people in attendance agreed with the scholar.

   (a) How many of those in attendance agreed with the scholar?

   (b) How many of those in attendance did not agree with the scholar?

8. Find the length of the segment

   (a) in millimeters.

   (b) in centimeters.

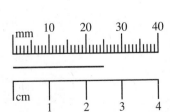

9. Write $6\dfrac{1}{6}$ as a decimal number.

10. Divide 6.3 by 11 and write the answer rounded to the nearest ten-thousandth.

11. Estimate the difference between 39,875 and 21,158 by rounding to the nearest thousand before subtracting.

12. Write 50,704 in expanded notation using powers of 10.

13. (a) What is the perimeter of this square?

    (b) What is the area of this square?

10 cm

14. We know that $10^2 = 100$ and $12^2 = 144$. If $N^2 = 225$, what is N?

15. (a) What fraction of this square is shaded?

    (b) What decimal part of this square is not shaded?

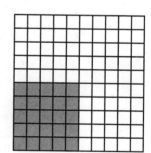

Solve:

16. $\dfrac{12}{100} = \dfrac{3}{d}$

17. $\dfrac{10}{y} = \dfrac{15}{27}$

18. $453 = 76 + a$

19. $c - 3.5 = 1.47$

Add, subtract, multiply, or divide, as indicated:

20. $8^2 + 6^2$

21. 4.7×10^4

22. $\dfrac{4}{5} + \dfrac{4}{4} + \dfrac{4}{3}$

23. $3\dfrac{1}{2} - \left(5 - 2\dfrac{3}{5}\right)$

24. $3\frac{1}{5} \times 3\frac{1}{8} \times 3$ **25.** $6\frac{2}{3} \div 3\frac{1}{8}$

26. $0.52 + 0.76 + 0.8 + 0.29 + 0.016$

27. $3.6 - (7 - 5.437)$ **28.** $(12)(1.2)(0.12)$

29. $0.05 \div 25$ **30.** $300 \div 0.015$

LESSON
55

Square Root

We remember that the exponent of an exponential expression tells how many times the base is to be used as a factor. Five squared means 5 is to be used as a factor twice.

$$5^2 = 25$$

We say, "Five squared is twenty-five."

The inverse operation of squaring a number is the operation that "undoes" squaring. This operation is called finding the **square root** of a number. We indicate square root with a **radical sign**.

$$\sqrt{}$$

To show the square root of 25, we write

$$\sqrt{25}$$

We say, "The square root of twenty-five."

The square root of 25 is the number which multiplied by itself produces 25. Since 5×5 equals 25, the square root of 25 is 5.

$$\sqrt{25} = 5$$

Example 1 Find the square root: (a) $\sqrt{16}$ (b) $\sqrt{100}$ (c) $\sqrt{625}$

Solution   (a)   Since 4 · 4 equals 16, $\sqrt{16}$ = **4**.

(b)   Since 10 · 10 equals 100, $\sqrt{100}$ = **10**.

(c)   Since 25 · 25 equals 625, $\sqrt{625}$ = **25**.

Example 2   Subtract: $\sqrt{25} - \sqrt{16}$

Solution   We must simplify radicals before we can subtract.

$$\sqrt{25} - \sqrt{16} \qquad \text{radicals}$$

$$5 - 4 = 1 \qquad \text{simplified}$$

The term "square root" comes from a geometric idea. The length of the side of a square is the square root of the area of the square.

Example 3   The area of this square is 36 in.²

(a)   What is the length of each side?

(b)   What is the perimeter?

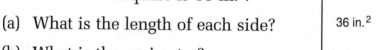

36 in.²

Solution   (a)   Since a square is a rectangle with sides of equal length, we can find the length of each side if we know the area of the square.

$$\text{Side} \times \text{side} = 36 \text{ in.}^2$$

Since each side is equal, each side must be 6 in.

$$6 \text{ in.} \times 6 \text{ in.} = 36 \text{ in.}^2$$

Each side equals the square root of 36 in.²

$$\sqrt{36 \text{ in.}^2} = \textbf{6 in.}$$

(b)   The perimeter of the square is the sum of its 4 sides.

$$\text{Perimeter} = 4 \times 6 \text{ in.}$$

$$= \textbf{24 in.}$$

Practice   Simplify:

a. $\sqrt{64}$ 　　　　　**b.** $\sqrt{121}$ 　　　　　**c.** $\sqrt{9} + \sqrt{16}$

d. What is the perimeter of a square whose area is 100 cm²?

Problem set 55

1. Alaska was purchased by the United States in 1867. Alaska became the forty-ninth state 92 years later. In what year did Alaska become a state?

2. Brand X costs $1.26 for 14 ounces. Brand Y costs $1.58 for 16 ounces. Which brand is the better buy?

3. The ratio of green beans to peas in the garden was 11 to 4. What was the ratio of peas to green beans?

4. During the month of February, Christy's weekly grocery bills were $110.47, $115.68, $96.40, and $120.10. Find her average weekly grocery bill in February to the nearest dollar.

5. Six and seven hundredths is how much less than eight? Write the answer in words.

6. Draw a diagram of this statement. Then answer the questions that follow.

   Seven twelfths of the 60 buttons in the box had 4 holes.

   (a) What fraction of the buttons did not have 4 holes?

   (b) How many buttons did not have 4 holes?

7. Replace the circle with the proper comparison symbol.

   4.06 $\bigcirc$ 4.060

8. Write each decimal as a fraction or as a mixed number.

   (a) 0.12   (b) 0.012

9. Divide 5.9 by 12 and write the answer with a bar over the repetend.

10. Find the length of the segment

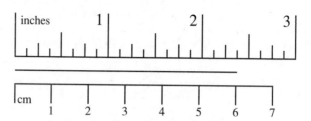

(a)   to the nearest centimeter.

(b)   to the nearest eighth of an inch.

11. If two million is the dividend and two hundred is the divisor, what is the quotient?

12. Simplify each of these numbers.

(a) $8\frac{20}{6}$

(b) $\frac{560}{640}$

13. Write 250,000 in expanded notation using exponents.

14. If the perimeter of a square is 36 inches, what is its area?

15. Draw segment AB to be 4 cm long. Next draw segment BC perpendicular to segment AB and 3 cm long. Then form a triangle by drawing segment AC. Measure to find the length of segment AC.

Solve:

16. $\frac{30}{25} = \frac{18}{f}$

17. $\frac{9}{75} = \frac{p}{100}$

18. $3w = 7.8$

19. $4 - m = 1.24$

Add, subtract, multiply, or divide, as indicated:

20. $9^2 - 3^4$

21. $\sqrt{25} - \sqrt{9}$

22. $26\frac{1}{3} + 15\frac{5}{8} + 8\frac{1}{2}$

23. $15\frac{7}{10} - 8\frac{3}{4}$

24. $7\frac{1}{2} \times 5\frac{1}{3} \times 1\frac{1}{10}$ **25.** $9\frac{3}{5} \div 5\frac{1}{3}$

26. $3.7 + 18.9 + 0.65 + (0.125 \times 10^2)$

27. $10 - (0.1 - 0.099)$ **28.** $0.1001 \div 13$

29. $1.3 \times 0.7 \times 1.1$ **30.** $7 \div 0.035$

LESSON 56

Rates

A **rate** is a ratio of two measurements. Either measurement can be on top. If Leo can walk 6 miles in 2 hours, we can write two rates. When writing rates, we simplify the fraction.

$$\frac{6 \text{ miles}}{2 \text{ hours}} = 3\frac{\text{miles}}{\text{hour}} \qquad \text{read ``3 miles per hour''}$$

$$\frac{2 \text{ hours}}{6 \text{ miles}} = \frac{1 \text{ hour}}{3 \text{ mile}} \qquad \text{read ``}\frac{1}{3}\text{ hour per mile''}$$

If cereal costs 27 cents for 2 ounces, we can write

$$\frac{27 \text{ cents}}{2 \text{ ounces}} = 13\frac{1}{2}\frac{\text{cents}}{\text{ounce}} \quad \text{or} \quad \frac{2 \text{ ounces}}{27 \text{ cents}} = \frac{2}{27}\frac{\text{ounces}}{\text{cent}}$$

If Jim can drive 100 miles on 3 gallons of gas, we can write

$$\frac{100 \text{ miles}}{3 \text{ gallons}} = 33\frac{1}{3}\frac{\text{miles}}{\text{gallon}} \quad \text{or} \quad \frac{3 \text{ gallons}}{100 \text{ miles}} = \frac{3}{100}\frac{\text{gallons}}{\text{mile}}$$

Some rates have special names.

The rate $\dfrac{\text{distance}}{\text{time}}$ is speed.

The rate $\dfrac{\text{miles}}{\text{gallon}}$ is mileage.

The rate $\dfrac{\text{francs}}{\text{dollar}}$ is a rate of exchange.

The rate $\dfrac{\text{dollar}}{\text{francs}}$ is also a rate of exchange.

Example   Edmund rode 24 miles in 3 hours.

(a) Write two rates for this statement.

(b) What was his speed?

Solution (a) $\dfrac{3 \text{ hours}}{24 \text{ miles}} = \dfrac{1 \text{ hour}}{8 \text{ mile}}$ or $\dfrac{24 \text{ miles}}{3 \text{ hours}} = 8 \dfrac{\text{miles}}{\text{hour}}$

(b) Speed has time as the denominator, so his speed is $8 \dfrac{\text{miles}}{\text{hour}}$.

Practice   a. When Monica landed in Belgium, she exchanged $40 for 1640 francs. What was the rate of exchange in francs per dollar? What was the rate of exchange in dollars per franc?

b. Their car traveled 322 miles on 14 gallons of gas. Write two rates for this statement. What was the mileage?

c. The Smiths drove 416 miles in 8 hours. Write two rates for this statement. What was their average speed?

Problem set 56   1. The train traveled 384 kilometers in 4 hours. Write the two rates for this statement. What was the train's average speed?

2. Twenty-four ounces of cereal cost $3.12. What is the unit price?

3. During 1 hour of television programming, there were 12 minutes of commercials.

(a) What fraction of the hour was commercial time?

(b) What was the ratio of commercial to noncommercial time?

4. What number is halfway between 6.7 and 11.9? (*Hint*: Find the average.)

5. Jason bought 2 pounds of meat for $1.85 per pound, 3 cans of tomato sauce for $0.39 per can, and a package of spaghetti for $1.49. What was the total cost of the items?

6. If the perimeter of a square is 1 foot, its area is how many square inches?

7. Draw a diagram of this statement. Then answer the questions that follow.

    It rained two fifths of the days in November.

    (a) How many days did it rain in November?

    (b) How many days did it not rain in November?

8. Find the reciprocal of $3\frac{2}{3}$.

9. Write $\frac{7}{8}$ as a decimal number.

10. Divide 123 by 4 and write the answer

    (a) with a remainder.

    (b) as a mixed number.

11. Round $5.\overline{45}$

    (a) to the nearest thousandth.

    (b) to the nearest hundredth.

12. Write three billion, two hundred million in expanded form using exponents.

13. (a) What fraction of this square is shaded?

    (b) What decimal part of this square is not shaded?

14. Copy this figure on your paper. Find the length of the unmarked sides and find the perimeter of the polygon. Dimensions are in centimeters. All angles are right angles.

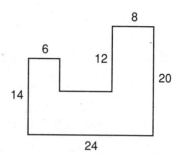

15. The moped traveled 78 miles on 1.2 gallons of gas. The moped averaged how many miles per gallon?

Solve:

16. $\dfrac{35}{60} = \dfrac{f}{24}$

17. $\dfrac{40}{100} = \dfrac{t}{15}$

18. $\dfrac{3}{4} + x = 2$

19. $58 = w - 467$

Add, subtract, multiply, or divide, as indicated:

20. $15^2 - 5^3$

21. $\sqrt{36} + \sqrt{64}$

22. $5\dfrac{5}{6} + 4\dfrac{1}{2} + 6\dfrac{7}{9}$

23. $5 - \left(4\dfrac{1}{4} - 3\dfrac{2}{3}\right)$

24. $4\dfrac{1}{6} \cdot 4 \cdot 3\dfrac{3}{4}$

25. $5\dfrac{5}{6} \div 7\dfrac{1}{2}$

26. $7.6 + 0.375 + 14.84 + 15 + 0.09$

27. $3 - (0.2 - 0.001)$

28. $2.4 \times 1.2 \times 10^3$

29. $0.1001 \div 110$

30. $\$14.52 \div 0.06$

LESSON 57

Percent

Percent is a Latin word that means **by the hundred**. Thus a percent is a fraction with a denominator of 100. The denominator of 100 is not written. The denominator of 100 is indicated by the word "percent" or by the symbol %.

1 percent	means	$\dfrac{1}{100}$
13%	means	$\dfrac{13}{100}$
130 percent	means	$\dfrac{130}{100}$
100%	means	$\dfrac{100}{100} = 1$

A percent describes a whole as though there were 100 parts, even though the whole may not actually contain 100 parts. For instance, we may say that 50 percent of this square is shaded because if this square were divided into 100 equal parts, 50 of the parts would be shaded.

Thus 50 percent is a way to describe $\frac{1}{2}$. Fifty percent is equivalent to $\frac{1}{2}$ because 50 percent means $\frac{50}{100}$, and $\frac{50}{100}$ is equivalent to $\frac{1}{2}$.

We note that 100 percent equals 1, so 100 percent of a number is the number. One hundred percent of 42 is 42. One hundred percent of 130.655 is 130.655.

We remember that we can multiply or divide a given number by 1 without changing the number.

$$50 \times 1 = 50$$

$$\frac{50}{1} = 50$$

Since 100 percent is equal to 1, we can multiply or divide a number by 100 percent without changing the number.

To write a given number as a percent, we multiply by 100 percent.

$$0.75 \times 100\% = 75\%$$

To write a given percent as a number, we divide by 100 percent.

$$\frac{75\%}{100\%} = \frac{75}{100} = \frac{3}{4} = 0.75$$

Example 1 Write $\frac{7}{10}$ as a percent.

Solution To change a number to a percent, we multiply the number by 100 percent.

$$\frac{7}{10} \times 100\% \qquad \text{multiplied by } 100\%$$

$$\frac{700}{10}\% \qquad \text{multiplied}$$

$$\textbf{70\%} \qquad \text{simplified}$$

Example 2 Write $\frac{2}{3}$ as a percent.

Solution We multiply by 100 percent.

$$\frac{2}{3} \times 100\% \qquad \text{multiplied by } 100\%$$

$$\frac{200}{3}\% \qquad \text{multiplied}$$

$$\mathbf{66\tfrac{2}{3}\%} \qquad \text{simplified}$$

Example 3 Write $1\frac{1}{4}$ as a percent.

Solution First we write $1\frac{1}{4}$ either as a fraction or as a decimal.

$$1\frac{1}{4} = \frac{5}{4} \qquad\qquad 1\frac{1}{4} = 1.25$$

Then we can change either of these numbers to a percent by multiplying by 100 percent.

$$\frac{5}{4} \times 100\% = \frac{500}{4}\% = \mathbf{125\%} \qquad 1.25 \times 100\% = \mathbf{125\%}$$

Example 4 Change 70% to a fraction.

Solution To remove the % symbol, we divide by 100%.

$$\frac{70\%}{100\%} = \frac{7}{10}$$

Thus 70 percent means the same thing as $\frac{7}{10}$.

Example 5 (a) What fraction of the square is shaded?

(b) What percent of the square is shaded?

(c) What percent of the square is not shaded?

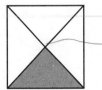

Solution (a) **One fourth** of the square is shaded.

(b) To change $\frac{1}{4}$ to a percent, we multiply by 100 percent.

$$\frac{1}{4} \times 100\% = \frac{100\%}{4} = 25\%$$

(c) The whole square equals 100%. Since 25% is shaded, **75%** is not shaded.

Practice Write each percent as a fraction or as a mixed number.

a. 30%   **b.** 120%

Write each number as a percent.

c. $\frac{3}{5}$ **d.** 1.5

Problem set 57

1. Sam pedaled hard. He traveled 80 kilometers in 2.5 hours. What was his average speed in kilometers per hour?

2. Write the prime factorization of 2016.

3. Write each percent as a fraction or as a mixed number.

(a) 8%   (b) 150%

The graph shows how one family spends their annual income. Use this graph to answer questions 4–6.

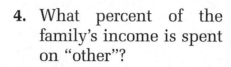

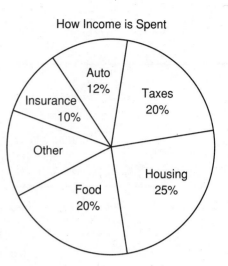

How Income is Spent

Auto 12%

Insurance 10%

Taxes 20%

Other

Housing 25%

Food 20%

4. What percent of the family's income is spent on "other"?

5. What fraction of the family's income is spent on food?

6. If $3200 is spent on insurance, how much is spent on taxes?

7. Draw a diagram of this statement. Then answer the questions that follow.

Van has read five eighths of the 336-page novel.

(a) How many pages has Van read?

(b) How many more pages are left to read?

8. Write each of these numbers as a percent.

(a) 0.25

(b) $1\frac{2}{5}$

9. Divide 2016 by 20 and write the answer as a decimal number.

10. Write $0.\overline{54}$ as a decimal number rounded to three decimal places.

11. A package contained 17 pieces and cost 85 cents. Write the two rates implied by this statement.

12. Write 623 in expanded notation using exponents.

13. What is the perimeter of this shape? Dimensions are in inches. All angles are right angles.

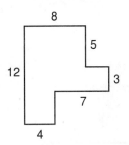

Refer to the rectangles in figure *ABCDEF* to answer questions 14 and 15.

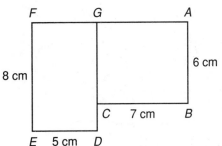

14. (a) What is the area of rectangle *ABCG*?

(b) What is the area of rectangle *DEFG*?

15. What is the perimeter of hexagon *ABCDEF*?

Solve:

16. $\dfrac{6}{40} = \dfrac{15}{w}$

17. $\dfrac{20}{x} = \dfrac{15}{12}$

18. $1.44 = 6m$

19. $\dfrac{1}{2} = \dfrac{1}{3} + f$

Add, subtract, multiply, or divide, as indicated:

20. $2^5 + 1^4 + 3^3$

21. $\sqrt{100} - \sqrt{36}$

22. $3\dfrac{5}{6} - \left(1\dfrac{1}{4} + 1\dfrac{1}{6}\right)$

23. $8\dfrac{3}{4} + \left(4 - \dfrac{2}{3}\right)$

24. $\dfrac{15}{16} \cdot \dfrac{24}{25} \cdot 1\dfrac{1}{9}$

25. $1\dfrac{1}{3} \div \left(2\dfrac{2}{3} \div 4\right)$

26. $8.7 + 9.64 + 25 + 1.456$

27. $10 - (0.9 - 0.876)$ **28.** $4 \times 3.16 \times 10^4$

29. $0.1 \div 25$ **30.** $\$13.93 \div 0.07$

LESSON 58

Mixed Measures • Adding Mixed Measures

Mixed measures   The measurement

$$18 \text{ inches}$$

can be changed into feet and inches. Since 12 inches equals 1 foot, we can divide 18 by 12 to find the number of feet in 18 inches. The remainder is the remaining number of inches.

$$
\begin{array}{r}
1 \text{ ft} \\
12\overline{)18} \\
\underline{12} \\
6 \text{ in.}
\end{array}
$$

$$18 \text{ in.} = 1 \text{ ft } 6 \text{ in.}$$

Example 1   Change 17 ft to yards and feet.

Solution   Since 3 ft equal 1 yd, we divide 17 by 3 to find the number of yards. The remainder is the remaining number of feet.

$$
\begin{array}{r}
5 \text{ yd} \\
3\overline{)17} \\
\underline{15} \\
2 \text{ ft}
\end{array}
$$

$$17 \text{ ft} = \textbf{5 yd 2 ft}$$

Example 2   Simplify: 1 ft 18 in.

Solution   The units are feet and inches. However, the number of inches is greater than the 12 in. that are in 1 ft. So we change 18 in.

to feet and inches. Then we find the total number of feet and inches.

$$1 \text{ ft} \qquad 18 \text{ in.}$$
$$\downarrow \qquad\quad \downarrow$$
$$1 \text{ ft} \quad 1 \text{ ft } 6 \text{ in.}$$

So 1 ft 18 in. = **2 ft 6 in.**

Adding mixed measures To add mixed measures, we align the numbers so that we add the same units. Then we simplify when possible.

Example 3 Add and simplify: 1 yd 2 ft 7 in. + 2 yd 2 ft 8 in.

Solution We add like units, and then we simplify from right to left.

$$
\begin{array}{r}
1 \text{ yd} \quad 2 \text{ ft} \quad 7 \text{ in.} \\
+\ 2 \text{ yd} \quad 2 \text{ ft} \quad 8 \text{ in.} \\
\hline
3 \text{ yd} \quad 4 \text{ ft} \quad 15 \text{ in.}
\end{array}
$$

We change 15 in. to 1 ft 3 in. and add to 4 ft. Now we have

3 yd 5 ft 3 in.

Then we change 5 ft to 1 yd 2 ft and add to 3 yd. Now we have

4 yd 2 ft 3 in.

Example 4 Add and simplify: 2 hr 40 min 35 sec
                                    + 1 hr 45 min 50 sec

Solution We add. Then we simplify from right to left.

$$
\begin{array}{r}
2 \text{ hr} \quad 40 \text{ min} \quad 35 \text{ sec} \\
+\ 1 \text{ hr} \quad 45 \text{ min} \quad 50 \text{ sec} \\
\hline
3 \text{ hr} \quad 85 \text{ min} \quad 85 \text{ sec}
\end{array}
$$

We change 85 sec to 1 min 25 sec and add to 85 min. Now we have

3 hr 86 min 25 sec

Then we simplify 86 min to 1 hr 26 min and combine hours.

4 hr 26 min 25 sec

Practice

a. Change 70 inches to feet and inches.

b. Simplify: 5 ft 20 in.

c. Add: 2 yd 1 ft 8 in. + 1 yd 2 ft 9 in.

d. Add: 5 hr 42 min 53 sec + 6 hr 17 min 27 sec

Problem set 58

1. What is the quotient when the sum of 0.2 and 0.05 is divided by the product of 0.2 and 0.05?

2. Darren carried the football 20 times and gained a total of 184 yards. What was the average number of yards he gained on each carry? Write the answer as a decimal number.

3. Robin bought two dozen arrows for six dollars. Write the two rates implied by this statement.

4. Jeffrey counted the sides on three octagons, two hexagons, a pentagon, and two quadrilaterals. Altogether, how many sides did he count?

5. What is the average of these numbers?

$$6.21, \ 4.38, \ 7.5, \ 6.3, \ 5.91, \ 8.04$$

6. Draw a diagram of this statement. Then answer the questions that follow.

Only two ninths of the 72 billy goats were gruff. The rest were cordial.

(a) How many of the billy goats were cordial?

(b) What was the ratio of gruff billy goats to cordial billy goats?

7. Arrange these numbers in order from least to greatest:

$$0.\overline{5}, \ 0.5, \ 0.\overline{54}$$

8. (a) Estimate the length of segment AB in inches.

(b)   Measure the length of segment *AB* to the nearest eighth of an inch.

9. Divide 365 by 12 and write the answer

(a)   with a remainder.

(b)   as a mixed number.

10. Write each percent as a fraction or as a mixed number.

(a)   3%   (b)   175%

11. Write each of these numbers as a percent.

(a)   0.1   (b) $1\frac{3}{5}$

12. Use exponents to write sixteen million in expanded notation.

13. Estimate the sum of 2,198,475 and 3,315,497 by rounding each number to the nearest hundred thousand before adding.

Refer to figure *ABDEFG* to answer questions 14 and 15.

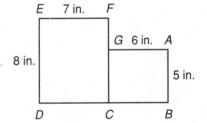

14. (a)   What is the area of rectangle *ABCG*?

(b)   What is the area of rectangle *DEFC*?

15. What is the perimeter of hexagon *ABDEFG*?

Solve:

16. $\dfrac{y}{18} = \dfrac{45}{15}$ 17. $\dfrac{35}{40} = \dfrac{14}{m}$

18. $\dfrac{1}{2} - n = \dfrac{1}{6}$ 19. $9d = 2.61$

Add, subtract, multiply, or divide, as indicated:

20. $\sqrt{100} + 4^3$ 21. 3.14×10^4

22. $3\frac{3}{4} + \left(4\frac{1}{6} - 2\frac{1}{2}\right)$

23. $6\frac{2}{3} \cdot \left(3\frac{3}{4} \div 1\frac{1}{2}\right)$

24.
$$\begin{array}{r} 3 \text{ days} \quad 8 \text{ hr} \quad 15 \text{ min} \\ + \ 2 \text{ days} \quad 15 \text{ hr} \quad 45 \text{ min} \\ \hline \end{array}$$

25.
$$\begin{array}{r} 1 \text{ yd} \quad 2 \text{ ft} \quad 6 \text{ in.} \\ + \ 2 \text{ yd} \quad 1 \text{ ft} \quad 9 \text{ in.} \\ \hline \end{array}$$

26. $25.875 + 4.36 + 19.9 + 15$

27. $0.8 - (7 - 6.543)$

28. $3.45 \times 0.2 \times 0.05$

29. $0.06 \times \$18.00$

30. $\$18.00 \div 0.06$

LESSON 59 — Multiplying Rates

There are two forms of every rate. To solve rate problems we simply multiply by the correct form of the rate. Consider the following statement.

There were 5 chairs in each row.

We can use this statement to write two rates.

(a) $\dfrac{5 \text{ chairs}}{1 \text{ row}}$ (b) $\dfrac{1 \text{ row}}{5 \text{ chairs}}$

If we multiply rate (a) by 6 rows, the rows will cancel and we will find the number of chairs in 6 rows.

$$\frac{5 \text{ chairs}}{1 \text{ row}} \times 6 \text{ rows} = 30 \text{ chairs}$$

If we multiply rate (b) by 20 chairs, the chairs will cancel and we will find the total number of rows that contain 20 chairs.

$$\frac{1 \text{ row}}{5 \text{ chairs}} \times 20 \text{ chairs} = 4 \text{ rows}$$

Example 1 Eight ounces of the solution cost 40 cents.

(a) Write the two rates given by this statement.

(b) Find the cost of 32 ounces of the solution.

(c) How many ounces can be purchased for $1.20?

Solution   (a)   The two rates are

$$(1) \quad \frac{8 \text{ oz}}{40 \text{ cents}} \qquad (2) \quad \frac{40 \text{ cents}}{8 \text{ oz}}$$

Rates have an original form and a reduced form. In Lesson 56 we learned to reduce rates to lowest terms. If we reduce these rates to lowest terms we get

$$(1) \quad \frac{1}{5}\frac{\text{oz}}{\text{cents}} \qquad (2) \quad 5\frac{\text{cents}}{\text{oz}}$$

This step is not necessary as the rates can be used without reducing them first as we will do in this problem. This saves a step.

(b)   **To find the cost, we use the rate that has money on top.**

$$\frac{40 \text{ cents}}{8 \text{ oz}} \times 32 \text{ oz} \qquad \text{canceled ounces}$$

$$\frac{1280}{8} \text{ cents} \qquad \text{multiplied}$$

160 cents $\qquad$ simplified

We usually write answers equal to a dollar or more by using a dollar sign. Thus the cost is **$1.60**.

(c)   Again we will not bother to reduce the rate before we use it. Using the rate in unreduced form is more convenient because it save a step. **To find the number of ounces, we use the rate that has ounces on top.**

$$\frac{8 \text{ oz}}{40 \text{ cents}} \times 120 \text{ cents} \qquad \text{canceled cents}$$

$$\frac{960}{40} \text{ oz} \qquad \text{multiplied}$$

24 oz $\qquad$ simplified

Example 2   Jennifer's speed was 60 miles per hour.

(a)   Write the two rates given by this statement.

(b)   How far did she drive in 5 hours?

(c)   How long would it take her to drive 300 miles?

Solution (a) The two rates are

$$(1) \ \frac{60 \text{ miles}}{1 \text{ hour}} \qquad (2) \ \frac{1 \text{ hour}}{60 \text{ miles}}$$

(b) To find how far, we use the rate with miles on top.

$$\frac{60 \text{ miles}}{1 \text{ hour}} \times 5 \text{ hours} = \textbf{300 miles}$$

(c) To find how much time, we use the rate with time on top.

$$\frac{1 \text{ hour}}{60 \text{ miles}} \times 300 \text{ miles} = \textbf{5 hours}$$

Practice In the lecture hall there were 18 rows. Fifteen chairs were in each row.

a. Write the two rates given by this statement.

b. Find the total number of chairs in the lecture hall.

A car could travel 24 miles on one gallon of gas.

c. Write the two rates given by this statement.

d. How many gallons would it take to go 160 miles?

Problem set 59

1. When the product of 3.5 and 0.4 is subtracted from the sum of 3.5 and 0.4, what is the difference?

2. (a) What fraction of this circle is marked with a 1?

(b) What percent of this circle is marked with a number greater than 1?

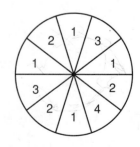

3. The 13-ounce box of cooked cereal costs $1.17, while the 18-ounce box costs $1.44. Find the unit cost for both sizes. Which size is the better buy?

4. Nelson covered the first 20 miles in $2\frac{1}{2}$ hours. What was his average speed in miles per hour?

5. The parking lot charges $2 for the first hour plus 50¢ for each additional half hour or part thereof. What is the total charge for parking in the lot for 3 hours 20 minutes?

6. The train traveled at an average speed of 60 miles per hour.

  (a) Write the two rates given by this statement.

  (b) How long did it take the train to go 420 miles?

7. Draw a diagram of this statement. Then answer the questions that follow.

> Two fifths of the 30 football players were endomorphic.

  (a) How many of the football players were endomorphic?

  (b) What percent of the football players were not endomorphic?

8. Which percent best identifies the shaded part of this circle?

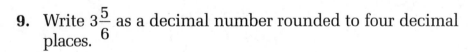

  (a) 25%   (b) 40%

  (c) 50%   (d) 60%

9. Write $3\frac{5}{6}$ as a decimal number rounded to four decimal places.

10. Write 250% as a mixed number.

11. Write $\frac{5}{6}$ as a percent.

12. Use exponents to write seventy-five thousand in expanded notation.

13. List the prime numbers between 90 and 100.

Refer to figure *ABCDEFGH* to answer questions 14 and 15. All angles are right angles. Dimensions are in centimeters.

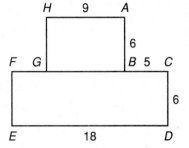

14. (a) What is the area of rectangle *ABGH*?

(b) What is the area of rectangle *CDEF*?

15. What is the perimeter of octagon *ABCDEFGH*?

Solve:

16. $\dfrac{10}{x} = \dfrac{7}{42}$

17. $\dfrac{1.5}{1} = \dfrac{w}{4}$

18. $3.56 = 5.6 - y$

19. $\dfrac{3}{4} = w + \dfrac{1}{8}$

Add, subtract, multiply, or divide, as indicated:

20. $12^2 - \sqrt{81}$

21. $4\,\text{tires} \cdot \dfrac{\$48.85}{1\,\text{tire}}$

22. $\begin{aligned} &5\ \text{hr}\ \ 48\ \text{min}\ \ 45\ \text{sec} \\ +\,&6\ \text{hr}\ \ 20\ \text{min}\ \ 20\ \text{sec} \end{aligned}$

23. $\begin{aligned} &4\ \text{yd}\ \ 2\ \text{ft}\ \ 7\ \text{in.} \\ +\,&3\ \text{yd}\ \ \ \ \ \ \ \ \ 5\ \text{in.} \end{aligned}$

24. $5\dfrac{1}{6} - \left(1\dfrac{3}{4} \div 2\dfrac{1}{3}\right)$

25. $3\dfrac{5}{7} + \left(3\dfrac{1}{8} \cdot 2\dfrac{2}{5}\right)$

26. $(3.26 \times 10^3) + (8.36 \times 10^2)$

27. $2 - (0.86 + 0.9)$

28. $0.625 \times 80 \times 0.02$

29. $1.44 \div 160$

30. $72 \div 0.018$

LESSON
60

Rectangular Area, Part 2

We have practiced finding the areas of rectangles. Sometimes we can find the area of more complex shapes by dividing the shape into rectangular parts. We find the area of each part and then add the areas of the parts to find the total area.

Example 1 Find the area of this figure. Dimensions are in centimeters. All angles are right angles.

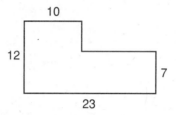

Solution We show two ways to solve this problem.

SOLUTION 1

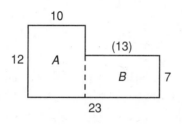

Total area = area A + area B

Area A = 10 cm · 12 cm = 120 cm²
+ Area B = 13 cm · 7 cm = 91 cm²
Total area = **211 cm²**

SOLUTION 2

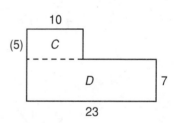

Total area = area C + area D

Area C = 10 cm · 5 cm = 50 cm²
+ Area D = 23 cm · 7 cm = 161 cm²
Total area = **211 cm²**

Example 2 Find the area of this figure. Dimensions are in inches. All angles are right angles.

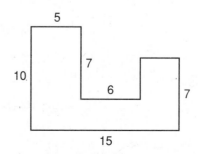

Solution   There are many ways to divide this figure into rectangles. We show just one way to find the answer.

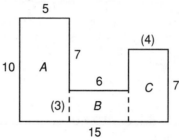

Total area = area *A* + area *B* + area *C*

Area *A* = 10 in. · 5 in. = 50 in.²
Area *B* = 6 in. · 3 in. = 18 in.²
+ Area *C* = 7 in. · 4 in. = 28 in.²
Total area   = **96 in.²**

Example 3   Find the area of this figure. Dimensions are in meters. All angles are right angles.

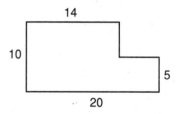

Solution   This time we will imagine this figure as a large rectangle with a small rectangular piece removed. If we find the area of the large rectangle and then subtract the area of the small rectangle, the answer will be the area of the figure shown above.

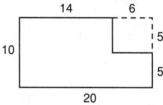

Area of figure = area of large rectangle − area of small rectangle

Area of large rectangle = 20 m · 10 m = 200 m²
− Area of small rectangle = 6 m · 5 m = 30 m²
Area of figure   = **170 m²**

We did not need to use subtraction to find this area. We could have added the areas of two smaller rectangles as we did in Example 1. However, sometimes subtraction is easier.

Practice   Find the area of each figure. Try finding the area of the figure in Problem **c** by subtracting. All dimensions are centimeters. All angles are right angles.

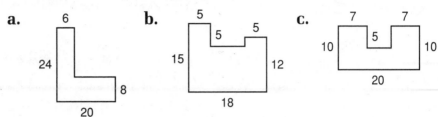

a.   **b.**   **c.**

Problem set 60   Refer to the graph to answer questions 1 and 2.

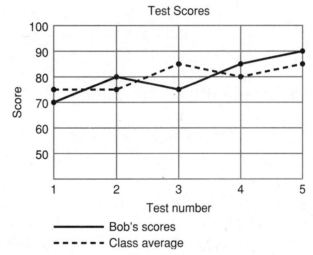

1. On how many tests was Bob's score better than the class average?

2. What was Bob's average score on these five tests?

3. Jim's car could go 75 miles on 3 gallons of gas. Write the two rates given by this statement. How far could his car go on 12 gallons of gas?

4. Fifty-two and one hundred eight thousandths is how much less than one hundred one and one hundredth?

5. When 9 squared is divided by the square root of 9, what is the quotient?

6. Draw a diagram of this statement. Then answer the questions that follow.

> Five ninths of the 3960 voters supported Mayor Cobb.

(a) How many voters did not support Mayor Cobb?

(b) What was the ratio of voters who supported the mayor to those who did not support the mayor?

7. (a) What percent of the circle is shaded?

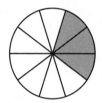

(b) What percent of the circle is not shaded?

8. Write 160% as a mixed number.

9. Write $\frac{2}{3}$ as a percent.

10. Write $0.0\overline{6}$ as a decimal number rounded to the nearest thousandth.

11. Write twelve billion in expanded notation using exponents.

12. Jose bunted the ball and ran 90 feet to first base. How many yards did he run?

13. Divide 365 by 7 and write the answer

(a) with a remainder.

(b) as a mixed number.

Solve:

14. $\dfrac{h}{10} = \dfrac{1.5}{2}$

15. $\dfrac{35}{50} = \dfrac{21}{k}$

16. $x + 4.35 = 7$

17. $m - \dfrac{2}{5} = \dfrac{1}{10}$

Refer to this hexagon to answer questions 18 and 19. Dimensions are in feet. All angles are right angles.

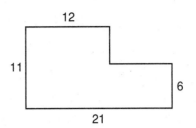

18. What is the perimeter of the hexagon?

19. What is the area of the hexagon?

Add, subtract, multiply, or divide, as indicated:

20. $1^3 + 2^3 + 3^3 - \sqrt{36}$

21. $48\dfrac{\text{miles}}{\text{hour}} \cdot 5 \text{ hours}$

22. $5\dfrac{1}{3} - \left(1\dfrac{3}{4} + 2\dfrac{7}{8}\right)$

23. $3\dfrac{1}{3} \div \left(2\dfrac{2}{3} \cdot 1\dfrac{1}{2}\right)$

24. $\begin{array}{r} 3 \text{ days } 15 \text{ hr } 17 \text{ min} \\ +\quad\quad\ 16 \text{ hr } 50 \text{ min} \\ \hline \end{array}$

25. $\begin{array}{r} 4 \text{ yd } 2 \text{ ft } 9 \text{ in.} \\ +\quad\quad 2 \text{ ft } 5 \text{ in.} \\ \hline \end{array}$

26. $4.296 + 15.47 + 0.0416 + 1$

27. $90 - (8.7 - 6.54)$

28. $0.04 \times 7.5 \times 10^6$

29. $0.06 \times \$24.00$

30. $\$24.00 \div 0.06$

LESSON 61

Scientific Notation for Large Numbers

We use scientific notation as an easy way to write large numbers. We use a decimal number followed by a power of 10 that indicates the true location of the decimal point. **The power of 10 tells us where the decimal point really should be.** Consider this notation.

$$4.62 \times 10^6$$

The power of 10 tells us that the decimal point **really should** be six places **to the right** of where it is written. We have to use zeros as placeholders. We get

$$4620000. \quad \longrightarrow \quad 4,620,000$$

To write a number in scientific notation, it is customary to place the decimal point to the right of the first nonzero digit. Then we use a power of 10 to tell us where the decimal point **really should be**. To write

$$405,700,000$$

in scientific notation, we begin by placing the decimal point to the right of 4 and counting the places to where the decimal point **really should be**.

$$4.05700000$$

8 places

We see that the decimal point **really should be** eight places to the right of where we put it. We omit the terminal zeros and write

$$4.057 \times 10^8$$

The 10^8 tells us the decimal point **really should be** eight places to the right of where it is written.

Example 1 Write 2.46×10^8 in standard form.

Solution The 10^8 tells us that the decimal point really should be eight places to the right of where it is written. We use zeros as placeholders and get

$$246000000. \quad \longrightarrow \quad \mathbf{246,000,000}$$

Example 2 Write 40720000 in scientific notation.

Solution We begin by placing the decimal point after the 4.

$$4.0720000$$

7 places

Now we discard the terminal zeros and write 10^7 to show that the decimal point really should be seven places to the right of where it is written. We get

$$\mathbf{4.072 \times 10^7}$$

Practice Write each number in scientific notation.

  a. 15,000,000   **b.** 400,000,000,000 **c.** 5,090,000

Write each number in standard form.

  d. 3.4×10^6 **e.** 5×10^8 **f.** 1×10^5

Problem set 61

1. Twenty-three billion, nine hundred fifty million is how much less than two hundred seven billion? Use words to write the answer.

2. In the pattern on a soccer ball, a regular hexagon and a regular pentagon share a common side. If the perimeter of the hexagon is 9 in., what is the perimeter of the pentagon?

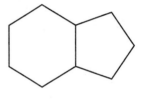

3. Five dozen apples cost $1.25. Write the two rates given by this statement. What would 7 dozen apples cost?

4. The store sold juice for 40¢ per can or 6 cans for $1.98. How much can be saved per can by buying 6 cans at the 6-can price?

5. Five sevenths of those people who saw the phenomenon were convinced.

  (a) What fraction of those who saw the phenomenon were unconvinced?

  (b) What was the ratio of the convinced to the unconvinced?

6. Write twelve million in scientific notation.

7. Write 1.2×10^4 in standard form.

8. Write $\dfrac{1}{8}$ as a decimal number.

9. Round to the nearest thousand.

   (a) 29,647   (b) 5280.08

10. Write 95% as a fraction.

11. Divide 96 by 5 and write the answer as a decimal number.

12. Consider the quadrilateral *WXYZ*. For each statement write true or false.

    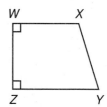

    (a) $\overline{WX} \perp \overline{WZ}$

    (b) $\overline{WX} \parallel \overline{YZ}$

    (c) $\angle WXY$ is a right angle.

Refer to this figure to answer questions 13 and 14. Dimensions are in meters. All angles are right angles.

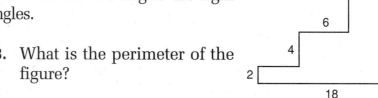

13. What is the perimeter of the figure?

14. What is the area of the figure?

15. What percent of this regular hexagon is shaded?

Solve:

16. $\dfrac{24}{x} = \dfrac{60}{25}$ 17. $\dfrac{6}{42} = \dfrac{n}{7}$

18. $5.37 + m = 8.4$ 19. $6.5 - y = 5.06$

Add, subtract, multiply, or divide, as indicated:

20. $5^2 + 3^3 + \sqrt{64}$

21. $16 \text{ cm} \cdot \dfrac{10 \text{ mm}}{1 \text{ cm}}$

22.
$$
\begin{array}{r}
5 \text{ days} \ \ 18 \text{ hr} \ \ 50 \text{ min} \\
+ \ 2 \text{ days} \ \ \ \ 8 \text{ hr} \ \ 25 \text{ min} \\
\hline
\end{array}
$$

23.
$$
\begin{array}{r}
3 \text{ yd} \ \ 2 \text{ ft} \ \ 5 \text{ in.} \\
+ \ 1 \text{ yd} \ \ \ \ \ \ \ \ \ \ 9 \text{ in.} \\
\hline
\end{array}
$$

24. $6\dfrac{2}{3} + \left(5\dfrac{1}{4} - 3\dfrac{7}{8}\right)$

25. $3\dfrac{1}{3} \cdot \left(2\dfrac{2}{3} \div 1\dfrac{1}{2}\right)$

26. $4.367 + 38.54 + 8.59 + 15$

27. $4.5 - (3 - 2.875)$

28. $\$40.00 \cdot 0.065$

29. $3.75 \div 75$

30. $3 \div 0.08$

LESSON 62

Order of Operations

The four basic operations of arithmetic are addition, subtraction, multiplication, and division. When more than one operation occurs in the same expression, we perform the operations in the order listed below.

ORDER OF OPERATIONS

1. Multiply and divide in order from left to right.

2. Then add and subtract in order from left to right.

Example 1 Simplify: $2 + 4 \times 3 \div 2 - 4$

Solution We multiply and divide in order from left to right before we add or subtract.

$$2 + 4 \times 3 \div 2 - 4 \qquad \text{problem}$$
$$2 + 12 \div 2 - 4 \qquad \text{multiplied } 4 \times 3$$
$$2 + 6 - 4 \qquad \text{divided 12 by 2}$$
$$4 \qquad \text{added and subtracted}$$

Example 2   Simplify: $10 \cdot 3 \div 5 \div 2 + 6(3)$

Solution   **We perform the multiplications and division first.**

$$30 \div 5 \div 2 + 6(3) \qquad \text{multiplied } 3 \cdot 10$$
$$6 \div 2 + 6(3) \qquad \text{divided 30 by 5}$$
$$3 + 6(3) \qquad \text{divided 6 by 2}$$
$$3 + 18 \qquad \text{multiplied}$$
$$\mathbf{21} \qquad \text{added}$$

Practice   Simplify:

a. $5 + 5 \cdot 5 - 5 \div 5$

b. $50 - 8 \cdot 5 + 6 \div 3$

c. $24 - 8 - 6 \cdot 2 \div 4$

d. $24 - 8 \div 4 \cdot 2 + (3)4$

Problem set 62

1. If the product of the first three prime numbers is divided by the sum of the first three prime numbers, what is the quotient?

2. Sean counted a total of 100 sides on the heptagons and nonagons. If there were 4 heptagons, how many nonagons were there?

3. Twenty-five and two hundred seventeen thousandths is how much less than two hundred two and two hundredths?

4. Albert bought a pack of 3 blank tapes for $5.95. What was the cost per tape to the nearest cent?

5. Ginger is starting a 330-page book. Suppose she reads for 4 hours and averages 35 pages per hour.

    (a) How many pages will she read in 4 hours?

    (b) After four hours, how many pages will she still have to read to finish the book?

6. Draw a diagram of this statement. Then answer the questions that follow.

    Three fourths of the 60 passengers disembarked at the terminal.

    (a) How many passengers disembarked at the terminal?

    (b) What percent of the passengers did not disembark at the terminal?

7. Write 3,750,000 in scientific notation.

8. Write 2.05×10^6 in standard form.

9. Write 7.6 as a mixed number.

10. Write $3.\overline{27}$ as a decimal number rounded to the nearest thousandth.

11. Write $2\frac{3}{4}$ as a percent.

12. Divide 70 by 9 and write the answer

    (a) as a decimal number with a bar over the repetend.

    (b) as a decimal number rounded to the nearest thousandth.

13. What decimal number names the point marked by the arrow?

0.9   1.0

Draw a rectangle that is 3 cm long and 2 cm wide. Then answer questions 14 and 15.

14. What is the perimeter of the rectangle in millimeters?

15. What is the area of the rectangle in square centimeters?

Solve:

16. $\dfrac{8}{f} = \dfrac{56}{105}$

17. $\dfrac{12}{15} = \dfrac{w}{2.5}$

18. $p + 6.8 = 20$

19. $q - 3.6 = 6.4$

Add, subtract, multiply, or divide, as indicated:

20. $5^3 - 10^2 - \sqrt{25}$

21. $4 + 4 \cdot 4 - 4 \div 4$

22. $\dfrac{24 \text{ mi}}{1 \text{ gal}} \cdot 5.5 \text{ gal}$

23.
$$\begin{array}{r} 5 \text{ hr } 45 \text{ min } 30 \text{ sec} \\ + \ 2 \text{ hr } 53 \text{ min } 55 \text{ sec} \\ \hline \end{array}$$

24. $6\dfrac{3}{4} + \left(5\dfrac{1}{3} \cdot 2\dfrac{1}{2} \right)$

25. $5\dfrac{1}{2} - \left(3\dfrac{3}{4} \div 2 \right)$

26. $8.575 + 12.625 + 8.4 + 70.4$

27. $4.26 - (9 - 5.74)$

28. $0.8 \times 1.25 \times 10^6$

29. $0.1001 \div 77$

30. $\$2.60 \div 0.065$

LESSON 63

Unit Multipliers • Unit Conversion

Let's take a moment to review the procedure for reducing a fraction. When we reduce a fraction, we remove pairs of numbers that appear as factors in both the numerator and denominator.

$$\frac{24}{36} \quad \rightarrow \quad \frac{\cancel{2} \cdot \cancel{2} \cdot 2 \cdot \cancel{3}}{\cancel{2} \cdot \cancel{2} \cdot 3 \cdot \cancel{3}} \quad = \quad \frac{2}{3}$$

We may reduce before we multiply. This is sometimes called **canceling**.

$$\frac{2}{\cancel{3}} \cdot \frac{\overset{1}{\cancel{3}}}{5} = \frac{2}{5}$$

We may apply this procedure to units as well. We may cancel units before we multiply.

$$5 \cancel{\text{ft}} \cdot \frac{12 \text{ in.}}{1 \cancel{\text{ft}}} = 60 \text{ in.}$$

We remember that we change the name of a number by multiplying by a fraction whose value equals 1. Here we have changed the name of 3 to $\frac{12}{4}$ by multiplying by $\frac{4}{4}$.

$$3 \cdot \frac{4}{4} = \frac{12}{4}$$

The fraction $\frac{12}{4}$ is another name for 3 because $12 \div 4$ equals 3.

Whenever the numerator and denominator of a fraction are equal (and are not zero), the fraction is equal to 1. There is an unlimited number of fractions that are equal to 1. A fraction equal to 1 may have units, such as

$$\frac{12 \text{ inches}}{12 \text{ inches}}$$

Since 12 inches equals 1 foot, we can write two more fractions that equal 1.

$$\frac{12 \text{ inches}}{1 \text{ foot}} \qquad \frac{1 \text{ foot}}{12 \text{ inches}}$$

Because these fractions have units and are equal to 1, we call them **unit multipliers**. Unit multipliers are very useful for converting from one unit of measure to another. For instance, if we want to convert 5 feet to inches, we can multiply 5 feet by a multiplier that has inches on top. The units cancel and we get 60 inches.

$$5 \cancel{ft} \cdot \frac{12 \text{ in.}}{1 \cancel{ft}} = 60 \text{ in.}$$

If we want to convert 96 inches to feet, we can multiply 96 inches by a multiplier that has feet on top. The units cancel and we get 8 feet.

$$96 \cancel{\text{in.}} \cdot \frac{1 \text{ ft}}{12 \cancel{\text{in.}}} = 8 \text{ ft}$$

Notice that we selected unit multipliers that canceled the unit we wanted to remove and kept the unit we wanted in the answer.

When we set up unit conversion problems, we will write the numbers involved in this order.

$$\boxed{\begin{array}{c} \text{Given} \\ \text{measure} \end{array}} \times \boxed{\begin{array}{c} \text{Unit} \\ \text{multiplier} \end{array}} = \boxed{\begin{array}{c} \text{Converted} \\ \text{measure} \end{array}}$$

Example 1   Write two unit multipliers for these equivalent measures.

$$3 \text{ ft} = 1 \text{ yd}$$

Solution   We write one measure as the numerator and its equivalent as the denominator.

$$\frac{\textbf{3 ft}}{\textbf{1 yd}} \quad \text{and} \quad \frac{\textbf{1 yd}}{\textbf{3 ft}}$$

Example 2   Use one of the unit multipliers from Example 1 to convert

(a)   240 yards to feet.

(b)   240 feet to yards.

Solution   (a)   We are given a measure in yards. We want the answer in feet. We write this down.

$$240 \text{ yd} \cdot \boxed{\begin{array}{c} \text{Unit} \\ \text{multiplier} \end{array}} = \quad \text{ft}$$

We want to cancel the unit "yd" and keep the unit "ft," so we select the unit multiplier that has ft on the top and yd below. Then we multiply and cancel units.

$$240 \text{ y̶d̶} \cdot \frac{3 \text{ ft}}{1 \text{ y̶d̶}} = \textbf{720 ft}$$

The answer is reasonable because feet are smaller units than yards, so it takes more feet than yards to measure the same distance.

(b) We are given the measure in feet, and we want the answer in yards. We choose the unit multiplier that has yd on the top.

$$240 \text{ f̶t̶} \cdot \frac{1 \text{ yd}}{3 \text{ f̶t̶}} = \textbf{80 yd}$$

The answer is reasonable because yards are longer units than feet, so it takes fewer yards than feet to measure the same distance.

Example 3 Convert 350 millimeters to centimeters (1 cm = 10 mm).

Solution We are given millimeters and are asked to convert to centimeters. We form a unit multiplier from the equivalence that has cm on the top.

$$350 \text{ m̶m̶} \cdot \frac{1 \text{ cm}}{10 \text{ m̶m̶}} = \textbf{35 cm}$$

Practice Write two unit multipliers for each pair of equivalent measures.

a. 1 yd = 36 in.

b. 100 cm = 1 m

c. 16 oz = 1 lb

Use unit multipliers to perform the following conversions.

d. Convert 10 yards to inches.

e. Twenty-four feet is how many yards? (1 yd = 3 ft)

f. In old England 12 pence equaled 1 shilling. Merlin had 24 shillings. This was the same as how many pence?

Problem set 63   Refer to this bar graph to answer questions 1–3.

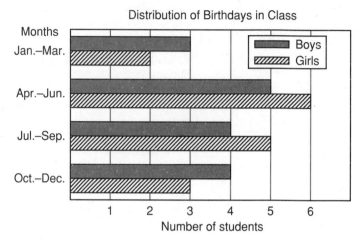

1. (a)   How many boys are in the class?

   (b)   How many girls are in the class?

2. What percent of the students have birthdays in January through June?

3. What fraction of the boys have birthdays in April through June?

4. At the book fair, Bill bought 4 books. One book cost $3.95. Another book cost $4.47. The other 2 books cost $4.95 each.

   (a)   Altogether, how much did Bill spend?

   (b)   What was the average price of the books?

5. Draw a diagram of this statement. Then answer the questions that follow.

   Seven twelfths of the 840 gerbils were hiding in their burrows.

   (a)   What fraction of the gerbils were not hiding in their burrows?

   (b)   How many gerbils were not hiding in their burrows?

6. Write one trillion in scientific notation.

7. Write 7×10^2 in standard form.

8. Use unit multipliers to perform the following conversions.

   (a)   35 yards to feet (3 ft = 1 yd)

   (b)   2000 cm to m (100 cm = 1 m)

9. Write the prime factorization of 10,000.

10. Estimate the difference of 19,827 and 12,092 by rounding to the nearest thousand before subtracting.

11. Write 140% as a mixed number.

12. Divide 430 by 20 and write the answer as a decimal number.

13. Big Bill is 2 m tall. Stephanie is 165 cm tall. Big Bill is how many centimeters taller than Stephanie?

Refer to this figure to answer questions 14 and 15. Dimensions are in feet. All angles are right angles.

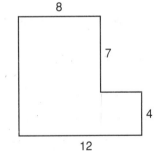

14. What is the area of the figure?

15. What is the perimeter of the figure?

Solve:

16. $\dfrac{18}{14} = \dfrac{90}{p}$

17. $\dfrac{6}{9} = \dfrac{t}{1.5}$

18. $8 = 7.25 + m$

19. $1.5 = 10 - n$

Add, subtract, multiply, or divide, as indicated:

20. $\sqrt{81} + 9^2 - 2^5$

21. $16 \div 4 \div 2 + 3 \times 4$

22. $84 \text{ in.} \cdot \dfrac{1 \text{ ft}}{12 \text{ in.}}$

23. $\begin{array}{r} 3 \text{ yd} \ \ 1 \text{ ft} \ \ 7\frac{1}{2} \text{ in.} \\ + \qquad 2 \text{ ft} \ \ 6\frac{1}{2} \text{ in.} \\ \hline \end{array}$

24. $12\dfrac{2}{3} + \left(5\dfrac{5}{6} \div 2\dfrac{1}{3}\right)$

25. $8\dfrac{3}{5} - \left(1\dfrac{1}{2} \cdot 3\dfrac{1}{5}\right)$

26. $10.6 + 4.2 + 16.4 + (3.875 \times 10^1)$

27. $4.06 - 3.975$

28. $0.065 \times \$12.00$

29. $5.4 \div 4.5$

30. $2.6 \div 0.052$

LESSON 64

Ratio Word Problems

In this lesson we will use proportions to solve ratio word problems. Consider the following ratio word problems.

> The ratio of parrots to macaws was 5 to 7. If there were 750 parrots, how many macaws were there?

In this problem there are two kinds of numbers, ratio numbers and actual count numbers. The ratio numbers are 5 and 7. The number 750 is an actual count of parrots. We will arrange these numbers into two columns to form a ratio box.

	RATIO	ACTUAL COUNT
Parrots	5	750
Macaws	7	M

We were not given the actual count of macaws, so we have used M to stand for the number of macaws.

The numbers in this ratio box can be used to write a proportion. By solving the proportion, we find the actual count of macaws.

	RATIO	ACTUAL COUNT
Parrots	5	750
Macaws	7	M

$$\rightarrow \quad \frac{5}{7} = \frac{750}{M}$$
$$5M = 5250$$
$$M = 1050$$

We find that the actual count of macaws was 1050.

Example The ratio of boys to girls was 5 to 4. If there were 200 girls in the auditorium, how many boys were there?

Solution We begin by making a ratio box.

	RATIO	ACTUAL COUNT
Boys	5	B
Girls	4	200

$$\rightarrow \quad \frac{5}{4} = \frac{B}{200}$$
$$4B = 1000$$
$$B = 250$$

We use the numbers in the ratio box to write a proportion. Then we solve the proportion and answer the question. There were **250 boys**.

Practice Solve each of these ratio word problems. Begin by making a ratio box.

a. The girl-boy ratio was 9 to 7. If 63 girls attended, how many boys attended?

b. The ratio of sparrows to bluejays in the yard was 5 to 3. If there were 15 bluejays in the yard, how many sparrows were in the yard?

c. The ratio of tagged fish to untagged fish was 2 to 9. Ninety fish were tagged. How many fish were untagged?

Problem set 64

1. Thomas Jefferson died on the fiftieth anniversary of the signing of the Declaration of Independence. He was born in 1743. The Declaration of Independence was signed in 1776. How many years did Thomas Jefferson live?

2. The heights of the five basketball players are 190 cm, 195 cm, 197 cm, 201 cm, and 203 cm. What is the average height of the players to the nearest centimeter?

3. Use a ratio box to solve this problem. The ratio of winners to losers was 5 to 4. If there were 1200 winners, how many losers were there?

4. What is the cost of 2.6 pounds of cheese at $1.75 per pound?

5. Maria shut the front door, but not before two hundred eighty-five thousand, six hundred slipped in. Meanwhile, another two million, fifteen thousand slipped in through the back door. How many slipped in altogether?

6. Draw a diagram of this statement. Then answer the questions that follow.

    Four fifths of the 80 trees were infested.

    (a) How many trees were infested?

    (b) How many trees were not infested?

7. Write 405,000 in scientific notation.

8. Write 0.04×10^5 in standard form.

9. Use unit multipliers to perform the following conversions.

    (a) 5280 ft to yards (3 ft = 1 yd)

    (b) 300 cm to millimeters (1 cm = 10 mm)

10. Write 3.1415926 as a decimal number rounded to four decimal places.

11. Write $2\frac{1}{3}$ as a percent.

12. Divide 100 by 6 and write the answer

    (a)   as a decimal number with a bar over the repetend.

    (b)   as a decimal number rounded to the nearest hundredth.

13. The positive square root of 100 is how much greater than 2 cubed?

Refer to this figure to answer questions 14 and 15. Dimensions are in centimeters. All angles are right angles.

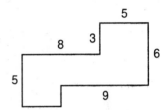

14. What is the perimeter of the figure?

15. What is the area of the figure?

Solve:

16. $6.2 = x + 4.1$

17. $1.2 = y - 0.21$

18. $\dfrac{24}{r} = \dfrac{36}{27}$

Add, subtract, multiply, or divide, as indicated:

19. $11^2 + 1^3 - \sqrt{121}$

20. $24 - 4 \times 5 \div 2 + 5$

21. $1000 \text{ cm} \cdot \dfrac{1 \text{ m}}{100 \text{ cm}}$

22.
$$\begin{array}{r} 1 \text{ week } 5 \text{ days } 14 \text{ hr} \\ + \ 2 \text{ week } 6 \text{ days } 10 \text{ hr} \\ \hline \end{array}$$

23. $3\dfrac{5}{10} + \left(9\dfrac{1}{2} - 6\dfrac{2}{3}\right)$

24. $7\dfrac{1}{3} \cdot \left(6 \div 3\dfrac{2}{3}\right)$

25. $3.47 + 6.3 + 12$

26. $23.6 - (10 - 8.91)$

27. $4.50 × 0.06   **28.** 6.25 × 0.16

29. 7.35 ÷ 70   **30.** 24 ÷ 0.016

LESSON 65   # Average, Part 2

If we know the average of a group of numbers and how many numbers are in the group, we can figure out the sum of the numbers.

Example 1   The average of three numbers is 17. What is their sum?

Solution   We are not told what the numbers are. We are only told their average. All of these sets of three numbers have an average of 17.

$$\frac{16 + 17 + 18}{3} = \frac{51}{3} = 17$$

$$\frac{10 + 11 + 30}{3} = \frac{51}{3} = 17$$

$$\frac{1 + 1 + 49}{3} = \frac{51}{3} = 17$$

Notice that for each set the sum of the three numbers is 51. Since average means what the numbers would be if they were "equalized," the sum is the same as if each of the three numbers is 17.

$$17 + 17 + 17 = \mathbf{51}$$

Thus the number of numbers times their average equals the sum of the numbers.

Example 2   The average of four numbers is 25. If three of the numbers are 16, 26, and 30, what is the fourth number?

Solution   If the average of four numbers is 25, the sum is the same as if all four numbers were 25.

$$25 + 25 + 25 + 25 = 100$$

Thus the sum of the four numbers is 100. We are given three of the numbers. The sum of these three numbers plus the fourth number must equal 100.

$$16 + 26 + 30 + N = 100$$

The sum of the first three numbers is 72. For the sum of the four numbers to total 100, the fourth number must be **28**.

$$16 + 26 + 30 + (28) = 100$$

$$100 \div 4 = 25 \qquad \text{check}$$

Example 3   After 4 tests, Annette's average score was 89. What score does Annette need on her fifth test to bring her average up to 90?

Solution   Although we do not know the specific scores on the first 4 tests, the total is the same as if each of the scores was 89. Thus the total after 4 tests is

$$4 \times 89 = 356$$

The total of her first 4 scores is 356. However, to have an average of 90 after 5 tests, she needs a 5-test total of 450.

$$5 \times 90 = 450$$

Therefore she needs to raise her total from 356 to 450 on the fifth test. To do this, she needs to score **94**.

$$
\begin{array}{rl}
356 & \leftarrow \quad \text{4-test total} \\
+\ 94 & \leftarrow \quad \text{fifth test} \\
\hline
450 & \leftarrow \quad \text{5-test total}
\end{array}
$$

Practice   a. Ralph scored an average of 18 points in each of his first 5 games. Altogether, how many points did Ralph score in the first 5 games?

b. The average of four numbers is 45. If three of the numbers are 24, 36, and 52, what is the fourth number?

c. After 5 tests, Mike's average score was 91. After 6 tests, his average score was 89. What was his score on the sixth test?

Problem set 65

1. Use a ratio box to solve this problem. The ratio of sailboats to rowboats in the bay was 7 to 4. If there were 56 sailboats in the bay, how many rowboats were there?

2. The average of four numbers is 85. If three of the numbers are 76, 78, and 81, what is the fourth number?

3. A one-quart container of oil costs 89¢. A case of 12 one-quart containers costs $8.64. How much is saved per container by buying the oil by the case?

4. Segment *BC* is how much longer than segment *AB*?

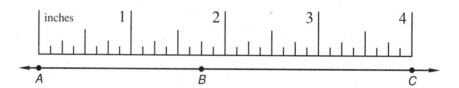

5. Draw a diagram of this statement. Then answer the questions that follow.

    Three tenths of the 30 students earned an A.

    (a) How many students earned an A?

    (b) What percent of the students earned an A?

6. Write 675,000,000 in scientific notation.

7. Write 1.86×10^5 in standard form.

8. Use unit multipliers to perform the following conversions.

    (a) 24 feet to inches

    (b) 500 mm to centimeters

9. Use digits and symbols to write "The product of two hundredths and twenty-five thousandths is five ten-thousandths."

10. Write 48% as a fraction.

11. Divide 75 by 8 and write the answer

    (a) with a remainder.

    (b) as a mixed number.

12. Refer to quadrilateral *ABCD* to answer the following questions.

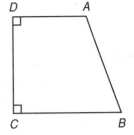

    (a) Which side is parallel to side *BC*?

    (b) Which side is perpendicular to side *BC*?

    (c) Which angle is an obtuse angle?

13. Don is 6 feet 2 inches tall. Bob is 68 inches tall. Don is how many inches taller than Bob?

Refer to this figure to answer questions 14 and 15. Dimensions are in inches. All angles are right angles.

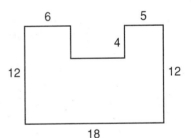

14. What is the area of the figure?

15. What is the perimeter of the figure?

Solve:

16. $4.56 + w = 10$

17. $\dfrac{a}{6} = \dfrac{35}{10}$

18. $4.7 - n = 4.7$

Add, subtract, multiply, or divide, as indicated:

19. $12^2 - 4^3 - 2^4 - \sqrt{144}$

20. $50 + 30 \div 5 \cdot 2 - 6$

21. $10 \text{ yd} \cdot \dfrac{36 \text{ in.}}{1 \text{ yd}}$

22. $\begin{array}{r} 8 \text{ yd } 2 \text{ ft } 7 \text{ in.} \\ + \underline{ 5 \text{ in.}} \end{array}$

23. $2\frac{1}{2} + 6\frac{5}{6} + 4\frac{7}{8}$

24. $6 - \left(7\frac{1}{3} - 4\frac{4}{5}\right)$

25. $6\frac{2}{3} \cdot 5\frac{1}{4} \cdot 2\frac{1}{10}$

26. $3\frac{1}{3} \div 3 \div 2\frac{1}{2}$

27. $3.47 + (6 - 1.359)$

28. $(0.6)(0.28)(0.01)$

29. $2.5 \div 1000$

30. $6.3 \div 0.018$

LESSON 66

Subtracting Mixed Measures

We have practiced adding mixed measures. In this lesson we will practice subtracting mixed measures. When subtracting mixed measures, we may need to borrow in order to subtract. When we borrow, we need to keep in mind the units in the problem. If we borrow 1 hour, it becomes 60 minutes. If we borrow 1 day, it becomes 24 hours.

Example 1 Subtract:

$$
\begin{array}{r}
5 \text{ days } \ 10 \text{ hr } \ 15 \text{ min} \\
- 1 \text{ day } \quad 15 \text{ hr } \ 40 \text{ min}
\end{array}
$$

Solution Before we can subtract minutes, we borrow 1 hour, which is 60 minutes. We combine 60 minutes and 15 minutes, making 75 minutes. Then we can subtract.

$$
\begin{array}{r}
\overset{9}{} \overset{(60\ min)}{} \\
5 \text{ days } \cancel{10} \text{ hr } 15 \text{ min} \\
- 1 \text{ day } \ 15 \text{ hr } 40 \text{ min}
\end{array}
\longrightarrow
\begin{array}{r}
\overset{9}{} \ \overset{75}{} \\
5 \text{ days } \cancel{10} \text{ hr } \cancel{15} \text{ min} \\
- 1 \text{ day } \ 15 \text{ hr } 40 \text{ min} \\
\hline
35 \text{ min}
\end{array}
$$

Next we borrow 1 day, which is 24 hours, and complete the subtraction.

$$
\longrightarrow
\begin{array}{r}
\overset{(24\ hr)}{} \\
\overset{4}{} \ \overset{9}{} \ \overset{75}{} \\
\cancel{5} \text{ days } \cancel{10} \text{ hr } \cancel{15} \text{ min} \\
- 1 \text{ day } \ 15 \text{ hr } 40 \text{ min} \\
\hline
35 \text{ min}
\end{array}
\longrightarrow
\begin{array}{r}
\overset{33}{} \\
\overset{4}{} \ \overset{\cancel{9}}{} \ \overset{75}{} \\
\cancel{5} \text{ days } \cancel{10} \text{ hr } \cancel{15} \text{ min} \\
- 1 \text{ day } \ 15 \text{ hr } 40 \text{ min} \\
\hline
\mathbf{3 \text{ days } 18 \text{ hr } 35 \text{ min}}
\end{array}
$$

Example 2 Subtract: 4 yd 3 in. – 2 yd 1 ft 8 in.

Solution We carefully align the numbers with like units. We borrow 1 yd, which equals 3 ft.

$$\begin{array}{r} \overset{3}{\cancel{4}}\text{ yd} \overset{\text{(3 ft)}}{\quad} 3\text{ in.} \\ -\ 2\text{ yd }1\text{ ft }8\text{ in.} \\ \hline \end{array}$$

Next we borrow 1 ft, which is 12 in. This combines with 3 in., making 15 in. Then we can subtract.

$$\begin{array}{r} \overset{3}{\cancel{4}}\text{ yd }\overset{2}{\cancel{3}}\text{ ft }\overset{15}{\cancel{3}}\text{ in.} \\ -\ 2\text{ yd }1\text{ ft }8\text{ in.} \\ \hline \mathbf{1\text{ yd }1\text{ ft }7\text{ in.}} \end{array}$$

Practice Subtract:

a. $\begin{array}{r}3\text{ hr}\qquad\quad 3\text{ sec}\\ -\ 1\text{ hr }15\text{ min }55\text{ sec}\\\hline\end{array}$
 b. $\begin{array}{r}8\text{ yd }1\text{ ft }5\text{ in.}\\ -\ 3\text{ yd }2\text{ ft }7\text{ in.}\\\hline\end{array}$

c. 2 days 3 hr 30 min – 1 day 8 hr 45 min

Problem set 66

1. Three hundred twenty-nine ten-thousandths is how much greater than thirty-two thousandths? Use words to write the answer.

2. Use a ratio box to solve this problem. The ratio of the length to the width of the rectangle is 4 to 3. If the length of the rectangle is 12 feet,

(a) what is its width?

(b) what is its perimeter?

3. The parking lot charges $2 for the first hour and 50¢ for each additional half hour or part thereof. What is the total charge for parking a car in the lot from 11:30 a.m. until 2:15 p.m.?

4. After four tests Trudy's average score was 85. If her score is 90 on the fifth test, what will be her average for all five tests?

5. Twelve ounces of Brand X costs $1.50. Sixteen ounces of Brand Y costs $1.92. Find the unit price of each. Which brand is the better buy?

6. Five eighths of the rocks in the box were metamorphic. The rest were igneous.

   (a) What fraction of the rocks were igneous?

   (b) What was the ratio of igneous to metamorphic rocks?

7. Write six hundred ten thousand in scientific notation.

8. Write 1.5×10^4 in standard form.

9. Use unit multipliers to perform the following conversions.

   (a) 216 hours to days

   (b) 5 minutes to seconds

10. Write $5\frac{1}{6}$ as a decimal number rounded to the nearest hundredth.

11. How many pennies equal one million dollars? Write the answer in scientific notation.

12. Write $\frac{1}{6}$ as a percent.

13. Which even two-digit number is a common multiple of 5 and 7?

Solve:

14. $\dfrac{3}{2.5} = \dfrac{48}{c}$

15. $x + 56 = 500$

16. $k - 0.75 = 0.75$

Refer to this figure to answer questions 17 and 18. Dimensions are in millimeters. All angles are right angles.

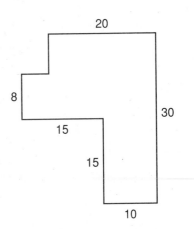

17. What is the perimeter of the figure?

18. What is the area of the figure?

Add, subtract, multiply, or divide, as indicated:

19. $15^2 - 5^3 - \sqrt{100}$

20. $6 + 12 \div 3 \cdot 2 - 3 \cdot 4$

21. $\quad$ 5 yd 2 ft 3 in.
$\quad$ + 2 yd 2 ft 9 in.

22. $\quad$ 5 yd 2 ft 3 in.
$\quad$ − 2 yd 2 ft 9 in.

23. $\dfrac{88 \text{ km}}{1 \text{ hr}} \cdot 4 \text{ hr}$

24. $2\dfrac{3}{4} + \left(5\dfrac{1}{6} - 1\dfrac{1}{4} \right)$

25. $3\dfrac{3}{4} \cdot 2\dfrac{1}{2} \div 3\dfrac{1}{8}$

26. $3\dfrac{3}{4} \div 2\dfrac{1}{2} \cdot 3\dfrac{1}{8}$

27. $4.87 + 12 - 7.363$

28. $\$24.50 \times 0.06$

29. $2.5 \times 4 \times 10^4$

30. $3.6 \div 10^3$

LESSON 67

Liquid Measure

The volume of a container tells us how much the container will hold. Thus, the volume of a container is a measure of the capacity of the container. To measure quantities of liquid, we use units of capacity. Units of capacity in the U.S. Customary System include ounces* (oz), pints (pt), quarts (qt), and gallons (gal). Units of capacity in the metric system include liters (L)

---

*The word "ounce" is used to describe a weight as well as an amount of liquid. An ounce of liquid is often called a **fluid ounce.** Although ounce has two meanings, a fluid ounce of water does weigh about 1 ounce.

and milliliters (mL). Equivalent measures are summarized in the tables below.

U.S. Customary
Equivalents

1 gal	= 4 qt
1 qt	= 2 pt
1 pt	= 2 cups
1 pt	= 16 oz
1 cup	= 8 oz

Metric
Equivalents

1 liter = 1000 mL

Note: 1 liter is a
little more
than 1 quart.

From these equivalents we can form unit multipliers to help us convert one measurement to another.

Example 1 Convert 3.5 liters to milliliters.

Solution From the equivalent 1 liter = 1000 mL we can form two unit multipliers:

$$\frac{1 \text{ liter}}{1000 \text{ mL}} \quad \textbf{or} \quad \frac{1000 \text{ mL}}{1 \text{ liter}}$$

To cancel liters and give us milliliters, we use the unit multiplier that has milliliters on top.

$$3.5 \text{ liters} \cdot \frac{1000 \text{ mL}}{1 \text{ liter}} = \textbf{3500 mL}$$

Example 2 (a) Convert 5 gallons to quarts.

(b) Convert 5 gallons to pints.

Solution (a) To convert gallons to quarts, we multiply by a unit multiplier that has quarts on top.

$$5 \text{ gal} \cdot \frac{4 \text{ qt}}{1 \text{ gal}} = \textbf{20 qt}$$

(b) To convert gallons to pints, we take two steps. For the first step we convert gallons to quarts. We did this in part (a). The second step is to convert quarts to pints. Since we showed the first step in (a), we will show only the second step here.

$$20 \text{ qt} \cdot \frac{2 \text{ pt}}{1 \text{ qt}} = \textbf{40 pt}$$

Example 3   Add: 1 qt 1 pt   7 oz
                 + 1 qt 1 pt 12 oz

Solution   First we add like units. Then we simplify from right to left.

$$\begin{array}{r} 1 \text{ qt } 1 \text{ pt } 7 \text{ oz} \\ + 1 \text{ qt } 1 \text{ pt } 12 \text{ oz} \\ \hline 2 \text{ qt } 2 \text{ pt } 19 \text{ oz} \end{array}$$ added

              2 qt 3 pt 3 oz   simplified ounces
                                 (19 oz = 1 pt 3 oz)

              3 qt 1 pt 3 oz   simplified
                                 (3 pt = 1 qt 1 pt)

Practice   **a.** Complete the chart.   **b.** Compare quantities.

1 gal = ___ qt
1 qt = ___ pt
1 pt = ___ oz

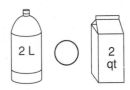

c. One liter equals how many milliliters?

Complete each unit conversion.

d. 3 gal = ___ pt   **e.** 2 qt = ___ oz

f. 2.5 L = ___ mL   **g.** 600 mL = ___ L
                                       (decimal answer)

h. Add: 1 gal 3 qt 1 pt 9 oz
             + 2 gal 1 qt 1 pt 8 oz

i. Subtract: 3 qt   5 oz
                  − 1 qt 1 pt 7 oz

Problem set   **1.** Find the average of these numbers:
67
                  6.38, 8.9, 7.14, 8, 9.32, 10.3

2. Use a ratio box to solve this problem. The ratio of tall ships to small ships floating in the bay was 2 to 7. If there were 28 tall ships, how many small ships were there?

3. Al can rent a chain saw for $35 per day or for $8.75 per hour. If he can finish the job and get the saw back in 3 hours, how much will he save by renting by the hour instead of by the day?

4. If lemonade costs 1.5¢ per ounce, what is the cost per pint?

5. Five and six hundredths is how much less than six and five thousandths? Use words to write the answer.

6. Draw a diagram of this statement. Then answer the questions that follow.

> Five ninths of the 720 students in the assembly were girls.

(a) How many boys were in the assembly?

(b) What was the boy-girl ratio in the assembly?

7. Write 300,000,000 in scientific notation.

8. Write 1×10^5 in standard form.

9. Use unit multipliers to change 5 liters to milliliters.

10. The velocity of light is 299,792,458 meters per second. Round this number to the nearest million.

11. Write 110% as a mixed number.

12. Divide 16.3 by 12 and write the answer as a decimal with a bar over the repetend.

13. Estimate the product of 586 and 716.

Refer to this figure to answer questions 14 and 15. Dimensions are in feet. All angles are right angles.

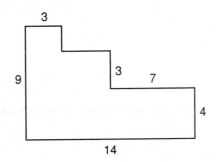

14. What is the area of the figure?

15. What is the perimeter of the figure?

Solve:

16. $1 = 0.01 + a$ **17.** $0.01 = 1 - c$

18. $\dfrac{3}{8} = \dfrac{x}{100}$

Add, subtract, multiply, or divide, as indicated:

19. $2^3 + 4^2 + 1^3 + \sqrt{49}$ **20.** $24 - 12 \div 6 \cdot 2 - 2(5)2$

21.   3 gal 2 qt 1 pt
      + 1 gal 3 qt 1 pt

22.   4 qt 1 pt 3 oz
      − 1 qt 1 pt 7 oz

23. $2.5 \text{ liters} \cdot \dfrac{1000 \text{ mL}}{1 \text{ liter}}$ **24.** $12\dfrac{1}{12} - \left(4\dfrac{1}{2} + 6\dfrac{2}{3}\right)$

25. $3\dfrac{3}{5} \cdot 2\dfrac{2}{3} \cdot 2\dfrac{1}{12}$ **26.** $3\dfrac{3}{5} \div \left(5 \div 1\dfrac{2}{3}\right)$

27. $12 + 0.8 + 1.46$ **28.** $4.37 - (2 - 0.416)$

29. $0.6 \times 3.5 \times 10^3$ **30.** $4.2 \div (6 \div 0.24)$

Scientific Notation for Small Numbers

LESSON 68

We have used scientific notation to write large numbers. We may also use scientific notation to write small numbers. When we write a number in scientific notation, the power of 10 tells us where the decimal point **really should be**. If the exponent is a positive number, the decimal point really should be that many places **to the right** of where it is written.

$$6.32 \times 10^7$$

The exponent is **positive seven**, so we know that the decimal point **really should be** seven places **to the right** of where it is written. This means the number is

63200000. → 63,200,000

7 places

If the exponent is a **negative number**, the decimal point **really should be** that many places **to the left** of where it is written.

$$6.32 \times 10^{-7}$$

The exponent is **negative seven**, so we know that the decimal point **really should be** seven places **to the left** of where it is written.

.000000632 → 0.000000632

7 places

We had to use zeros as placeholders.

Example 1 Write 4.63×10^{-8} in standard notation.

Solution The negative exponent tells us that the decimal point really should be eight places to the left of where it is written. We have to insert zeros as placeholders.

.0000000463 → **0.0000000463**

8 places

Example 2 Write 0.0000033 in scientific notation.

Solution We place the decimal point to the right of the first digit that is not zero.

$$0\underset{\sim}{000000}3.3$$

6 places

The decimal point really should be six places to the left of where we have placed it. So we write

$$\mathbf{3.3 \times 10^{-6}}$$

Practice Write each number in scientific notation.

a. 0.00000025   **b.** 0.000000001   **c.** 0.000105

Write each number in standard form.

d. 4.5×10^{-7} **e.** 1×10^{-3} **f.** 1.25×10^{-5}

Problem set 68

1. Make a ratio box to solve this problem. The ratio of walkers to riders was 5 to 3. If 315 were walkers, how many were riders?

2. After five tests Allison's average score was 88. After six tests her average score had increased to 90. What was her score on the sixth test?

3. When Richard rented a car, he paid $34.95 per day plus 18¢ per mile. If he rented the car for 2 days and drove 300 miles, how much did he pay?

4. If lemonade costs $0.52 per quart, then what is the cost per pint?

5. Draw a diagram of this statement. Then answer the questions that follow.

    Jason finished his math homework in two fifths of an hour.

    (a) How many minutes did it take Jason to finish his math homework?

    (b) What percent of an hour did it take for Jason to finish his math homework?

6. Write each number in scientific notation.

   (a) 186,000   (b) 0.00004

7. Write each number in standard form.

   (a) 3.25×10^1 (b) 1.5×10^{-6}

8. Use unit multipliers to convert 2000 mL to liters.

9. Write $\frac{1}{7}$ as a decimal number rounded to five decimal places.

10. Write $\frac{8}{5}$ as a percent.

11. Divide 330 by 24 and write the answer as a decimal number.

12. (a) What decimal part of the square is shaded?

    (b) What percent of the square is not shaded?

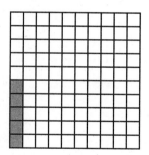

13. Compare: $2.5 \times 10^{-2} \bigcirc 2.5 \div 10^2$

Refer to this figure to answer questions 14 and 15. Dimensions are in yards. All angles are right angles.

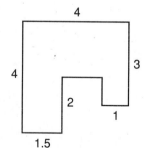

14. What is the perimeter of the figure?

15. What is the area of the figure?

Solve:

16. $17.3 = d + 5.7$

17. $\dfrac{3.5}{w} = \dfrac{28}{20}$

18. $8.4 = q - 3.6$

Add, subtract, multiply, or divide, as indicated:

19. $20^2 + 10^3 - \sqrt{36}$

20. $48 \div 12 \div 2 + 2(3)$

21.
$$\begin{array}{r} 3 \text{ yd } 2 \text{ ft } 1 \text{ in.} \\ - 1 \text{ yd } 2 \text{ ft } 3 \text{ in.} \\ \hline \end{array}$$

22.
$$\begin{array}{r} 4 \text{ gal } 3 \text{ qt } 1 \text{ pt } 6 \text{ oz} \\ - 1 \text{ gal } 2 \text{ qt } 1 \text{ pt } 5 \text{ oz} \\ \hline \end{array}$$

23. $48 \text{ oz} \cdot \dfrac{1 \text{ pt}}{16 \text{ oz}}$

24. $5\dfrac{1}{3} \cdot \left(7 \div 1\dfrac{3}{4}\right)$

25. $5\dfrac{1}{6} + 3\dfrac{5}{8} + 2\dfrac{7}{12}$

26. $5\dfrac{1}{2} - 3\dfrac{3}{5}$

27. $(4.6 \times 10^{-2}) + 0.46$

28. $10 - (2.3 - 0.575)$

29. $0.24 \times 0.15 \times 0.05$

30. $10 \div (0.14 \div 70)$

LESSON
69

Classifying Quadrilaterals

We remember from Lesson 19 that a four-sided polygon is called a quadrilateral. Here we show several kinds of quadrilaterals.

| Trapezium | Trapezoid | Parallelogram | Parallelogram Rectangle | Parallelogram Rectangle Square |

Quadrilaterals are classified by certain characteristics. The chart below names the five illustrated quadrilaterals and identifies some characteristics of each one.

Type of Quadrilateral	Characteristics
Trapezium	No parallel sides
Trapezoid	Exactly one pair of parallel sides
Parallelogram	Two pairs of parallel sides
Rectangle	A parallelogram with four right angles
Square	A rectangle with all sides equal in length

Notice that a square is a special kind of rectangle and that a rectangle is a special kind of parallelogram.

Example 1 Answer true or false:

(a) A square is a rectangle.

(b) All rectangles are parallelograms.

(c) Some squares are trapezoids.

Solution (a) **True**, a square is a parallelogram with four right angles.

(b) **True**, all rectangles have two pairs of parallel sides.

(c) **False**, all squares have two pairs of parallel sides. Trapezoids have only one pair of parallel sides.

Example 2 Sketch a trapezoid.

Solution We begin by drawing two parallel segments of different lengths.

Then we connect the endpoints of the segments to form a quadrilateral.

Practice Answer true or false:

   a. All rectangles are squares.

   b. Some parallelograms are rectangles.

   c. No trapezoid is a parallelogram.

Sketch the following figures.

   d. A parallelogram that is not a rectangle

   e. A trapezoid

In this figure, quadrilateral *ABDE* is a rectangle and $\overline{AC} \parallel \overline{FD}$. Classify the following quadrilaterals.

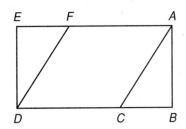

   f. Quadrilateral *ACDF*

   g. Quadrilateral *ABDF*

Problem set 69

   1. The bag contained red marbles, white marbles, and blue marbles. If half the marbles were red and one third of the marbles were white, what fraction of the marbles were blue?

   2. One hundred twenty-five millionths is how much less than two thousandths? Use words to write the answer.

   3. What is the cost of $3\frac{1}{2}$ pounds of bananas at 38¢ per pound?

4. Use a ratio box to solve this problem. The ratio of numismatists to philatelists at the auction was 7 to 5. If there were 140 philatelists, how many numismatists were there?

5. Jenny's average score for six tests is 89. The teacher said that the lowest score would not be included in the average. If her lowest score was 79, what was the average of the remaining scores?

6. Compare: 4 liters $\bigcirc$ 4000 mL

7. The survey found that 4 out of 5 teenagers prefer the taste of Brand A.

   (a) According to the survey, what percent of teenagers prefer the taste of Brand A?

   (b) According to the survey, what fraction of teenagers do not prefer the taste of Brand A?

8. Write each number in scientific notation.

   (a) 30 trillion   (b) 0.0000037

9. Write each number in standard form.

   (a) 8×10^4 (b) 4×10^{-6}

10. Use unit multipliers to convert 16 pints to quarts.

11. Which of the following is not a parallelogram?

   (a) square   (b) trapezoid   (c) rectangle

12. Write 2% as a fraction.

13. Divide 5.43 by 0.11 and write the answer as a decimal with a bar over the repetend.

Refer to this figure to answer questions 14 and 15. Dimensions are in meters. All angles are right angles.

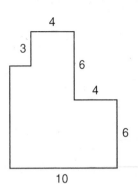

14. What is the perimeter of the figure?

15. What is the area of the figure?

Solve:

16. $101 + f = 1000$

17. $10 - h = 9.7$

18. $\dfrac{y}{4} = \dfrac{150}{20}$

Add, subtract, multiply, or divide, as indicated:

19. $10^2 - \sqrt{100} - 3^4$

20. $48 - 12 \div 2 - 2(3)$

21.   3 wk 4 days   5 hr
     + 1 wk 6 days 21 hr

22.   3 gal
     − 1 gal 2 qt 1 pt

23. $8 \text{ gal} \cdot \dfrac{4 \text{ qt}}{1 \text{ gal}}$

24. $4\dfrac{3}{5} + \left(5\dfrac{1}{6} - 3\dfrac{2}{3}\right)$

25. $4\dfrac{1}{6} \cdot 3\dfrac{1}{5} \cdot 30$

26. $4\dfrac{1}{2} \div \left(3 \div 4\dfrac{1}{2}\right)$

27. $4.9 - (5 - 0.101)$

28. $(0.37)(5.1)(10^3)$

29. $0.24 \div 15$

30. $\$175 \div 0.25$

LESSON 70

Area of a Parallelogram

As we saw in the preceding lesson, a parallelogram is a quadrilateral in which both pairs of opposite sides are parallel.

Parallelogram

Parallelogram

Not a parallelogram

We also saw that a rectangle is a special kind of parallelogram. For several lessons we have practiced finding the areas of rectangles. In this lesson we will practice finding the area of a parallelogram. We may use a paper parallelogram and scissors to help us understand the concept.

Project

> Cut a piece of paper to form a parallelogram as shown.
>
>
>
> Next, cut the paper into two pieces along a line shown by the dotted line below.
>
>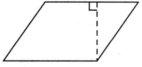
>
> Finally, move the triangular piece as shown to form a rectangle. We see that the area of the parallelogram equals the area of this rectangle.
>
>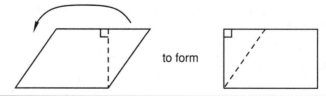
>
> to form

We find the area of a rectangle by multiplying the length times the width. When describing a parallelogram we do not use the words "length" and "width." Instead we use the words **base** and **height**.

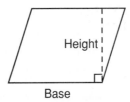

Notice that the height is not one of the sides of the parallelogram (unless the parallelogram is a rectangle). Instead, **the height is perpendicular to the base**. Multiplying the base and height gives us the area of a rectangle. However, as we saw in the project, the area of the rectangle equals the area of the parallelogram we are considering. Thus, we find the area of a parallelogram by multiplying its base and height.

> **Area of a parallelogram = base · height**

Example Find (a) the perimeter and (b) the area of this parallelogram. Dimensions are in inches.

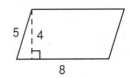

Solution (a) We find the perimeter by adding the lengths of the sides. The opposite sides of a parallelogram are equal in length. So the perimeter is

$$5 \text{ in.} + 8 \text{ in.} + 5 \text{ in.} + 8 \text{ in.} = \textbf{26 in.}$$

(b) We find the area of a parallelogram by multiplying the base and the height. The base is 8 in. and the height is 4 in. So the area is

$$(8 \text{ in.})(4 \text{ in.}) = \textbf{32 in.}^2$$

Practice Find the perimeter and area of each parallelogram. Dimensions are in centimeters.

a.

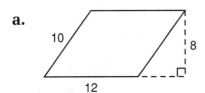

b.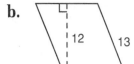

c.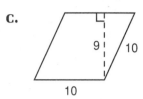

Problem set 70

1. If $\frac{1}{2}$ gallon of milk costs $1.12, what is the cost per pint?

2. Use a ratio box to solve this problem. The cookie recipe called for oatmeal and brown sugar in the ratio of 2 to 1. If 3 cups of oatmeal were called for, how many cups of brown sugar were needed?

3. Matt ran the 400-meter race 3 times. His fastest time was 54.3 seconds. His slowest time was 56.1 seconds. If his average time was 55.0 seconds, what was his time for the third race?

4. It is $4\frac{1}{2}$ miles to the end of the trail. If Paul runs to the end of the trail and back in 60 minutes, what is his average speed in miles per hour?

5. Sixty-three million, one hundred thousand is how much greater than seven million, sixty-five thousand? Write the answer in words.

6. Only three tenths of the print area of the newspaper carried news. The rest of the area was filled with advertisements.

   (a) What percent of the print area was filled with advertisements?

   (b) What was the ratio of news area to advertisement area?

7. Write 0.00105 in scientific notation.

8. Write 3.02×10^5 in standard form.

9. Use unit multipliers to convert 1760 yards to feet.

Quadrilateral *ABDE* is a rectangle and $\overline{EC} \parallel \overline{FB}$. Refer to this figure in Problems 10–12.

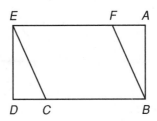

10. Classify each of the following quadrilaterals.

(a) *ECBF*

(b) *ECBA*

11. In the figure above, if *AB* = 4 cm, *BC* = 6 cm, and *BD* = 8 cm, then what is the area of quadrilateral *BCEF*?

12. Classify each of the following angles.

(a) ∠*ECB*   (b) ∠*EDC*   (c) ∠*FBA*

13. Write $\frac{5}{3}$ as a percent.

Refer to this parallelogram to answer questions 14 and 15.

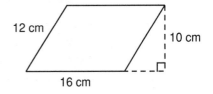

14. What is the perimeter of this parallelogram?

15. What is the area of this parallelogram?

Solve:

16. $\dfrac{4}{1.5} = \dfrac{n}{21}$

17. $p + 3.6 = 5$

18. $r - 15 = 1.5$

Add, subtract, multiply, or divide, as indicated:

19. $10 + 10 \times 10 - 10 \div 10$ **20.** $10^4 - \sqrt{81} + 2^3$

21.
$$\begin{array}{r} 3\text{ gal } 3\text{ qt } 1\text{ pt } 9\text{ oz} \\ +\ \underline{\hspace{3cm} 7\text{ oz}} \end{array}$$

22.
$$\begin{array}{r} 3\text{ yd} \\ -\ \underline{1\text{ yd } 2\text{ ft } 7\text{ in.}} \end{array}$$

23. $2.75 \text{ liters} \cdot \dfrac{1000 \text{ mL}}{1 \text{ liter}}$ **24.** $5\dfrac{7}{8} + \left(3\dfrac{1}{3} - 1\dfrac{1}{2}\right)$

25. $4\dfrac{4}{5} \cdot 1\dfrac{1}{9} \cdot 1\dfrac{7}{8}$ **26.** $6\dfrac{2}{3} \div \left(3\dfrac{1}{5} \div 8\right)$

27. $12 - (0.8 + 0.97)$ **28.** $(2.4)(0.05)(0.005)$

29. $0.2 \div (4 \times 10^2)$ **30.** $0.36 \div (4 \div 0.25)$

LESSON 71

Fraction-Decimal-Percent Equivalents

We remember that if we multiply a number by another number whose value is 1, we do not change the value of the number. We just change its name. All numbers have many fraction names and many decimal names, but only one percent name. **To find the percent name for a number, we multiply the number by 100 percent.**

We can write 4.3 as a percent by multiplying by 100 percent because 100 percent is another way to write 1.

$$4.3 \times 100\% = \mathbf{430\%}$$

We can write one fifth as a percent by multiplying by 100 percent.

$$\frac{1}{5} \times 100\% = \frac{100\%}{5} = 20\%$$

To write any percent as a fraction, we divide by 100 percent. We can change 53 percent to a fraction by dividing by 100 percent.

$$53\% = \frac{53\%}{100\%} = \frac{53}{100}$$

And of course this fraction equals the decimal number 0.53.

$$\frac{53}{100} = 0.53$$

Future problem sets will contain problems that allow us to practice changing from percents to fractions to decimal numbers. The problems will require that we complete a table as we show in the following example.

Example Complete the table.

FRACTION	DECIMAL	PERCENT
$\frac{1}{3}$	(a)	(b)
(c)	1.5	(d)
(e)	(f)	60%

Solution For (a) and (b) we find the decimal and percent equal to $\frac{1}{3}$.

(a)
$$3\overline{\smash{)}\,1.00}\quad\overset{0.\overline{3}}{}$$

(b) $\frac{1}{3} \times 100\% = \frac{100\%}{3} = \mathbf{33\frac{1}{3}\%}$

For (c) and (d) we find a fraction (or a mixed number) and a percent equal to 1.5.

(c) $1.5 = 1\frac{5}{10} = \mathbf{1\frac{1}{2}}$

(d) $1.5 \times 100\% = \mathbf{150\%}$

For (e) and (f) we find a fraction and decimal number for 60%.

(e) $60\% = \frac{60}{100} = \mathbf{\frac{3}{5}}$

(f) $60\% = \frac{60}{100} = \mathbf{0.6}$

Practice   Complete the table.

Fraction	Decimal	Percent
$\frac{2}{3}$	a.	b.
c.	1.1	d.
e.	f.	4%

Problem set 71

1. At the post office Eva bought forty 25-cent stamps, thirty 20-cent stamps, and twenty 15-cent stamps. If Eva paid for the stamps with a $20 bill, how much did she get back in change?

2. When Jim is resting, his heart beats 70 times per minute. When Jim is jogging, his heart beats 150 times per minute. During a half hour of jogging, Jim's heart beats how many more times than it would if he were resting?

3. The product of the number N and 12 is 288. What number is N?

4. Use a ratio box to solve this problem. The ratio of brachiopods to trilobites in the fossil find was 2 to 9. If 720 trilobites were found, how many brachiopods were found?

5. In her first 5 basketball games Sherry scored a total of 72 points. What was the average number of points Sherry scored per game?

6. Draw a diagram of this statement. Then answer the questions that follow.

   The survey found that 3 out of 4 doctors recommend Brand X.

   (a) According to the survey, what percent of the doctors recommend Brand X?

   (b) According to the survey, what fraction of the doctors do not recommend Brand X?

7. Use a unit multiplier to convert 20 centimeters to meters.

8. Compare: $30(150 - 70) \bigcirc 30 \cdot 150 - 30 \cdot 70$

9. Light travels at a speed of about three hundred thousand kilometers per second. Write that number in scientific notation.

10. Write each number in standard form.

   (a) 6.05×10^6 (b) 4×10^{-5}

11. Divide 1000 by 32 and write the answer as a decimal number.

12. Complete the table.

Fraction	Decimal	Percent
(a)	1.2	(b)

13. Write the prime factorization of 5000.

Refer to this figure to answer questions 14 and 15. Dimensions are in inches.

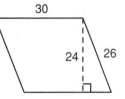

14. What is the perimeter of this parallelogram?

15. What is the area of this parallelogram?

Solve:

16. $5a = 1.7$ **17.** $8.1 + x = 12$

Add, subtract, multiply, or divide, as indicated:

18. $7 + 7 \cdot 7 - 7 \div 7$ **19.** $25^2 - 5^3 - \sqrt{100}$

20. $5.8 - (6 - 3.17)$ **21.** $(2.4)(0.75)(0.05)$

22.
$$
\begin{array}{r}
\text{1 yd} \ \text{1 ft} \ \text{1 in.} \\
- \qquad \text{2 ft} \ \text{2 in.} \\
\hline
\end{array}
$$

23.
$$
\begin{array}{r}
\text{1 hr} \ \text{15 min} \ \text{45 sec} \\
+ \qquad \text{45 min} \ \text{15 sec} \\
\hline
\end{array}
$$

24. $12 \text{ gal} \cdot \dfrac{24 \text{ mi}}{1 \text{ gal}}$

25. $3\dfrac{5}{8} + 5\dfrac{1}{6} + 2\dfrac{1}{2}$

26. $3\dfrac{1}{3} - \left(2\dfrac{1}{2} \div 3\right)$

27. $6\dfrac{1}{4} \cdot 5\dfrac{1}{3} \cdot 3\dfrac{1}{3}$

28. $0.8 \div 32$

29. $\$15.00 \div 0.75$

30. $6.4 + 5.88 + 15.7 + 24 + 0.09 + 23.86$

LESSON 72

Sequences • Functions, Part 1

Sequences   A **sequence** is an ordered list of numbers that are arranged by following a certain rule. To find the next number in a sequence, we first determine the rule for the sequence. Then we use the rule to find the next number.

Example 1   Find the next three numbers in this sequence.

$$5, 20, 35, 50, \ldots$$

Solution   As we inspect the numbers in the sequence, we see that each number is 15 greater than the preceding number. This is an addition sequence.

$$
\underbrace{5}_{5,} \qquad \underbrace{5 + 15}_{20,} \qquad \underbrace{5 + 15 + 15}_{35,} \qquad \underbrace{5 + 15 + 15 + 15}_{50}
$$

We continue this pattern and find that the next three numbers in the sequence are

65, 80, 95

Example 2   Find the next three numbers in this sequence.

$$5, 10, 20, 40, \ldots$$

Solution   As we inspect the numbers in the sequence, we see that each number is twice the preceding number. This is a multiplication sequence.

$$\underbrace{5}_{5,} \qquad \underbrace{5 \times 2}_{10,} \qquad \underbrace{5 \times 2 \times 2}_{20,} \qquad \underbrace{5 \times 2 \times 2 \times 2}_{40}$$

We continue this pattern and find that the next three numbers in the sequence are

80, 160, 320

Functions   A **function** is a set of number pairs that are related by a certain rule. We will be finding a missing number from a number pair. The thinking we use for these problems is similar to the thinking we use for sequence problems. We study pairs of numbers to determine a rule for the function. Then we use the rule to find the missing number.

Example 3   Find the missing number.

IN | FUNCTION | OUT
3 → 9
5 → 15
7 → ▢
10 → 30

Solution   We study each "in-out" number pair to determine the rule for the function. We see that for each complete pair, if the number "in" is multiplied by 3, it equals the number "out." Thus the rule of the function is "multiply by 3."

We use this rule to find the missing number. We multiply 7 by 3 and find that the missing number is **21**.

Practice   Find the next three numbers in each sequence.

a. 5, 11, 17, 23, . . .   **b.** 3, 9, 27, 81, . . .

Find the missing number in each diagram.

c. IN | FUNCTION | OUT
4 → 20
3 → 15
7 → ▢
9 → 45

d. IN | FUNCTION | OUT
0 → 4
1 → ▢
3 → 7
5 → 9

Problem set 72

1. It is 1.4 kilometers from Jim's house to school. How far does Jim walk going to and from school every day for 5 days?

2. The parking lot charges 25¢ for each half hour or part of a half hour. If Edie parks her car in the lot from 10:45 a.m. until 1:05 p.m., how much money will she pay?

3. If the product of the number N and 17 is 340, what is the sum of the number N and 17?

4. The football team won 3 of their 12 games but lost the rest.

   (a) What was the team's won-lost ratio?

   (b) What fraction of the games did the team lose?

   (c) What percent of the games did the team win?

5. Will's bowling average after 5 games was 120. In his next 3 games Will scored 118, 124, and 142. What was Will's bowling average after 8 games?

6. Draw a diagram of this statement. Then answer the questions that follow.

   Three fifths of the 60 questions on the test were multiple-choice.

   (a) How many of the 60 questions were multiple-choice?

   (b) What percent of the 60 questions were not multiple-choice?

7. Write 0.0000001 in scientific notation.

8. Write 1.5×10^7 in standard form.

9. Compare: 20 qt $\bigcirc$ 5 gal

10. Divide 3.45 by 0.18 and write the answer rounded to the nearest whole number.

11. Find the next three numbers in this sequence.

20, 15, 10, . . .

12. Complete the table.

FRACTION	DECIMAL	PERCENT
$\frac{1}{6}$	(a)	(b)

13. Find the missing number.

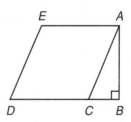

Refer to this figure to answer questions 14 and 15.

14. In this figure, $\overline{AE} \parallel \overline{CD}$ and $\overline{AC} \parallel \overline{ED}$.

   (a) Classify quadrilateral $ACDE$.

   (b) Classify quadrilateral $ABDE$.

15. If $AB = 24$ cm, $AC = 25$ cm, and $AE = 30$ cm, what is the area of quadrilateral $ACDE$?

Solve:

16. $\dfrac{1.5}{2} = \dfrac{7.5}{w}$

17. $1.7 - y = 0.17$

18. $3x = 0.45$

Add, subract, multiply, or divide, as indicated:

19. $10^3 - 10^2 + 10 - 1$ **20.** $6 + 3(2) - 4 - (5 + 3)$

21. $\begin{array}{r} 1\text{ gal } 2\text{ qt } 1\text{ pt} \\ + 1\text{ gal } 2\text{ qt } 1\text{ pt} \\ \hline \end{array}$ **22.** $\begin{array}{r} 1\text{ day } 3\text{ hr } 15\text{ min} \\ - \quad\quad 8\text{ hr } 30\text{ min} \\ \hline \end{array}$

23. $2\text{ mi} \cdot \dfrac{5280\text{ ft}}{1\text{ mi}}$ **24.** $10 - \left(5\dfrac{3}{4} - 1\dfrac{5}{6}\right)$

25. $\left(2\dfrac{1}{5} + 5\dfrac{1}{2}\right) \div 2\dfrac{1}{5}$ **26.** $3\dfrac{3}{4} \cdot \left(6 \div 4\dfrac{1}{2}\right)$

27. $5 - (4.3 - 0.021)$ **28.** $(3.6)(2.5)(10^2)$

29. $4.6 \div 80$ **30.** $15 \div 0.015$

LESSON 73

Adding Integers on the Number Line

Integers The **whole numbers** are zero and the counting numbers. All the numbers in this sequence are whole numbers.

$$0, 1, 2, 3, \ldots$$

Integers include all the whole numbers and also the opposites of the positives integers, that is, their negatives. All the numbers in this sequence are integers.

$$\ldots, -3, -2, -1, 0, 1, 2, 3, \ldots$$

The dots on this number line mark the integers from -5 through $+5$.

Notice that the numbers between the whole numbers, such as $3\frac{1}{2}$ and 1.3, are not integers.

All numbers except zero are **signed numbers**, either positive or negative. Zero is neither positive nor negative. Positive and negative numbers have a sign and a value, which is called **absolute value**.

NUMERAL	NUMBER	SIGN	ABSOLUTE VALUE
+3	Positive three	+	3
−3	Negative three	−	3

The absolute value of both +3 and −3 is 3. Notice on the number line that +3 and −3 are both 3 units from zero. Absolute value can be represented by distance, whereas the sign can be represented by direction. Thus positive and negative numbers are sometimes called **directed numbers** because the sign of the number (+ or −) can be thought of as indicating direction.

When we add, subtract, multiply, or divide signed numbers, we need to pay attention to the signs as well as to the absolute values of the numbers. In this lesson we will practice adding positive and negative numbers.

Number line model for addition   A number line may be used to illustrate the addition of signed numbers. A positive 3 is indicated by a 3-unit arrow that points to the right. A negative 3 is indicated by a 3-unit arrow that points to the left.

To show the addition of +3 and −3, we begin at zero on the number line and draw the +3 arrow. From this arrowhead we draw the −3 arrow. The sum of +3 and −3 is found at the point on the number line that corresponds to the second arrowhead.

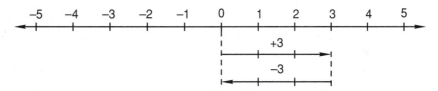

We see that the sum of +3 and −3 is 0. We find that the sum of two opposites is always zero.

Example 1   Sketch a number line to show each addition problem.

(a) $(-3) + (+5)$ (b) $(-4) + (-2)$

Solution (a)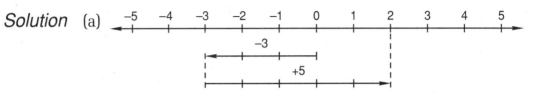

We begin at zero and draw an arrow 3 units long that points to the left. From this arrowhead we draw an arrow 5 units long that points to the right. We see that the sum of -3 and $+5$ is **2**.

(b)

We used the arrows to show that the sum of -4 and -2 is **−6**.

Example 2   Sketch a number line to show this addition problem.

$$(-2) + (+5) + (-4)$$

Solution This time we draw three arrows. We always begin the first arrow at zero. We begin each remaining arrow at the arrowhead of the previous arrow.

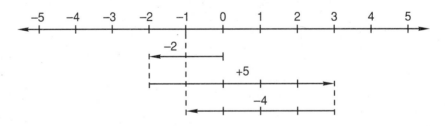

The last arrowhead corresponds to -1 on the number line, so the sum of -2 and $+5$ and -4 is **−1**.

Practice   Sketch a number line to show each addition problem.

a. $(-2) + (-3)$

b. $(+4) + (+2)$

   c. $(-5) + (+2)$

   d. $(+5) + (-2)$

   e. $(-4) + (+4)$

   f. $(-3) + (+6) + (-1)$

Problem set 73

1. School pictures cost $4.25 for an 8 by 10 print. They cost $2.35 for a 5 by 7 print, and 60¢ for each wallet-size print. What is the total cost of two 5 by 7 prints and six wallet-size prints?

Refer to this graph to answer questions 2 and 3.

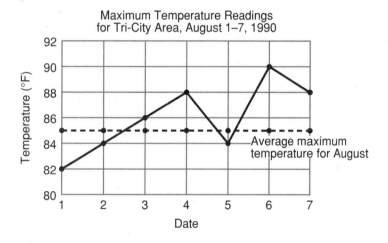

2. The highest temperature reading on August 6, 1990, was how much greater than the average maximum temperature for August?

3. What was the average maximum temperature during the first seven days of August 1990?

4. The sum of the number n and 12 is 30. What is the product of n and 12?

5. Use a ratio box to solve this problem. The ratio of sonorous to discordant voices in the crowd was 7 to 4. If 56 voices were discordant, how many voices were sonorous?

6. Draw a diagram of this statement. Then answer the questions that follow.

   The Celts won three fourths of their first 20 games.

   (a) How many of their first 20 games did the Celts win?

   (b) What percent of their first 20 games did the Celts fail to win?

7. Write 4,000,000,000,000 in scientific notation.

8. Pluto's average distance from the sun is 3.67×10^9 miles. Write that number in standard form.

9. A micron is 1×10^{-6} meter. Write that number in standard form.

10. Use a unit multiplier to convert 300 mm to centimeters.

11. Complete the table.

FRACTION	DECIMAL	PERCENT
(a)	(b)	12%

12. Sketch a number line to show each addition problem.

    (a) $(-5) + (+2)$

    (b) $(+5) + (-2)$

13. Find the missing number.

IN	FUNCTION	OUT
2	→	14
12	→	24
8	→	☐
0	→	12

Solve:

14. $4.4 = 8w$

15. $\dfrac{0.8}{1} = \dfrac{x}{1.5}$

16. $n + \dfrac{2}{3} = \dfrac{3}{4}$

Refer to this figure to answer questions 17 and 18. Dimensions are in millimeters. All angles are right angles.

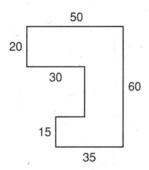

17. What is the perimeter of this figure?

18. What is the area of this figure?

Add, subtract, multiply, or divide, as indicated:

19. $4 + 5(6) - (7 + 8) - 9$ **20.** $\sqrt{64} - 2^3 + \sqrt{1}$

21.   3 yd 2 ft $7\frac{1}{2}$ in.
     + 1 yd $5\frac{1}{2}$ in.

22.   1 qt 1 pt   6 oz
    −   1 pt 12 oz

23. $2\dfrac{1}{2}$ hr $\cdot \dfrac{50 \text{ mi}}{1 \text{ hr}}$

24. $\left(\dfrac{5}{9} \cdot 12\right) \div 6\dfrac{2}{3}$

25. $3\dfrac{5}{6} - \left(4 - 1\dfrac{1}{9}\right)$

26. $\left(5\dfrac{5}{8} + 6\dfrac{1}{4}\right) \div 6\dfrac{1}{4}$

27. $0.1 - (0.2)(0.3)$

28. $0.065 \times \$18.00$

29. $0.364 \div 7$

30. $3 \div (30 \div 0.03)$

LESSON
74

Fractional Part of a Number and Decimal Part of a Number, Part 1

Problems about parts of a number are stated by using decimal fractions or common fractions. The procedure for solving both kinds of problems is the same. We can solve fractional-part-of-a-number problems by translating the question into an equation. Then we solve the equation. To translate,

We replace **is** with =

We replace **of** with ×

Example 1   What number is 0.6 of 31?

Solution   This problem uses a decimal number to ask the question. We will use W_N to represent *what number*. We will translate *is* by writing an equals sign. We will translate *of* by writing a multiplication symbol.

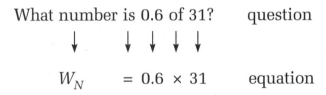

$$W_N = 0.6 \times 31 \qquad \text{equation}$$

To find the answer, we multiply.

$$W_N = 18.6 \qquad \text{multiplied}$$

Example 2   Three fifths of 120 is what number?

Solution   This time the question is phrased by using a common fraction. The procedure is the same. We translate directly.

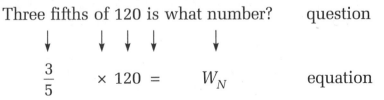

$$\frac{3}{5} \times 120 = W_N \qquad \text{equation}$$

To find the answer, we multiply.

$$\frac{360}{5} = W_N \qquad \text{multiplied}$$

$$72 = W_N \qquad \text{simplified}$$

Practice Write equations to solve each problem.

a. What number is $\frac{4}{5}$ of 71?

b. Three eighths of $3\frac{3}{7}$ is what number?

c. What number is 0.6 of 145?

d. Seventy-five hundredths of 14.4 is what number?

Problem set 74

1. Five and seven hundred eighty-four thousandths is how much less than seven and twenty-one ten-thousandths?

2. Cynthia was paid 20¢ per board for painting the fence. If she was paid $10 for painting half the boards, how many boards were there?

3. When 72 is divided by n, the quotient is 12. What is the product when 72 is multiplied by n?

4. Four fifths of the students passed the test.

   (a) What percent of the students did not pass the test?

   (b) What was the ratio of students who passed to students who did not pass?

5. The average height of the five players on the basketball team was 77 inches. One player was 71 inches tall. Another was 74 inches tall, and two were 78 inches tall. How tall was the tallest player on the team?

6. Write each number in scientific notation.

   (a) 0.00000008   (b) 67,500,000,000

7. Draw a diagram of this statement. Then answer the questions that follow.

Two thirds of the 96 members approved of the plan.

(a) How many of the 96 members approved of the plan?

(b) What percent of the members did not approve of the plan?

Write equations to solve problems 8 and 9.

8. What number is $\frac{3}{4}$ of 17?

9. What number is 0.7 of 6.5?

10. Compare: $\frac{1}{3}$ ◯ 0.33

11. Complete the table.

FRACTION	DECIMAL	PERCENT
$\frac{1}{8}$	(a)	(b)

12. Sketch a number line to show the addition $(-3) + (-1)$.

13. Find the next three numbers in this sequence.

$$1, 4, 9, 16, 25, \ldots$$

14. Write the prime factorization of 360.

Solve:

15. $p - \dfrac{1}{3} = \dfrac{1}{2}$ 16. $9m = 0.117$

17. $\dfrac{48}{y} = \dfrac{32}{20}$

Refer to this figure to answer questions **18** and **19**. Dimensions are in feet.

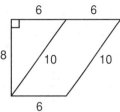

18. What is the area of the parallelogram?

19. What is the perimeter of the trapezoid?

Add, subract, multiply, or divide, as indicated:

20. $3^2 + 4(3 + 2) - 8 \div 4 + \sqrt{36}$

21. $\begin{array}{l} \text{3 days}\quad\text{16 hr}\quad\text{48 min} \\ +\ \text{1 day}\quad\text{15 hr}\quad\text{54 min} \\ \hline \end{array}$

22. $1\,\text{m} - 20\,\text{cm} = ?\,\text{cm}$

23. $5\,\text{sec} \cdot \dfrac{344\,\text{m}}{1\,\text{sec}}$

24. $19\dfrac{3}{4} + 27\dfrac{7}{8} + 24\dfrac{5}{6}$

25. $3\dfrac{3}{5} - \left(\dfrac{5}{6} \cdot 4\right)$

26. $\left(1\dfrac{1}{4} \div \dfrac{5}{12}\right) \div 24$

27. $4.6 + 0.375 + 12$

28. $6.5 - (0.65 - 0.065)$

29. $(0.4)(0.3)(0.2)(0.1)$

30. $0.3 \div (3 \div 0.03)$

LESSON 75

Variables and Evaluation

In algebra we often use letters in place of numbers. The expression

$$x + y$$

means that two numbers are added. The value of the expression depends upon the numbers we choose for x and y. Since the value of a letter may vary, a letter is called a **variable**.

In arithmetic we use the symbols $+$, $-$, $\times$, and $\div$ to indicate the operations of arithmetic. In algebra we use the $+$ and $-$ signs, but sometimes we do not use the $\times$ and $\div$ signs. To show multiplication of variables, we can write the variables together with no sign between. Thus, xy means that x and y are multiplied. To show division, we can use a division line. The chart below summarizes how the operations of arithmetic can be indicated in algebra.

EXPRESSION	MEANING
$a + b$	b is added to a
$a - b$	b is subtracted from a
ab	b is multiplied by a
$\dfrac{a}{b}$	a is divided by b

The value of an expression depends upon the numbers that are substituted for the variables. We **evaluate** an expression by finding its value.

Example 1   Evaluate: $a + ab$ if $a = 3$ and $b = 4$

Solution   We will begin by writing parentheses in place of each variable. This step may seem unnecessary, but many errors can be avoided if this step is the first step.

$$a + ab$$

$$( ) + ( )( ) \qquad \text{parentheses}$$

Then we replace a with 3 and b with 4.

$$a \; + \; ab$$

$$(3) + (3)(4) \qquad \text{substituted}$$

We follow the order of operations by multiplying first. Then we add.

$$(3) + (3)(4) \qquad \text{problem}$$

$$3 \; + \; 12 \qquad \text{multiplied}$$

$$\mathbf{15} \qquad \text{added}$$

Example 2 Evaluate: $xy - \dfrac{x}{2}$ if $x = 9$ and $y = \dfrac{2}{3}$

Solution First we replace each letter with parentheses.

$$xy \; - \; \frac{x}{2}$$

$$(\;)(\;) - \frac{(\;)}{2} \qquad \text{parentheses}$$

Then we write 9 in place of x and $\dfrac{2}{3}$ in place of y.

$$xy \; - \; \frac{x}{2}$$

$$(9)\left(\frac{2}{3}\right) - \frac{(9)}{2} \qquad \text{substituted}$$

We follow the order of operations by multiplying and dividing before we subtract.

$$(9)\left(\frac{2}{3}\right) - \frac{(9)}{2}$$

$$6 \; - \; 4\frac{1}{2} \qquad \text{multiplied and divided}$$

$$1\frac{1}{2} \qquad \text{subtracted}$$

Practice Evaluate:

    a. $ab - bc$ if $a = 5$, $b = 3$, and $c = 4$

b. $ab + \dfrac{a}{c}$ if $a = 6$, $b = 4$, and $c = 2$

c. $x - xy$ if $x = \dfrac{2}{3}$ and $y = \dfrac{3}{4}$

Problem set 75

1. At 1:30 p.m. David found a parking meter that still had 10 minutes until it expired. He put 2 dimes into the meter and went to his meeting. If 5 cents buys 15 minutes of parking time, at what time will the meter expire?

Use the information in the next paragraph to answer questions 2 and 3.

The Barkers started their trip with a full tank of gas and a total of 39,872 miles on their car. They stopped and filled the gas tank 4 hours later with 8.0 gallons of gas. At that time the car's total mileage was 40,060.

2. How far did they travel in 4 hours?

3. The Barkers' car traveled an average of how many miles per gallon during the first 4 hours of the trip?

4. When 24 is multiplied by w, the product is 288. What is the quotient when 24 is divided by w?

5. Use a ratio box to solve this problem. There were 144 Bolsheviks in the crowd. If the ratio of Bolsheviks to czarists was 9 to 8, how many czarists were in the crowd?

6. Draw a diagram of this statement. Then answer the questions that follow.

Exit polls showed that 7 out of 10 voters cast their ballot for the incumbent.

(a) According to the exit polls, what percent of the voters cast their ballot for the incumbent?

(b) According to the exit polls, what fraction of the voters did not cast their ballot for the incumbent?

Write equations to solve Problems 7 and 8.

7. What number is $\dfrac{5}{6}$ of $3\dfrac{1}{3}$?

8. What number is 0.06 of 23.5?

9. Write 1.6×10^7 in standard form. Then use words to write the number.

10. Use a unit multiplier to convert 1.2 liters to milliliters.

11. Sketch a number line to show this addition problem.

$$(-3) + (+4)$$

12. Complete the table.

Fraction	Decimal	Percent
(a)	0.9	(b)

13. Find the missing number.

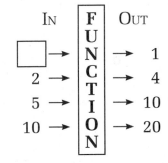

14. Evaluate: $ab - a + bc$ if $a = 5$, $b = 3$, and $c = 4$

Refer to this figure to answer questions 15 and 16. Dimensions are in inches. All angles are right angles.

15. What is the perimeter of the figure?

16. What is the area of the figure?

Solve:

17. $\dfrac{12}{16} = \dfrac{9}{m}$ 　　　　　　　　**18.** $5.4 = 1 + x$

Add, subtract, multiply, or divide, as indicated:

19. $10 - (9 - 8) - 6 \div 3 + 1$

20. $1^4 + 2^3 + 3^2 + 4^1$

21. $1\,\text{m} - 20\,\text{mm} = \underline{\ \ }\,\text{mm}$

22.　　$\begin{array}{l} 3\text{ gal }\ 2\text{ qt }\ 1\text{ pt} \\ -\ 1\text{ gal }\ 2\text{ qt }\ 1\text{ pt }\ 2\text{ oz} \\ \hline \end{array}$

23. $\dfrac{344\text{ m}}{1\text{ sec}} \cdot \dfrac{60\text{ sec}}{1\text{ min}}$ 　　　**24.** $2\dfrac{1}{2} + \left(3 - 2\dfrac{1}{6}\right)$

25. $10\dfrac{1}{2} - \left(3\dfrac{1}{3} \div 2\dfrac{1}{2}\right)$ 　**26.** $4\dfrac{4}{5} \cdot 1\dfrac{7}{8} \cdot 1\dfrac{1}{9}$

27. $\$20 - (0.25 \times \$20)$ 　　**28.** $4.36 \div 400$

29. $(0.1)(0.01)(0.001)$ 　　　**30.** $0.1 \div (0.001 \div 0.01)$

LESSON 76

Classifying Triangles

Certain kinds of triangles have special names. These triangles are classified according to the relative lengths of the sides or the measures of the angles.

When we classify triangles by the lengths of the sides, we determine whether two or more sides are equal in length.

 **Scalene** triangles have three unequal sides and three unequal angles.

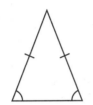 **Isosceles** triangles have at least two equal sides and at least two equal angles. Marks may be used to show sides of equal length and angles of equal measure.

 Equilateral triangles have three equal sides and three equal angles. An equilateral triangle is a **regular triangle**.

When we classify triangles by their angle measures, we determine whether the largest angle of the triangle is acute, right, or obtuse. We use the same words to describe the triangles that contain these angles.

 Acute triangles have three acute angles.

 Right triangles have a right angle.

 Obtuse triangles have an obtuse angle.

Example 1 The perimeter of an equilateral triangle is 2 feet. How many inches long is each side?

Solution All three sides of an equilateral triangle are equal in length. Since 2 feet equals 24 inches, we divide 24 inches by 3 and find that the length of each side is **8 inches.**

Example 2 Sketch an isosceles right triangle.

Solution   "Isosceles" means the triangle has at least two equal sides. "Right" means the triangle contains a right angle. We sketch a right angle, making both segments equal in length. Then we complete the triangle.

Practice   Classify each triangle by its angles.

a.   b.   c.

Classify each triangle by its sides.

d.   e.   f.

g. If we know that two sides of an isosceles triangle are 3 cm and 4 cm and that its perimeter is not 10 cm, then what is its perimeter?

Problem set 76

1. Dan spent a total of 9 hours doing homework during the first 4 nights of the week. What was the average number of minutes he spent on homework each night?

2. Eva has just enough money to buy thirty 25-cent stamps. If Eva bought 15-cent stamps instead of 25-cent stamps, how many stamps could she buy?

3. The earliest Indian head penny was minted in 1859. The latest Indian head penny was minted in 1909. For how many years were the Indian head pennies minted? (Be careful.)

4. The product of *y* and 15 is 600. What is the sum of *y* and 15?

5. Thirty percent of those gathered agreed that the king should abdicate his throne. All the rest disagreed.

    (a) What fraction of those gathered disagreed?

    (b) What was the ratio of those who agreed to those who disagreed?

6. Draw a diagram of this statement. Then answer the questions that follow.

       Five eighths of the 600 trees in the forest were deciduous.

    (a) How many of the trees in the forest were deciduous?

    (b) What percent of the trees in the forest were not deciduous?

7. Write an equation to solve this problem. What number is $\frac{3}{8}$ of 510?

8. Write twenty-five thousandths in scientific notation.

9. Write 1.86×10^5 in standard form. Then use words to write this number.

10. Compare: 1 qt $\bigcirc$ 1 liter

11. Sketch a number line to show $(-3) + (+4) + (-2)$.

12. Complete the table.

Fraction	Decimal	Percent
$\frac{5}{8}$	(a)	(b)

13. Evaluate: $x + \dfrac{x}{y} - y$ if $x = 12$ and $y = 3$

14. Find the next three numbers in this sequence.

$$\frac{1}{8}, \frac{1}{4}, \frac{3}{8}, \frac{1}{2}, \frac{5}{8}, \ldots$$

15. This figure contains an acute triangle, a right triangle, and an obtuse triangle.

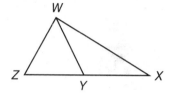

    (a) Which triangle is an acute triangle? __

    (b) Which triangle is an obtuse triangle?

    (c) Which triangle is a right triangle?

Solve:

16. $7q = 1.428$

17. $\dfrac{x}{1} = \dfrac{15}{6}$

18. $3.4 = 5 - w$

Add, subtract, multiply, or divide, as indicated:

19. $5^2 + 2^5 - \sqrt{49}$

20. $3(8) - (5)(2) + 10 \div 2$

21. 1 yd 2 ft $3\frac{3}{4}$ in.
  +   2 ft $6\frac{1}{2}$ in.

22. $1\,\text{L} - 50\,\text{mL} = ?\,\text{mL}$

23. $\dfrac{60 \text{ mi}}{1 \text{ hr}} \cdot \dfrac{1 \text{ hr}}{60 \text{ min}}$

24. $4\dfrac{1}{5} + 7\dfrac{3}{4} + 3\dfrac{1}{2}$

25. $2\dfrac{2}{5} \div \left(4\dfrac{1}{5} \div 1\dfrac{3}{4}\right)$

26. $20 - \left(7\dfrac{1}{2} \div \dfrac{2}{3}\right)$

27. $6.7 + 0.98 + 12 + 9.9$ **28.** $\$30 - (0.3 \times \$30)$

29. $4.72 - (6 - 1.375)$ **30.** $5 \div (2.5 \div 0.05)$

LESSON 77

Symbols of Inclusion

Parentheses, brackets, and braces

Parentheses are called **symbols of inclusion**. We have used parentheses to show which operation to perform first. To simplify the following expression, we add 5 and 7 before subtracting from 15.

$$15 - (5 + 7)$$

Brackets [ ] and braces { } are also symbols of inclusion. When an expression contains multiple symbols of inclusion, we simplify within the innermost symbols first.

To simplify the expression

$$20 - [15 - (5 + 7)]$$

we simplify within the parentheses first.

$20 - [15 - (12)]$ simplified within parentheses

Next we simplify within the brackets. Then we subtract.

$20 - [3]$ simplified within brackets

17 subtracted

Example 1 Simplify: $50 - [20 + (10 - 5)]$

Solution First we simplify within the parentheses.

$50 - [20 + (5)]$ simplified within parentheses

$50 - [25]$ simplified within brackets

25   subtracted

Division line   A division line also serves as a symbol of inclusion. We simplify above and below the division line before we divide. We follow the order of operations within the symbol of inclusion.

Example 2   Simplify: $\dfrac{4 + 5 \times 6 - 7}{10 - (9 - 8)}$

Solution   We will simplify above and below the line before we divide. Above the line we multiply first. Below the line we simplify within the parentheses first. This gives us

$$\frac{4 + 30 - 7}{10 - (1)}$$

We continue by simplifying above and below the division line.

$$\frac{27}{9}$$

Now we divide and get

$$3$$

Practice   Simplify:

   a. $30 - [40 - (10 - 2)]$ **b.** $100 - 3[2(6 - 2)]$

   c. $\dfrac{10 + 9 \cdot 8 - 7}{6 \cdot 5 - 4 - 3 + 2}$ **d.** $\dfrac{1 + 2(3 + 4) - 5}{10 - 9(8 - 7)}$

Problem set 77

1. Jennifer and Jason each earn $5 per hour doing yard work. On one job Jennifer worked 3 hours and Jason worked $2\frac{1}{2}$ hours. Altogether, how much money were they paid?

2. If p is a two-digit prime number greater than 90, what prime number is 90 less than p?

3. Use a ratio box to solve this problem. The ratio of favorable to unfavorable outcomes was 3 to 5. If 45 unfavorable outcomes were possible, how many favorable outcomes were possible?

4. During the first 5 days of the journey, the wagon train averaged 18 miles per day. During the next 2 days the wagon train traveled 16 miles and 21 miles, respectively. If the total journey is 1017 miles, how much farther does the wagon train have to travel?

5. Write an equation to solve this problem. What number is 0.35 of 840?

6. The average distance from the Earth to the Sun is 1.496×10^8 km. Use words to write that number.

7. Draw a diagram of this statement. Then answer the questions that follow.

   Three tenths of the 40 cars pulled by the locomotive were tankers.

   (a) How many of the cars were tankers?

   (b) What percent of the cars were not tankers?

8. The top speed of Dan's pet snail is 2×10^{-3} mile per hour. Use words to write that number.

9. Use a unit multiplier to convert 3000 sec to minutes.

10. Divide 4.36 by 0.012 and write the answer with a bar over the repetend.

11. Sketch a number line and draw arrows to show $(-3) + (+5) + (-2)$.

12. Complete the table.

FRACTION	DECIMAL	PERCENT
(a)	(b)	125%

13. Find the missing number.

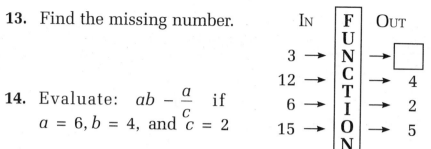

14. Evaluate: $ab - \dfrac{a}{c}$ if $a = 6, b = 4,$ and $c = 2$

15. In this figure, $AB = AD = BD = CD = 5$ cm.

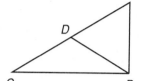

(a) Classify $\triangle BCD$ by its sides.

(b) What is the perimeter of the equilateral triangle?

(c) Which triangle appears to be a right triangle?

Solve:

16. $\dfrac{3}{4} = x + \dfrac{1}{3}$

17. $2 = 0.4p$

18. $\dfrac{14}{20} = \dfrac{n}{5}$

Add, subtract, multiply, or divide, as indicated:

19. $3[24 - (8 + 3 \cdot 2)] - \dfrac{6 + 4}{2}$

20. $3^3 - \sqrt{3^2 + 4^2}$

21. 1 m $- 95$ cm $= ?$ cm

22.
$$\begin{array}{r} 1 \text{ week } 2 \text{ days } 7 \text{ hr} \\ -\qquad\quad 5 \text{ days } 9 \text{ hr} \\ \hline \end{array}$$

23. $\dfrac{20 \text{ mi}}{1 \text{ gal}} \cdot \dfrac{1 \text{ gal}}{4 \text{ qt}}$

24. $4\dfrac{2}{3} + 3\dfrac{5}{6} + 2\dfrac{5}{9}$

25. $12\dfrac{1}{2} \cdot 4\dfrac{4}{5} \cdot 3\dfrac{1}{3}$

26. $6\dfrac{1}{3} - \left(1\dfrac{2}{3} \div 3\right)$

27. $4 \div (0.08 \div 16)$

28. $(0.25)(0.08)(0.05)(10^4)$

29. $10 - (0.1 - 0.001)$

30. $\$1.56 \div 0.06$

LESSON 78

Adding Signed Numbers

In this lesson we will summarize what we have learned about adding signed numbers.

From our practice on the number line we have seen that when we add two negative numbers, the sum is a negative number. When we add two positive numbers, the sum is a positive number.

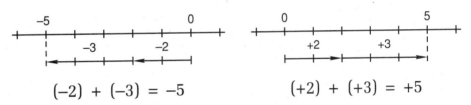

$$(-2) + (-3) = -5 \qquad (+2) + (+3) = +5$$

We have also seen that when we add a positive number and a negative number, the sum is positive or negative or zero depending upon which, if either, of the numbers has the greater absolute value.

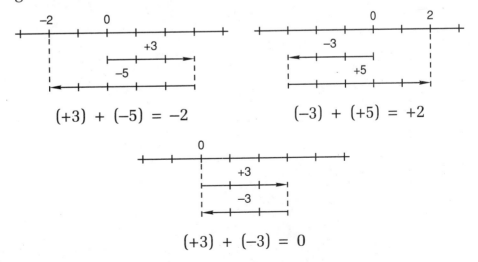

$$(+3) + (-5) = -2 \qquad (-3) + (+5) = +2$$

$$(+3) + (-3) = 0$$

We can summarize these observations with the following statements.

---

1. The sum of two numbers with the same sign is the sum of their absolute values. The sign of the sum is the same as the sign of the numbers.

2. The sum of two numbers with opposite signs is the difference of their absolute values. The sign of the difference is the same as the sign of the number with the greater absolute value.

3. The sum of two opposites is zero.

We can use these observations to help us add signed numbers without drawing a number line.

Example 1   Find each sum:

(a) $(-54) + (-78)$ (b) $(+45) + (-67)$ (c) $(-92) + (+92)$

Solution   (a) Since the signs are the same, we add the absolute values and use the same sign for the sum.

$$(-54) + (-78) = -132$$

(b) Since the signs are **different**, we **find the difference of** the absolute values and keep the sign of -67 because 67 is greater than 45.

$$(+45) + (-67) = -22$$

(c) The difference of the absolute values of -92 and 92 is zero. **Zero has no sign.** The sum of two opposites is zero.

$$(-92) + (+92) = 0$$

Example 2   Find the sum: $(-3) + (-2) + (+7) + (-4)$

Solution   We will show two methods.

Method 1. Add the first two numbers. Then add the third number. Then add the fourth number.

$$[(-3) + (-2)] + (+7) + (-4) \qquad \text{problem}$$

$$[(-5) + (+7)] + (-4) \qquad \text{added } -3 \text{ and } -2$$

$$(+2) + (-4) \qquad \text{added } -5 \text{ and } +7$$

$$-2 \qquad \text{added } +2 \text{ and } -4$$

Method 2. Rearrange the terms and add all numbers with the same sign first.

$$[(-3) + (-2) + (-4)] + (+7) \qquad \text{rearranged}$$

$$(-9) + (+7) \qquad \text{added}$$

$$-2 \qquad \text{added}$$

Example 3 Find each sum:

(a) $\left(-2\frac{1}{2}\right) + \left(-3\frac{1}{3}\right)$ (b) $(+4.3) + (-7.24)$

Solution These numbers are not integers, but the method for adding these signed numbers is the same as the method for adding integers.

(a) The signs are both negative. We add the absolute values and keep the same sign.

$$\begin{aligned} 2\frac{1}{2} &= 2\frac{3}{6} \\ + 3\frac{1}{3} &= 3\frac{2}{6} \\ \hline &\quad\, 5\frac{5}{6} \end{aligned}$$

$$\left(-2\frac{1}{2}\right) + \left(-3\frac{1}{3}\right) = -5\frac{5}{6}$$

(b) The signs are different. We find the difference of the absolute values and keep the sign of -7.24.

$$\begin{aligned} \overset{6}{\cancel{7}}.\overset{1}{2}4 \\ -\,4.3 \\ \hline 2.94 \end{aligned}$$

$$(+4.3) + (-7.24) = -2.94$$

Practice   Find each sum:

    a. $(-56) + (+96)$ **b.** $(-28) + (-145)$

    c. $(-3) + (-8) + (+15)$ **d.** $(-5) + (+7) + (+9) + (-3)$

    e. $\left(-3\dfrac{5}{6}\right) + \left(+5\dfrac{1}{3}\right)$ **f.** $(-1.6) + (-11.47)$

Problem set 78

1. Two trillion is how much more than seven hundred fifty billion? Write the answer in scientific notation.

2. The taxi cost $2.25 for the first mile and 15¢ for each additional tenth of a mile. For a 5.2-mile trip Eric paid $10 and told the driver to keep the change. How much was the driver's tip?

3. The product of x and 12 is 84. The product of y and 12 is 48. What is the product of x and y?

4. Three hundred twenty boys and four hundred girls crowded into the assembly.

    (a) What fraction of the students in the assembly were boys?

    (b) What was the ratio of girls to boys in the assembly?

5. What is the average of 1.74, 2.8, 3.4, 0.96, 2, and 1.22?

6. Draw a diagram of this statement. Then answer the questions that follow.

    The viceroy conscripted two fifths of the 1200 serfs in the province.

    (a) How many of the serfs in the province were conscripted?

    (b) What percent of the serfs in the province were not conscripted?

7. Write an equation to solve this problem. What number is $\frac{5}{9}$ of 100?

8. The temperature at the center of the sun is about 1.6×10^7 degrees Celsius. Use words to write that number.

9. A red blood cell is about 7×10^{-6} meter in diameter. Use words to write that number.

10. Divide 456 by 28 and write the answer

    (a) as a mixed number.

    (b) as a decimal rounded to two decimal places.

    (c) rounded to the nearest whole number.

11. Find each sum:

    (a) $(-63) + (-14)$

    (b) $(-16) + (+20) + (-32)$

12. Complete the table.

FRACTION	DECIMAL	PERCENT
(a)	2.5	(b)

13. Find the next three numbers in this sequence.

    48, 24, 12, 6, . . .

14. Evaluate: $x + xy$ if $x = \frac{2}{3}$ and $y = \frac{3}{4}$

Refer to this hexagon to answer questions 15 and 16. Dimensions are in meters. All angles are right angles.

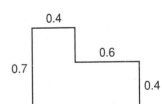

15. What is the perimeter of the hexagon?

16. What is the area of the hexagon?

Solve:

17. $\dfrac{4}{9} = y - \dfrac{2}{9}$ **18.** $25x = 10$

Add, subtract, multiply, or divide, as indicated:

19. $\dfrac{3^2 + 4^2}{\sqrt{3^2 + 4^2}}$

20. $100 - [20 + 5(4 + 3(2 + 1))]$

21.   5 gal 2 qt 1 pt 7 oz   **22.** $2\,\text{m} - 800\,\text{mm} = \_\_\,\text{mm}$
        + 1 gal 1 qt 1 pt 9 oz

23. $\dfrac{1088\text{ ft}}{1\text{ sec}} \cdot \dfrac{60\text{ sec}}{1\text{ min}}$ **24.** $3\dfrac{1}{5} \cdot 15 \cdot \dfrac{3}{8}$

25. $2\dfrac{4}{5} \div \left(6 \div 2\dfrac{1}{2} \right)$ **26.** $5\dfrac{1}{6} - \left(4 - 2\dfrac{1}{3} \right)$

27. $0.1 - (0.01 - 0.001)$ **28.** $0.1 + 0.2 + 0.3 + 0.4$

29. $\$15 + (0.06 \times \$15)$ **30.** $5.1 \div (5.1 \div 1.5)$

LESSON 79   Area of a Triangle

We have practiced finding the area of a rectangle and the area of a parallelogram. In this lesson we will practice finding the area of a triangle. We may cut out some triangles to help us understand the concept of this lesson.

Project

Fold a sheet of paper in half. Then use a pencil and a ruler to draw a few triangles on the paper that are large enough to cut out. Draw a right triangle, an acute triangle, and an obtuse triangle.

While the paper is still folded, cut out each triangle so that by cutting the folded paper you cut out a pair of identical (congruent) triangles at the same time. Then fit each pair of triangles together to form a quadrilateral with that pair.

- What kind of quadrilateral is formed by the two right triangles?

- What kind of quadrilateral is formed by the two acute triangles?

- What kind of quadrilateral is formed by the two obtuse triangles?

In all three cases above, we can arrange the two triangles to form a parallelogram. Thus, one triangle is half of a parallelogram. This fact gives us a clue for finding the area of a triangle.

 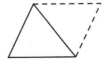 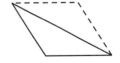

We remember that the area of a parallelogram equals the product of the base and the height. Since the area of a triangle is half the area of a parallelogram, to find the area of a triangle we multiply the base by the height and then divide the product by 2.

$$\text{Area of a triangle} = \frac{\text{base} \times \text{height}}{2}$$

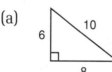 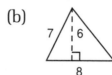

We notice that in a right triangle the height may be one side of the triangle, whereas in other triangles the height may fall inside or outside the triangle. In every case, the height is perpendicular to the base.

Example Find the area of each triangle. Dimensions are in inches.

(a)   (b)   (c)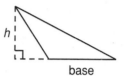

Solution The base of each triangle is 8 inches, and the height of each triangle is 6 inches. Thus the area of each triangle is the same.

$$\text{Area of a triangle} = \frac{\text{base} \times \text{height}}{2}$$

$$\text{Area} = \frac{8 \text{ in.} \cdot 6 \text{ in.}}{2}$$

$$\text{Area} = \textbf{24 in.}^2$$

Practice Find the area of each triangle. Dimensions are in centimeters.

a. 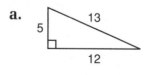   b. 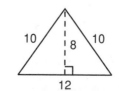   c.

**Problem set
79**   Refer to this table to answer questions 1 and 2.

NUMBER OF LINCOLN
HEAD PENNIES MINTED

DATE	NUMBER MINTED
1909–S	1,750,000
1909–S–VDB	484,000
1914–D	1,193,000
1931–S	866,000

1. The 1909-S and 1909-S-VDB were both minted in San Francisco. Together, how many Lincoln head pennies were minted in San Francisco in 1909?

2. How many more 1914-D pennies were minted than 1931-S pennies?

3. Gilbert wanted to buy packages of crackers and cheese from the vending machine. Each package cost 35¢. Gilbert had 5 quarters, 3 dimes, and 2 nickels. How many packages of crackers and cheese could he buy?

4. The two prime numbers p and m are between 50 and 60. Their difference is 6. What is their sum?

5. Use a ratio box to solve this problem. The ratio of flies to ants at the picnic was 2 to 15. If there were 60 flies, how many ants were there?

6. Draw a diagram of this statement. Then answer the questions that follow.

    The survey found that only 2 out of 5 Lilliputians believe in giants.

    (a) According to the survey, what fraction of the Lilliputians do not believe in giants?

    (b) If 60 Lilliputians were selected for the survey, how many of them believe in giants?

Write equations to solve Problems 7 and 8.

7. Three fifths of $7\frac{1}{2}$ is what number?

8. What number is 0.95 of 3500?

9. Write each number in scientific notation.

   (a) 400,000,000   (b) 0.00000078

10. Write 1.7×10^9 in standard form. Then use words to write this number.

11. Use a unit multiplier to convert $2\frac{1}{2}$ hr to minutes.

12. Complete the table.

Fraction	Decimal	Percent
(a)	(b)	85%

13. Find the sum.

$$(-48) + (+17) + (+31)$$

14. Write the prime factorization of 4900.

15. Evaluate: $xy - \dfrac{x}{y}$ if $x = 0.2$ and $y = 2$

16. In this figure the quadrilateral is a parallelogram. Dimensions are in meters.

    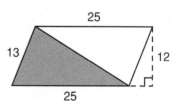

    (a) What is the area of the parallelogram?

    (b) What is the area of the shaded triangle?

Solve:

17. $3.14 = 10r$ 18. $\dfrac{36}{16} = \dfrac{2\frac{1}{4}}{q}$ 19. $436 + w = 600$

Add, subtract, multiply, or divide, as indicated:

20. $200 - \{100 - [3(4 + 5) + 5(3 + 4)]\}$

21. $2\,L - 380\,mL = ?\,mL$ **22.**   1 yd 2 ft   3 in.
$$-\qquad\quad\text{2 ft }5\tfrac{1}{4}\text{ in.}$$

23. $\dfrac{720\text{ mi}}{1\text{ hr}} \cdot \dfrac{1\text{ hr}}{60\text{ min}}$ **24.** $8\dfrac{1}{6} + \left(7 - 6\dfrac{1}{8}\right)$

25. $\left(8\dfrac{1}{3} \cdot 1\dfrac{4}{5}\right) \div 30$ **26.** $6\dfrac{2}{3} \div \left(24 \div 4\dfrac{1}{2}\right)$

27. $2.5 - (2.05 - 2.005)$ **28.** $[(0.6)(0.5)(0.4)] \div 0.3$

29. $\$37.50 \div 0.75$ **30.** $\$24 + (0.05 \times \$24)$

LESSON
80

Percent of a Number

We remember that percent means "by the hundred."

$$40\text{ percent means }\frac{40}{100}$$

$$6\text{ percent means }\frac{6}{100}$$

$$140\text{ percent means }\frac{140}{100}$$

We can translate percent problems into equations the same way we translate fractional-part-of-a-number problems. We remember to

replace **is** with $=$

replace **of** with $\times$

replace **percent** with $\dfrac{\textbf{percent}}{\textbf{100}}$

Example 1 What number is 40 percent of 75?

Solution We translate directly.

What number is 40 percent of 75?   question

$$W_N \qquad = \qquad \frac{40}{100} \quad \times \ 75 \qquad \text{equation}$$

To find the answer, we multiply.

$$W_N = \frac{40 \times 75}{100} \qquad \text{multiplied}$$

$$W_N = 30 \qquad \text{simplified}$$

Example 2 Eight percent of 36 is what number?

Solution We translate directly.

Eight percent of 36 is what number?   question

$$\frac{8}{100} \quad \times \ 36 \ = \qquad W_N \qquad \text{equation}$$

To solve, we multiply.

$$\frac{8 \times 36}{100} = W_N \qquad \text{multiplied}$$

$$2.88 = W_N \qquad \text{simplified}$$

Practice Write an equation to solve each problem.

a. What number is 50 percent of 150?

b. Three percent of 39 is what number?

c. What number is 25 percent of 64?

Problem set 80

1. Eight hundred seventy-six ten-thousandths is how much more than seventy-nine thousandths? Write the answer in words.

2. Use a ratio box to solve this problem. The ratio of princes to knights at the tournament was 4 to 27. If 108 knights attended, how many princes were at the tournament?

3. One eighth of the possible outcomes were favorable, while the rest of the possible outcomes were unfavorable.

    (a) What fraction of the possible outcomes were unfavorable?

    (b) What was the ratio of favorable outcomes to unfavorable outcomes?

4. What is the average of $2\frac{1}{2}$, $3\frac{2}{3}$, 4, and $4\frac{5}{6}$?

5. Write an equation to solve this problem. What number is 6 percent of 350?

6. Draw a diagram of this statement. Then answer the questions that follow.

    Diane gave $\frac{1}{3}$ of her 234 baseball cards to her brother.

    (a) What percent of her baseball cards did Diane give to her brother?

    (b) How many baseball cards did Diane have left?

7. Write each number in scientific notation.

    (a) One hundred-thousandth

    (b) One hundred thousand

8. Write 1.5×10^{-5} standard form. Then use words to write the number.

9. Compare: $\frac{2}{3}$ ◯ 0.667

10. Evaluate: $ab + a + b$ if $a = \frac{1}{2}$ and $b = \frac{1}{4}$

11. Divide 5 by 0.24 and write the answer

  (a) as a decimal with a bar over the repetend.

  (b) rounded to the nearest whole number.

12. Find each sum:

  (a) $(-148) + (-52)$

  (b) $(+7) + (-12) + (+6) + (-8)$

13. Complete the table.

FRACTION	DECIMAL	PERCENT
$\frac{2}{9}$	(a)	(b)

14. Find the missing number.

IN	F U N C T I O N	OUT
5 →		→ 25
☐ →		→ 16
8 →		→ 64
11 →		→ 121

15. Find the area of each triangle. Dimensions are in centimeters.

(a) 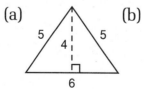 (b) 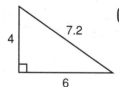 (c)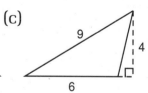

Solve:

16. $1.5c = 0.345$

17. $\dfrac{9}{10} - x = \dfrac{1}{2}$ **18.** $\dfrac{a}{14} = \dfrac{15}{35}$

Add, subtract, multiply, or divide, as indicated:

19. $\dfrac{10(10 + 10) - (10 \cdot 10 + 10)}{10}$

20. $6^2 + 8^2 - 10^2$ **21.** $2\,m - 25\,cm = ?\,cm$

22.   1 day
    $-$ 6 hr 15 min 45 sec

23. $\dfrac{1\,mi}{5\,min} \cdot \dfrac{60\,min}{1\,hr}$ **24.** $3\dfrac{5}{12} + \left(5\dfrac{1}{6} - 1\dfrac{1}{4}\right)$

25. $4\dfrac{1}{2} \cdot 36 \cdot \dfrac{5}{6}$ **26.** $6\dfrac{2}{3} \div \left(6 \div 1\dfrac{1}{5}\right)$

27. $42.3 - 5.787$ **28.** $(0.7)(1.1)(1.3)$

29. $1.02 \div 0.12$ **30.** $0.1 \div (5 \div 0.02)$

LESSON
81

Ratio Problems Involving Totals

Some ratio problems require that we use the total to solve the problem. Consider the following problem.

> The ratio of boys to girls was 5 to 4. If there were 180 students in the assembly, how many girls were there?

We begin by making a ratio box. This time we add a third row for the total number of students.

	Ratio	Actual Count
Boys	5	B
Girls	4	G
Total	9	180

In the ratio column we wrote 5 for boys and 4 for girls, then **added these to get 9 for the total ratio number.** We were given 180 as the actual count of students. This is a total. We can use two rows from this table to write a proportion. Since we were asked to find the number of girls, we will use the "girls" row. Since we know both "total" numbers, we will also use the "total" row. Then we solve the proportion.

	RATIO	ACTUAL COUNT
Boys	5	B
Girls	4	G
Total	9	180

$$\longrightarrow \quad \frac{4}{9} = \frac{G}{180}$$

$$9G = 720$$

$$G = 80$$

We find that there were 80 girls. We can use this answer to complete the ratio box.

	RATIO	ACTUAL COUNT
Boys	5	100
Girls	4	80
Total	9	180

Example The ratio of football players to soccer players in the room was 5 to 7. If 48 players were in the room, how many were football players?

Solution We use the information in the problem to form a table. We include a row for total. The total ratio number is 12.

	RATIO	ACTUAL COUNT
Football players	5	F
Soccer players	7	S
Total players	12	48

$$\frac{5}{12} = \frac{F}{48}$$

$$12F = 240$$

$$F = 20$$

To find the number of football players, we wrote a proportion from the "football players" row and the "total players" row. We solved the proportion to find that there were **20 football players** in the room. From this information we can complete the ratio box.

	RATIO	ACTUAL COUNT
Football players	5	20
Soccer players	7	28
Total players	12	48

Practice   Solve these problems. Begin by drawing a ratio box.

a. Acrobats and clowns converged on the center ring in the ratio of 3 to 5. If 72 entertainers performed in the center ring, how many were clowns?

b. The ratio of young men to young women at the prom was 8 to 9. If 240 young men were in attendance, how many young people attended in all?

Problem set 81

1. If 5 pounds of apples cost $2.40, then

    (a) what is the price per pound?

    (b) what is the cost for 8 pounds of apples?

2. What is the sum of ten million, fifty-five thousand, two hundred one and six million, seven hundred eight thousand, four hundred eighty?

3. Use a ratio box to solve this problem. The ratio of big fish to little fish in the pond was 4 to 11. If there were 1320 fish in the pond, how many big fish were there?

4. The car traveled 350 miles on 15 gallons of gasoline. The car averaged how many miles per gallon? Round the answer to the nearest tenth.

5. If the average of three numbers is 1440, what is the sum of the three numbers?

6. Write twelve billion in scientific notation.

7. Draw a diagram of this statement. Then answer the questions that follow.

    One sixth of the five dozen eggs were cracked.

    (a) How many eggs were not cracked?

    (b) What was the ratio of eggs that were cracked to eggs that were not cracked?

8. (a) Draw segment *AB*. Draw segment *DC* parallel to segment *AB* but not the same length. Draw segments between the endpoints of segments *AB* and *DC* to form a quadrilateral.

   (b) What type of quadrilateral was formed in part (a)?

9. Refer to the figure to answer the questions. Dimensions are in inches. Corners that look square are square.

   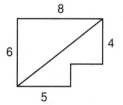

   (a) What is the perimeter of the hexagon?

   (b) What is the area of the hexagon?

10. What is the average of the two numbers indicated by arrows on this number line?

   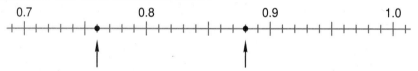

Write equations to solve Problems 11 and 12.

11. What number is 75 percent of 64?

12. What number is 0.3 of 7.4?

13. Find each sum:

   (a) $(-3) + (-8)$

   (b) $(+3) + (-8)$

   (c) $(-3) + (+8) + (-5)$

14. Complete the table.

Fraction	Decimal	Percent
(a)	0.35	(b)

15. Use unit multipliers to convert 0.95 L to milliliters.

16. Evaluate: $ab + a + \dfrac{a}{b}$ if $a = 5$ and $b = 0.2$

17. Find the missing number.

18. Divide 366 by 7 and write the answer as a mixed number.

IN | FUNCTION | OUT
$5 \rightarrow$ | | $\rightarrow 14$
$8 \rightarrow$ | | $\rightarrow 17$
$\square \rightarrow$ | | $\rightarrow 32$
$11 \rightarrow$ | | $\rightarrow 20$

Solve:

19. $\dfrac{x}{6} = \dfrac{21}{9}$

20. $4.8 + p = 7$

21. $10n = 240$

Add, subtract, multiply, or divide, as indicated:

22.   1 yd 2 ft 7 in.
     $+$ 1 ft 9 in.

23.   4 days 5 hr 15 min
     $-$ 1 days 7 hr 50 min

24. $4.5 \div (0.4 + 0.5)$

25. $\dfrac{3 + 0.6}{3 - 0.6}$

26. $3\dfrac{3}{5} + 5\dfrac{1}{4} + 2\dfrac{1}{2}$

27. $5\dfrac{1}{6} - \left(4 - 1\dfrac{2}{3}\right)$

28. $4\dfrac{1}{5} \div \left(1\dfrac{1}{6} \cdot 3\right)$

29. $3^2 + \sqrt{4 \cdot 7 - 3}$

30. $3 + 4[(5 - 2)(3 + 1)]$

LESSON
82

Geometric Solids

Geometric solids are shapes that take up space. Below we show a few geometric solids.

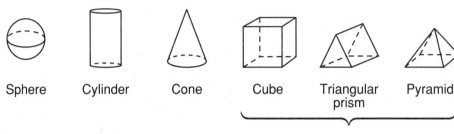

Sphere Cylinder Cone Cube Triangular prism Pyramid

Polyhedrons

Some geometric solids, such as spheres, cylinders, and cones, have one or more curved surfaces. If a solid has only flat surfaces that are polygons, the solid is called a **polyhedron**. Cubes, triangular prisms, and pyramids are examples of polyhedrons. When describing a polyhedron, we may refer to its faces, edges, or vertices. A **face** is one of the flat surfaces. An **edge** is formed where two faces meet. A **vertex** (plural, vertices) is formed where three or more edges meet.

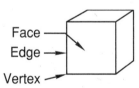

A **prism** is a special kind of polyhedron. A prism has a polygon of a constant size "running through" the prism. A rectangular prism has a rectangle of a constant size running through it. A triangular prism has a triangle of a constant size running through it. Thus, at least two faces of a prism are identical and are parallel.

To draw a prism, we draw two identical, parallel polygons, as shown below. Then we draw lines connecting corresponding vertices. We may use broken lines to indicate edges hidden from view.

Rectangular prism: Draw the same size rectangle twice.

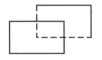

Then connect corresponding vertices. (Use broken lines for hidden edges.)

Triangular prism: Draw the same size triangle twice.

Connect corresponding vertices.

Example 1   Use the name of a geometric solid to describe the shape of each object.

(a)   Basketball

(b)   Shoe box

(c)   Can of beans

Solution   (a)   **Sphere**

(b)   **Rectangular prism**

(c)   **Cylinder**

Example 2   A cube has how many (a) faces, (b) edges, and (c) vertices?

Solution   (a)   **6 faces**

(b)   **12 edges**

(c)   **8 vertices**

Example 3   Draw a cube.

Solution   A cube is a special kind of rectangular prism. A cube has a square running through it. All faces are squares.

Practice Use the name of a geometric solid to describe each shape.

a.
Tent

b.
Cone

c.
Box

A triangular prism has how many of each?

d. Faces **e.** Edges **f.** Vertices

Draw a representation of each shape.

g. Sphere **h.** Rectangular prism **i.** Cylinder

Problem set 82

1. The bag contained 24 red marbles, 30 white marbles, and 40 blue marbles. What was the ratio of

   (a) red marbles to blue marbles?

   (b) white marbles to red marbles?

2. When the product of $\frac{1}{3}$ and $\frac{1}{2}$ is subtracted from the sum of $\frac{1}{3}$ and $\frac{1}{2}$, what is the difference?

3. If the cost of calling Albuquerque is 76¢ for the first minute and 48¢ for each additional minute, what is the cost of a 10-minute call?

4. On his first 5 tests Cliff averaged 92 points. On his next 3 tests Cliff scored 94 points, 85 points, and 85 points.

   (a) What was his average for his last 3 tests?

   (b) What was his average for all 8 tests?

5. Use a ratio box to solve this problem. The jeweler's tray was filled with diamonds and rubies in the ratio of 5 to 2. If 210 gems filled the tray, how many were diamonds?

6. Draw a diagram of this statement. Then answer the questions that follow.

   Four fifths of the 360 dolls were sold during November.

(a) How many of the dolls were sold during November?

(b) What percent of the dolls were not sold during November?

7. Draw a rectangular prism, and then answer these questions. A rectangular prism has how many

(a) edges?

(b) faces?

(c) vertices?

8. Refer to these triangles to answer the questions. Dimensions are in meters.

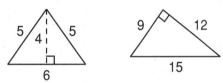

(a) What is the area of the scalene triangle?

(b) What is the perimeter of the isosceles triangle?

9. Find the length of segments (a) *AB*, (b) *AC*, and (c) *BC*.

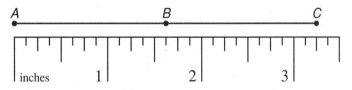

10. Write twenty-five ten-thousandths in scientific notation.

Write equations to solve Problems 11 and 12.

11. What number is 24 percent of 75?

12. What number is 1.2 of 12?

13. Find each sum:

(a) $(-2) + (-3) + (-4)$ (b) $(+2) + (-3) + (+4)$

14. Complete the table.

Fraction	Decimal	Percent
(a)	(b)	4%

15. Use a unit multiplier to convert 700 mm to centimeters.

16. Evaluate: $a^2 - \sqrt{a}$ if $a = 9$

17. Find the missing number.

18. Round 7856.427 to the nearest hundred.

In	FUNCTION	Out
5 →		→ 35
0 →		→ 0
11 →		→ 77
1 →		→ ☐

Solve:

19. $p - 5\frac{1}{2} = 7\frac{1}{2}$

20. $\dfrac{2.5}{w} = \dfrac{15}{12}$

21. $3600 = 9y$

Add, subtract, multiply, or divide, as indicated:

22. $9 + 8\{7 \cdot 6 - 5[4 + (3 - 2 \cdot 1)]\}$

23.
$$\begin{array}{r} 1 \text{ yd} \\ - \quad 1 \text{ ft } 3 \text{ in.} \\ \hline \end{array}$$

24. $2\frac{1}{2} \text{ hr} \cdot \dfrac{4 \text{ mi}}{1 \text{ hr}}$

25. $6.4 - (0.6 - 0.04)$

26. $\dfrac{3 + 0.6}{(3)(0.6)}$

27. $1\frac{2}{3} + 3\frac{1}{4} - 1\frac{5}{6}$

28. $\dfrac{3}{5} \div 3\frac{1}{5} \cdot 5\frac{1}{3} \cdot 1$

29. $3\frac{3}{4} \div \left(3 \div 1\frac{2}{3}\right)$

30. $5^2 - \sqrt{16} + 2^3$

LESSON 83

Weight

Physical objects are composed of matter. The attraction between the earth and a physical object is the weight of the object. In the metric system we use kilograms (kg), grams (g), and milligrams (mg) to measure the weight of an object.* The approximate weights of some familiar objects are shown here.

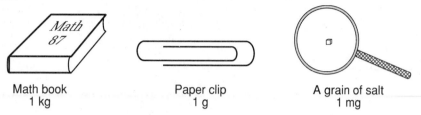

Math book
1 kg

Paper clip
1 g

A grain of salt
1 mg

1 kilogram = 1000 grams   1 gram = 1000 milligrams

To measure the weight of an object, we compare its weight to a known weight. The apple balances 112 grams so its weight is 112 g.

In the U.S. customary system we use tons (t), pounds (lb), and ounces (oz) to measure weight.

1 ton = 2000 pounds   1 pound = 16 ounces

The spring scale indicates that the apple weighs 4 ounces.

If a 1-kilogram weight was placed on a grocer's scale, the scale would read about 2.2 pounds.

---

*A scientist uses kilograms, grams, and milligrams to describe the masses of objects. A scientist uses a unit called a newton to describe weight. In the marketplaces of countries that use the metric system merchants use grams and kilograms as units of weight and the scientists in the same countries use grams and kilograms as units of mass. It can be confusing at times.

Example 1   Use a unit multiplier to perform each conversion.

(a)   1.2 kg to grams   (b)   250 mg to grams

Solution   (a)   We form a unit multiplier from the equivalence 1 kg = 1000 g that has grams on top. Then we multiply to cancel kilograms.

$$1.2 \, \cancel{kg} \cdot \frac{1000 \text{ g}}{1 \, \cancel{kg}} = \textbf{1200 g}$$

(b)   We multiply by a unit multiplier that has grams on top. Then we cancel milligrams.

$$250 \, \cancel{mg} \cdot \frac{1 \text{ g}}{1000 \, \cancel{mg}} = \frac{250}{1000} \text{ g} = \textbf{0.25 g}$$

Example 2   Simplify: (a)   5 lb 8 oz   (b)   5 lb 8 oz
                               + 3 lb 9 oz   − 3 lb 9 oz

Solution   (a)   5 lb   8 oz   (b)   4 $^{(16 \text{ oz})}$ 24
                     + 3 lb   9 oz $\cancel{5}$ lb $\cancel{8}$ oz
                     ――――――――――   − 3 lb   9 oz
                       8 lb 17 oz → **9 lb 1 oz**   ――――――――――
                                                            1 lb 15 oz

Practice   Simplify:

a.   10 lb   7 oz   b.   9 lb 3 oz
     + 4 lb 12 oz   − 1 lb 7 oz

Use unit multipliers to perform each conversion.

c.   32 oz to pounds   d.   2 tons to pounds

e.   2.4 kg to grams   f.   4000 mg to grams

Problem set 83   1.   With the baby in his arms Papa weighed 180 pounds. Without the baby in his arms Papa weighed $165\frac{1}{2}$ pounds. How much did the baby weigh?

2. Tim ran 15 miles in 2 hours.

    (a) What was his average speed in miles per hour?

    (b) What was the average number of minutes it took Tim to run each mile?

3. If 3 notebooks cost $8.91, then

    (a) what is the cost of 1 notebook?

    (b) what is the cost of 5 notebooks?

4. Use a ratio box to solve this problem. The ratio of black sheep to white sheep in the flock was 2 to 17. If there were 34 black sheep in the flock, how many sheep were in the flock in all?

Write equations to solve Problems 5 and 6.

5. What number is 40 percent of 65?

6. What number is 0.075 of 600?

7. Three fifths of the earth's surface is covered with water.

    (a) What percent of the earth's surface is covered with water?

    (b) What is the ratio of the area of land to the area of water on the earth's surface?

8. Write 5.2×10^6 in standard form.

9. Use the name of a geometric solid to describe the shape of each object shown.

    (a)   (b)   (c)

Refer to the figure to answer questions 10 – 12. Dimensions are in millimeters. Corners that look square are square.

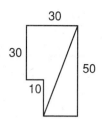

10. What is the perimeter of the hexagon?

11. What is the area of the hexagon?

12. What is the area of the triangle?

13. What is the average of the two numbers marked by arrows on this number line?

14. Complete the table.

Fraction	Decimal	Percent
$\frac{1}{25}$	(a)	(b)

15. Use a unit multiplier to convert 10 kg to grams.

Find each sum:

16. $(-6) + (-2) + (+5)$ **17.** $(-7) + (-8) + (-6) + (-1)$

18. Evaluate: $\dfrac{x + xy}{x}$ if $x = 7$ and $y = 8$

Solve:

19. $245 - x = 179$ **20.** $3n = 2.31$

21. $4\frac{1}{2} + w = 8$ **22.** $\dfrac{14}{18} = \dfrac{m}{45}$

Add, subtract, multiply, or divide, as indicated:

23.   8 lb   8 oz
       + 9 lb 11 oz

24.   5 gal 2 qt
       −   3 qt 1 pt

25. $16 \div 0.8 \div 0.04$ **26.** $0.4 + (0.5)(0.6) - 0.7$

27. $\dfrac{5}{6} + \dfrac{4}{5} + \dfrac{2}{3} - \dfrac{1}{2}$ **28.** $3\dfrac{3}{4} \cdot 5\dfrac{1}{3} \cdot 1\dfrac{1}{3} \div 4$

29. $3^2 - 2^3 + \sqrt{2 \cdot 18}$ **30.** $\dfrac{28 - 3[8 - (4 \cdot 2 - 1)]}{3 \cdot 4 - (3 + 4)}$

LESSON 84

Circles • Investigating Circumference

Circles A **circle** is a smooth curve, every point of which is the same distance from the **center**. The distance from the center to the circle is the **radius**. The plural of radius is **radii**. The distance across a circle through the center is the **diameter**. The distance around a circle is the **circumference**.

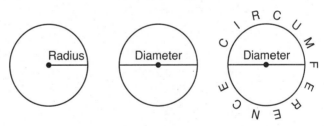

We see that the diameter of a circle is twice the radius of the circle.

Example 1 The diameter of this circle is 2 m. How many centimeters is its radius?

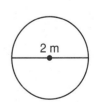

Solution The radius of a circle is half the length of the diameter. So the radius of this circle must be 1 m. We are asked to express the radius in centimeters. The radius equals **100 cm**.

Example 2 The radius of a circle is 18 inches. How many feet is its diameter?

Solution The diameter of a circle is twice the length of the radius. So the diameter of the circle is

$$2 \times 18 \text{ in.} = 36 \text{ in.}$$

We are asked to express the diameter in feet.

$$36 \text{ in.} = \textbf{3 ft}$$

The circumference of a circle is related to the diameter of the circle in a special way. The following investigation will explore that relationship.

Investigating circumference

This investigation requires a tape measure (preferably metric) and a number of circular objects. A calculator may also be useful.

Select a circular object and measure its circumference and its diameter as precisely as you can. To calculate the number of diameters that equal the circumference, divide the circumference by the diameter. Round the quotient to two decimal places. Then repeat the investigation with another circular object of a different size. Compare your results with the results of other students in the class. You will find that for every circular object the circumference is about 3.14 times as long as the diameter.

Practice

a. If the diameter of a circle is 3 meters, its radius is how many centimeters?

b. If the radius of a circle is 30 inches, its diameter is how many feet?

c. Find a circular object that you can measure, and record this information:
   1. The name of the object
   2. Its diameter
   3. Its circumference
   4. The result when the circumference is divided by the diameter (round to two decimal places)

**Problem set
84**

1. According to this graph, what percent of Dan's income was spent on items other than food and housing? If his income was $25,000, how much did he spend on food?

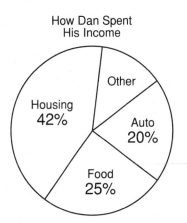

How Dan Spent His Income

2. It is $1\frac{1}{4}$ miles from Tim's house to school. How far does Tim travel in 5 days walking to school and back?

3. When the sum of 1.9 and 2.2 is subtracted from the product of 1.9 and 2.2, what is the difference?

4. Use a ratio box to solve this problem. There was a total of 520 dimes and quarters in the soda machine. If the ratio of dimes to quarters was 5 to 8, how many dimes were there?

5. Saturn is 900 million miles from the sun. Write that number in scientific notation.

6. Draw a diagram of this statement. Then answer the questions that follow.

   Three tenths of the 400 acres were planted with alfalfa.

   (a) What percent of the land was planted with alfalfa?

   (b) How many of the 400 acres were not planted with alfalfa?

7. Forty percent of the 30 students earned an A on the test. How many students earned an A on the test?

8. Draw a triangular prism so that the triangular bases are equilateral. The prism has how many faces?

9. If the radius of a circle is 50 millimeters, its diameter is how many centimeters?

Refer to the figure to answer questions 10 and 11. Dimensions are in centimeters.

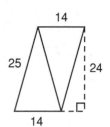

10. (a) What is the area of the parallelogram?

    (b) What is the area of each triangle?

11. Each triangle is isosceles. What is the perimeter of one of the triangles?

Write equations to solve Problems 12 and 13.

12. What number is 90 percent of 3500?

13. What number is $\frac{5}{6}$ of $2\frac{2}{5}$?

14. Complete the table.

FRACTION	DECIMAL	PERCENT
(a)	0.45	(b)

15. Find each sum:

    (a) (5) + (−4) + (6) + (−1)

    (b) 3 + (−5) + (+4) + (−2)

16. Use unit multipliers to convert 1.4 kg to grams.

17. Find the missing number.

IN	FUNCTION	OUT
26 →		→ 13
7 →		→ $3\frac{1}{2}$
16 →		→ 8
☐ →		→ $\frac{1}{2}$

18. Estimate this product by rounding each number to one nonzero digit before multiplying.

(2876)(513)(18)

Solve:

19. $5.6 = 7x$

20. $654 - p = 456$

21. $\dfrac{0.9}{1.5} = \dfrac{12}{n}$

22. $\dfrac{2}{3} + w = \dfrac{11}{12}$

Add, subtract, multiply, or divide, as indicated:

23. $\quad \begin{array}{r} 4 \text{ lb} \ \ 12 \text{ oz} \\ + \ 1 \text{ lb} \ \ \ \ 7 \text{ oz} \\ \hline \end{array}$

24. $\dfrac{3 \text{ ft}}{1 \text{ yd}} \cdot \dfrac{12 \text{ in.}}{1 \text{ ft}}$

25. $16 \div (0.8 \div 0.04)$

26. $0.4[0.5 - (0.6)(0.7)]$

27. $\dfrac{3}{8} \cdot 1\dfrac{2}{3} \cdot 4 \div 1\dfrac{2}{3}$

28. $6\dfrac{2}{3} + 2\dfrac{1}{2} + 1\dfrac{5}{6}$

29. $\sqrt{9} + \sqrt{16} - \sqrt{9 + 16}$

30. $30 - 5[4 + (3)(2) - 5]$

LESSON
85

Circumference and Pi

In the preceding lesson we investigated the circumference of a circle. We measured both the circumference and the diameter of a circle. Then we divided the circumference by the diameter to find the number of diameters that equal a circumference. How many diameters equal a circumference? This question has been asked by people for thousands of years. They found that the answer did not depend on the size of the circle. The circumference of a circle is slightly more than three diameters.

Another way to illustrate this fact is to cut a length of string equal to the diameter of a particular circle and find how many of these lengths are needed to reach around the circle. No matter what the size of the circle, it takes three diameters plus a little extra to equal the circumference.

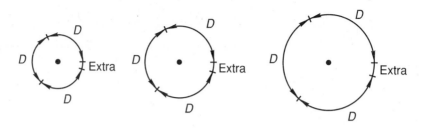

The extra amount needed is about, but not exactly, one seventh of a diameter. Thus the number of diameters needed to equal the circumference of a circle is about

$$3\frac{1}{7} \quad \text{or} \quad \frac{22}{7} \quad \text{or} \quad 3.14$$

Neither $3\frac{1}{7}$ nor 3.14 is exact. They are approximations. There is no fraction or decimal number that exactly states the number of diameters in a circumference. (Some computers have calculated the number to more than 1 million decimal places.) We use the symbol π, which is the Greek letter **pi** (pronounced "pie"), to stand for this number.

The circumference of a circle is π times the diameter of the circle. This idea is expressed by the formula

$$C = \pi d$$

To perform calculations with π, we can use an approximation. The commonly used approximations for π are

$$3.14 \quad \text{and} \quad \frac{22}{7}$$

For calculations that require great accuracy, more accurate approximations for π may be used, such as

3.14159265359

Sometimes the calculation is performed leaving π as π. Unless directed to use another approximation, we will use 3.14 for π to perform the calculations in this book.

Example 1 The radius of a circle is 10 cm. What is the circumference?

Solution If the radius is 10 cm, the diameter is 20 cm.

$$\text{Circumference} = \pi \cdot \text{diameter}$$
$$\approx 3.14 \cdot 20 \text{ cm}$$
$$\approx 62.8 \text{ cm}$$

The circumference is about **62.8 cm**.

Example 2 Find the circumference of each circle.

(a)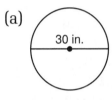

Use 3.14 for π

(b)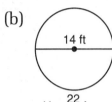

Use $\frac{22}{7}$ for π

(c)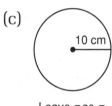

Leave π as π

Solution (a) $C = \pi d$

$C \approx 3.14(30 \text{ in.})$

$C \approx \textbf{94.2 in.}$

(b) $C = \pi d$

$C \approx \frac{22}{7}(14 \text{ ft})$

$C \approx \textbf{44 ft}$

(c) $C = \pi d$

$C = \pi (20 \text{ cm})$

$C = \textbf{20}\boldsymbol{\pi}\textbf{ cm}$

Note the form of answer (c): first 20 times π, then the unit of measure.

Practice Find the circumference of each circle.

a.

Use 3.14 for π

b.

Use $\frac{22}{7}$ for π

c.

Leave π as π

d. Sylvia used a compass to draw a circle. If the point of the compass was 3 inches from the point of the pencil, what was the circumference of the circle? (Use 3.14 for π.)

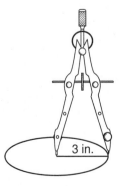

**Problem set
85**

1. Hilda ran 8 laps of the track at a steady speed. If it took $4\frac{1}{2}$ minutes to run the first 3 laps, how long did it take her to run all 8 laps?

2. The average of three numbers is 2. If the greatest is 2.8 and the least is 1.5, what is the third number?

3. Use a ratio box to solve this problem. The ratio of princes to paupers in the kingdom was 1 to 24. If the total number in both categories was 4800, how many princes were in the kingdom?

4. How far will a migrating duck fly in 8 hours at an average speed of 24 miles per hour?

5. James has read $\frac{5}{8}$ of the 320 pages in the book. How many pages are left to read?

6. If the diameter of a circle is 1 meter, its radius is how many millimeters?

7. (a) Draw a prism with bases that are right triangles.

   (b) A triangular prism has how many more edges than vertices?

8. Find the circumference of each circle.

   (a)
   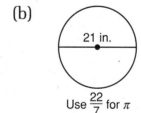
   21 in.

   Use 3.14 for π

   (b)

   21 in.

   Use $\frac{22}{7}$ for π

9. Refer to the figure to answer the questions. Dimensions are in millimeters.

   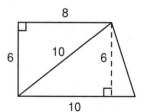

   (a) What is the area of the right triangle?

   (b) What is the area of the isosceles triangle?

10. What is the perimeter of this equilateral triangle?

11. Write 32,500,000,000 in scientific notation.

Write equations to solve Problems 12 and 13.

12. What number is 5 percent of 1000?

13. What number is 0.015 of 600?

14. Complete the table.

Fraction	Decimal	Percent
(a)	(b)	7.5%

15. Find each sum:

    (a) $6 + (-1) + (2) + (4)$

    (b) $(-5) + (-7) + (+6) + (-3)$

16. Use a unit multiplier to convert 3 g to milligrams.

17. Evaluate: $\dfrac{x + y}{xy}$ if $x = \dfrac{1}{4}$ and $y = \dfrac{1}{2}$

18. Divide 2.4 by 0.018 and write the quotient

    (a)   as a decimal with a bar over the repetend.

    (b)   rounded to the nearest whole number.

Solve:

19. $\dfrac{3}{4} + n = 1\dfrac{1}{2}$

20. $\dfrac{y}{14} = \dfrac{1.2}{0.8}$

21. $f - 479 = 563$

22. $25m = 225$

Add, subtract, multiply, or divide, as indicated:

23. $1 \text{ kg} - 350 \text{ g} = ? \text{ g}$

24. $\dfrac{2000 \text{ lb}}{1 \text{ ton}} \cdot \dfrac{16 \text{ oz}}{1 \text{ lb}}$

25. $16 \div 0.04 \div 0.8$

26. $10 - 0.1 - (0.01)(0.1)$

27. $\dfrac{3}{4} \cdot \dfrac{4}{5} \cdot \dfrac{5}{8} \div 3$ **28.** $3\dfrac{3}{4} + \left(8\dfrac{2}{3} - 5\dfrac{1}{6}\right)$

29. $4^2 - \sqrt{5^2 - 3^2}$ **30.** $\dfrac{3 \cdot 2 + 5 \cdot 6 - 1^2}{1 + 2 \cdot 3}$

LESSON **86**	# The Opposite of the Opposite • Algebraic Addition

The opposite of the opposite

The graphs of -3 and 3 are the same distance from zero on the number line. The graphs are on the opposite sides of zero.

We say that 3 and -3 are the **opposites** of each other.

3 is the opposite of -3

-3 is the opposite of 3

We can read -3 as **the opposite of 3**. Then $-(-3)$ can be read as the **opposite of the opposite of 3**. This means that $-(-3)$ is another way to write 3.

Algebraic addition

There are two ways to simplify this expression.

$7 - 3$

The first way is to let the minus sign mean to subtract. If we subtract 3 from 7, the answer is 4.

$7 - 3 = 4$

The second way is to use the thought process of **algebraic addition**. To use algebraic addition, we let the minus sign mean that -3 is a negative number and treat the problem as an addition problem. This is what we think.

$7 + (-3) = 4$

We get the same answer both ways. The only difference is in the way we think.

We can also use algebraic addition to simplify this expression.
$$7 - (-3)$$
We use an addition thought and think that 7 is added to $-(-3)$. This is what we think.
$$7 + [-(-3)]$$
But the opposite of the opposite of 3 is another name for 3, so we can write
$$7 + [3] = 10$$

We will practice using the thought process of algebraic addition because algebraic addition can be used to simplify expressions that would be very difficult to simplify if we used the thought process of subtraction.

Example 1   Simplify: $-3 - (-2)$

Solution   We think addition. We think we are to **add** -3 and $-(-2)$. This is what we think.
$$(-3) + [-(-2)]$$
But the opposite of the opposite of 2 is 2 itself. So we have
$$(-3) + [2] = \mathbf{-1}$$

Example 2   Simplify: $-(-2) - 5 - (-6)$

Solution   We see three numbers. **We think addition**. We think
$$[-(-2)] + (-5) + [-(-6)]$$
We simplify the first and third numbers and get
$$[+2] + (-5) + [+6] = \mathbf{3}$$

Practice   Use algebraic addition to find these sums.

   a. $(-3) - (+2)$ **b.** $(-3) - (-2)$

   c. $(+3) - (2)$ **d.** $(-3) - (+2) - (-4)$

   e. $(-8) + (-3) - (+2)$

Problem set 86

1. The weight of the beaker and the liquid was 1037 g. The weight of the empty beaker was 350 g. What was the weight of the liquid?

2. Use a ratio box to solve this problem. Jenny's soccer ball is covered with a pattern of pentagons and hexagons in the ratio of 3 to 5. If there are 12 pentagons, how many hexagons are in the pattern?

3. When the sum of $\frac{1}{4}$ and $\frac{1}{2}$ is divided by the product of $\frac{1}{4}$ and $\frac{1}{2}$, what is the quotient?

4. Pens were on sale 4 for $1.24.

   (a) What was the price per pen?

   (b) How much would 100 pens cost?

5. Christy rode her bike 60 miles in 5 hours.

   (a) What was her average speed in miles per hour?

   (b) What was the average number of minutes it took to ride each mile?

6. Sound travels about 331 meters per second in air. About how many seconds does it take sound to travel a kilometer?

7. The following scores were made on a test:

   72, 80, 84, 88, 100, 88, and 76

   (a) Which score was made most often?

   (b) If the scores were listed in order from the least to the greatest, what would be the middle score?

   (c) What is the average of all the scores?

8. What is the average of the two numbers marked by arrows on this number line?

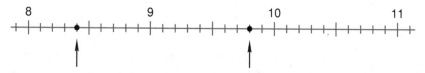

9. This rectangular shape is two cubes high and two cubes deep.

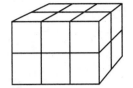

(a) How many cubes were used to build this shape?

(b) What is the name of this shape?

10. Find the circumference of each circle.

(a)   (b)

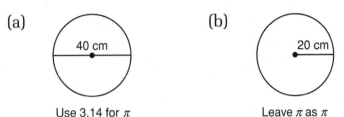

Use 3.14 for π Leave π as π

11. Draw a right triangle that is also an isosceles triangle.

12. Multiply twenty thousand and thirty thousand, and write the product in scientific notation.

Write equations to solve Problems 13 and 14.

13. What number is 75 percent of 400?

14. What number is $1\frac{1}{3}$ of $1\frac{1}{2}$?

15. Simplify:

(a) $(-4) - (-6)$ (b) $(-4) - (+6)$

16. Complete the table.

Fraction	Decimal	Percent
$\frac{3}{25}$	(a)	(b)

17. Use a unit multiplier to convert 72 qt to gallons.

18. Evaluate: $x^2 + xy + y^2$ if $x = 4$ and $y = 5$

Solve:

19. $a + 3.7 = 4.09$

20. $290 - k = 29$

21. $\dfrac{4}{c} = \dfrac{3}{7\frac{1}{2}}$

22. $2.25 = 15w$

Add, subtract, multiply, or divide, as indicated:

23.
$$\begin{array}{r} 1 \text{ gal} \\ -\quad \underline{\quad 1 \text{ qt} \ \ 1 \text{ pt} \ \ 1 \text{ oz}} \end{array}$$

24. $\dfrac{\$12.00}{1 \text{ hr}} \cdot \dfrac{1 \text{ hr}}{60 \text{ min}}$

25. $16 \div (0.04 \div 0.8)$

26. $10 - [0.1 - (0.01)(0.1)]$

27. $\dfrac{5}{8} + \dfrac{2}{3} \cdot \dfrac{3}{4} - \dfrac{3}{4}$

28. $4\dfrac{1}{2} \cdot 3\dfrac{3}{4} \div 1\dfrac{2}{3}$

29. $\sqrt{5^2 - 2^4}$

30. $3 + 6[10 - (3 \cdot 4 - 5)]$

LESSON 87

Operations with Fractions and Decimals

Sometimes we encounter expressions that contain both fractions and decimals, such as this expression.

$$\frac{3}{4} - 0.4$$

Before we simplify, we rewrite the expression so that both numbers are fractions or both numbers are decimals.

<table>
<tr><th>FRACTIONS</th><th>DECIMALS</th></tr>
<tr><td>$\dfrac{3}{4} - 0.4$</td><td>$\dfrac{3}{4} - 0.4$</td></tr>
<tr><td>$\left(0.4 = \dfrac{2}{5}\right)$</td><td>$\left(\dfrac{3}{4} = 0.75\right)$</td></tr>
<tr><td>$\dfrac{3}{4} - \dfrac{2}{5} = \dfrac{15}{20} - \dfrac{8}{20} = \mathbf{\dfrac{7}{20}}$</td><td>$0.75 - 0.4 = \mathbf{0.35}$</td></tr>
</table>

Both answers are correct because $\frac{7}{20}$ equals 0.35.

The problems we will see in this book will ask for an answer in one form or the other form.

Example   Simplify each expression to the form indicated.

(a) $3\frac{3}{4} + 3.4$ (decimal)

(b) $4.5 - \frac{4}{5}$ (fraction)

Solution   (a) We change $3\frac{3}{4}$ to the decimal number 3.75 and then add.

$$3\frac{3}{4} + 3.4 \qquad \text{problem}$$

$$3.75 + 3.4 \qquad \text{changed } 3\frac{3}{4} \text{ to 3.75}$$

$$\textbf{7.15} \qquad \text{added}$$

(b) We change 4.5 to the mixed number $4\frac{1}{2}$. Then we subtract.

$$4.5 - \frac{4}{5} \qquad \text{problem}$$

$$4\frac{1}{2} - \frac{4}{5} \qquad \text{changed 4.5 to } 4\frac{1}{2}$$

$$4\frac{5}{10} - \frac{8}{10} \qquad \text{common denominator}$$

$$\textbf{3}\frac{\textbf{7}}{\textbf{10}} \qquad \text{subtracted}$$

Practice   Simplify each problem to the form indicated.

a. $3.8 + \frac{3}{8}$ (decimal)   **b.** $\frac{1}{3} + 0.5$ (fraction)

c. $\frac{4}{5} - 0.45$ (decimal)   **d.** $2.3 - 1\frac{2}{5}$ (fraction)

Problem set
87

1. The five judges awarded scores of 8.7, 8.2, 8.1, 8.5, and 8.5 to the contestant. The highest and lowest scores were not counted. What was the average of the 3 middle scores?

2. Use a ratio box to solve this problem. The ratio of lords to ladies at the palace ball was 5 to 7. If 420 lords and ladies attended, how many lords were there?

3. Seven dictionaries were stacked on the shelf. If 3 of the dictionaries weigh a total of 96 ounces, all 7 dictionaries weigh how many pounds?

4. The diameter of Debbie's bicycle tire is 24 in. What is the circumference of the tire to the nearest inch?

96 oz

5. Five ninths of the 360 students were girls. How many of the students were boys?

6. Find the product of two thousandths and three thousandths. Then write the answer in scientific notation.

7. Use the name of a geometric solid to describe each object.

(a)    (b)    (c)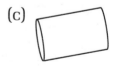

Write equations to solve Problems 8 and 9.

8. What number is 35 percent of 800?

9. What number is 2.5 of 40?

10. Find the circumference of each circle.

(a)
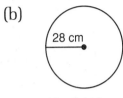

Use $\frac{22}{7}$ for π

(b)

Use 3.14 for π

Refer to the figure to do Problems 11 and 12. Dimensions are in centimeters. The larger shape is a rectangle.

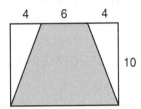

11. (a) What is the area of the rectangle?

(b) What is the area of each triangle?

12. Use the answers to Problem 11 to find the area of the shaded region.

13. Simplify:

(a) $(-5) - (-5)$

(b) $(-8) - (+7) - (-6) + (-4)$

14. Complete the table.

FRACTION	DECIMAL	PERCENT
(a)	1.75	(b)

15. Use a unit multiplier to convert 0.25 kg to grams.

16. Evaluate: $mn + \sqrt{m} - n^2$ if $m = 9$ and $n = 9$

17. Round $7.\overline{27}$ to the nearest thousandth.

18. Find the missing number. (*Hint:* The rule is to multiply by a number and then add another number.)

$$
\begin{array}{ccc}
\text{IN} & \boxed{\begin{array}{c}\textbf{F}\\\textbf{U}\\\textbf{N}\\\textbf{C}\\\textbf{T}\\\textbf{I}\\\textbf{O}\\\textbf{N}\end{array}} & \text{OUT}\\
2 \rightarrow & \rightarrow & 5\\
4 \rightarrow & \rightarrow & 9\\
3 \rightarrow & \rightarrow & 7\\
6 \rightarrow & \rightarrow & \square
\end{array}
$$

Solve:

19. $y - 46 = 217$ **20.** $3c = 5.64$

21. $4.54 + m = 9.2$ **22.** $\dfrac{15}{25} = \dfrac{x}{15}$

Add, subtract, multiply, or divide, as indicated:

23.   6 lb 10 oz
    $+$ 1 lb   9 oz

24. $\dfrac{\$12}{1 \text{ hr}} \cdot \dfrac{40 \text{ hr}}{1 \text{ week}} \cdot \dfrac{4 \text{ weeks}}{1 \text{ month}}$

25. $\dfrac{5}{8} + 0.58$ (decimal)   **26.** $4.5 \div \dfrac{4}{5}$ (fraction)

27. $(7 - 0.6) \div 0.05$ **28.** $4.25 - 4 + (0.2)(5)$

29. $3\dfrac{5}{8} - \left(5\dfrac{1}{3} - 1\dfrac{3}{4}\right)$ **30.** $4\dfrac{1}{5} \div \left(1\dfrac{1}{2} \div 3\dfrac{1}{3}\right)$

The Addition Rule for Equations

LESSON 88

In this lesson we will begin using a new method for solving equations. The method presented is used in algebra. This method uses the addition rule for equations.

ADDITION RULE FOR EQUATIONS

> If the same number is added to both sides of an equation, the solution (answer) is not changed.

Here's how the addition rule works. Consider the following equation whose solution is 5.

$$x + 3 = 8 \qquad \text{(solution is 5)}$$

The addition rule says that if we add the same number to both sides of this equation, the solution to the new equation will also be 5. To illustrate, we will add 2 to both sides of the equation.

$$
\begin{array}{ll}
x + 3 = 8 & \text{original equation (solution is 5)} \\
\underline{+2 = +2} & \text{add 2 to both sides} \\
x + 5 = 10 & \text{new equation (solution is 5)}
\end{array}
$$

We see that the solution to the new equation is also 5.

Adding 2 to both sides of the equation did not change the solution, but it did not help us find the solution either. However, if we carefully select the number to add to both sides of the equation, the solution to the original equation will appear in the new equation. We will show this by adding -3 to both sides of the equation.

$$
\begin{array}{ll}
x + 3 = 8 & \text{original equation} \\
\underline{-3 = -3} & \text{add } -3 \text{ to both sides} \\
x + 0 = 5 & \text{new equation}
\end{array}
$$

We see that the new equation, $x + 0 = 5$, *shows us* the solution to the original equation. Since $x + 0$ is x, the new equation is $x = 5$, which *is* the solution of the original equation.

Solving an equation by the addition rule gets the variable

by itself on one side of the equals sign. To use this method we follow these steps.

1. Find the variable in the equation and determine which side of the equals sign the variable occupies.

2. Find the number on the same side of the equals sign that is added to the variable.

3. Add the **opposite of this number** to both sides of the equals sign.

Example 1 Solve and check: $m - 248 = 352$

Solution **We want to get m all by itself on one side of the equals sign.** The number on the same side as m is -248. To remove -248 we add $+248$ to both sides of the equation.

$$
\begin{array}{ll}
m - 248 = 352 & \text{original equation} \\
\underline{\quad +248 = +248} & \text{add } +248 \text{ to both sides} \\
m + \quad 0 = 600 & \text{new equation} \\
\\
\mathbf{m = 600} & \text{solution}
\end{array}
$$

Now we check the solution.

$$
\begin{array}{ll}
m - 248 = 352 & \text{original equation} \\
(600) - 248 = 352 & \text{substituted 600 for } m \\
352 = 352 & \text{simplified; solution checks}
\end{array}
$$

Example 2 Solve and check: $263 = x + 47$

Solution **We want to get x all by itself on one side of the equals sign.** The variable x is on the right-hand side with 47. If we add -47 to both sides of the equation, we will get x all by itself.

$$
\begin{array}{ll}
263 = x + 47 & \text{original equation} \\
\underline{-47 = \quad -47} & \text{add } -47 \text{ to both sides} \\
216 = x + 0 & \text{new equation} \\
\\
\mathbf{216 = x} & \text{solution}
\end{array}
$$

Now we check the solution.

$$263 = x + 47 \qquad \text{original equation}$$

$$263 = (216) + 47 \qquad \text{substituted 216 for } x$$

$$263 = 263 \qquad \text{simplified; solution checks}$$

Example 3 Solve and check: $y - 4.7 = 5.79$

Solution **We want to get y all by itself.** We can do this if we add $+4.7$ to both sides of the equation.

$$
\begin{array}{ll}
y - 4.7 = 5.79 & \text{original equation} \\
\underline{+4.7 \quad +4.7} & \text{add 4.7 to both sides} \\
y + 0 = 10.49 & \text{new equation} \\
y = \mathbf{10.49} & \text{solution}
\end{array}
$$

Now we check the solution.

$$
\begin{array}{ll}
y - 4.7 = 5.79 & \text{original equation} \\
(10.49) - 4.7 = 5.79 & \text{substituted 10.49 for } y \\
5.79 = 5.79 & \text{simplified; solution checks}
\end{array}
$$

Example 4 Solve and check: $w + \dfrac{3}{4} = \dfrac{5}{6}$

Solution **We want to get w all by itself.** We can do this if we add $-\dfrac{3}{4}$ to both sides of the equation.

$$
\begin{array}{ll}
w + \dfrac{3}{4} = \dfrac{5}{6} & \text{original equation} \\[2mm]
\underline{-\dfrac{3}{4} = -\dfrac{3}{4}} & \text{add } -\dfrac{3}{4} \text{ to both sides} \\[2mm]
w + 0 = \dfrac{5}{6} - \dfrac{3}{4} & \text{new equation}
\end{array}
$$

To add $\dfrac{5}{6}$ and $-\dfrac{3}{4}$, we need to use a common denominator.

$$
\begin{array}{ll}
\dfrac{5}{6} - \dfrac{3}{4} & \text{problem} \\[2mm]
\dfrac{10}{12} - \dfrac{9}{12} = \dfrac{1}{12} & \text{used common denominators}
\end{array}
$$

Now we can write the solution.

$$w = \frac{1}{12}$$

Now we check our work by using $\frac{1}{12}$ for w in the original equation.

$$w + \frac{3}{4} = \frac{5}{6} \qquad \text{original equation}$$

$$\left(\frac{1}{12}\right) + \frac{3}{4} = \frac{5}{6} \qquad \text{substituted } \frac{1}{12} \text{ for } w$$

$$\frac{1}{12} + \frac{9}{12} = \frac{5}{6} \qquad \text{common denominators}$$

$$\frac{10}{12} = \frac{5}{6} \qquad \text{simplified}$$

$$\frac{5}{6} = \frac{5}{6} \qquad \text{simplified; solution checks}$$

Practice Use the addition rule to solve each equation. Show your work. Check each answer.

a. $463 = m - 281$

b. $p + 56 = 203$

c. $n - 1.3 = 12.28$

d. $5.7 = x + 1.35$

Problem set 88

1. Fully dressed, O'Riley weighs 123 lb. How much does O'Riley actually weigh if the weight of his clothes is 80 oz?

2. When the sum of $\frac{1}{3}$ and $\frac{1}{2}$ is divided by the product of $\frac{1}{3}$ and $\frac{1}{2}$, what is the quotient?

3. Nine seconds elapsed from the time Mark saw the lightning until he heard the thunder. The lightning was about how many kilometers from Mark? (Sound travels about 331 meters per second in air.)

4. Use a ratio box to solve this problem. The ratio of left-handed students to right-handed students in the math class was 2 to 3. If 18 of the students were right-handed, how many students were there in the math class?

5. It was estimated that $1\frac{1}{2}$ million people lined the parade route. Write that number in scientific notation.

6. Five out of every fifty fans cheered for the visiting team.

   (a) What percent of the fans cheered for the visiting team?

   (b) What fraction of the fans did not cheer for the visiting team?

7. The average weight of the three big men was 288 pounds. Two of the big men weighed 252 and 261 pounds, respectively. What was the weight of the other big man?

8. How many cubes were used to build this rectangular prism?

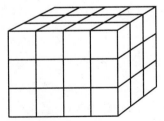

9. Find the circumference of each circle.

   (a)

   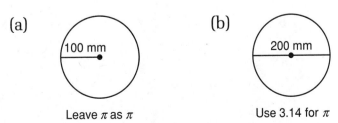

   100 mm

   Leave π as π

   (b)

   200 mm

   Use 3.14 for π

10. Refer to the figure to answer the questions. Dimensions are in centimeters.

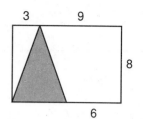

(a) What is the area of the rectangle?

(b) What is the area of the shaded triangle?

(c) What fraction of the rectangle is shaded?

Write equations to solve Problems 11 and 12.

11. What number is 80 percent of 500?

12. What number is $\frac{2}{3}$ of 100?

13. Simplify:

(a) $(-8) + (-5) - (-9)$

(b) $(+2) - (-7) + (-3) - (+4)$

14. Complete the table.

Fraction	Decimal	Percent
(a)	(b)	1%

15. Use a unit multiplier to convert $2\frac{1}{2}$ tons to pounds.

16. Evaluate: $x - \dfrac{x - y}{y}$ if $x = 12$ and $y = 3$

17. Find the next three numbers in this sequence.

10000, 1000, 100, ___, ___, ___

Solve and check. Show your work.

18. $x - 47 = 360$

19. $y + 1.4 = 5.17$

Solve:

20. $\dfrac{4}{6} = \dfrac{6}{n}$

21. $10p = 12.5$

Add, subtract, multipy, or divide, as indicated:

22. $\begin{array}{cccc} & 2\ hr & 16\ min & 7\ sec \\ - & 1\ hr & 20\ min & 15\ sec \end{array}$

23. $\dfrac{4\ qt}{1\ gal} \cdot \dfrac{2\ pt}{1\ qt} \cdot \dfrac{16\ oz}{1\ pt}$

24. $\dfrac{2}{5} - 0.025$ (decimal)

25. $2\dfrac{2}{3} \times 0.9$ (fraction)

26. $7 - (0.6)(0.05)$

27. $4.25 - [4 + (0.2 \div 5)]$

28. $3\dfrac{5}{8} + \left(3 - 1\dfrac{5}{6}\right)$

29. $\left(6\dfrac{1}{4}\right)\left(3\dfrac{1}{5} \div 1\dfrac{1}{3}\right)$

30. $\dfrac{20 + \{45 - 5[8 - 2(8 - 3 \cdot 2)]\}}{5^2 - 4^2}$

More on Scientific Notation

LESSON
89

When we write a number in scientific notation, we usually put the decimal point just to the right of the first digit that is not zero.

To write $\qquad 4600 \times 10^5$

in scientific notation, we will use two steps. First we will write 4600 in scientific notation. In place of 4600 we will write 4.6×10^3. Now we have

$$(4.6 \times 10^3) \times 10^5$$

For the second step we change the two powers of 10 into one power of 10. We recall that 10^3 means the decimal point is 3 places to the right and 10^5 means the decimal point is 5 places to the right. Since 3 places to the right and 5 places to the right is 8 places to the right, the power of 10 is 10^8.

$$4.6 \times 10^8$$

To perform the exercises in this lesson, first change the decimal number to scientific notation. Then change the two powers of 10 to one power of 10.

Example 1   Write 25×10^{-5} in scientific notation.

Solution   We write 25 in scientific notation.

$$(2.5 \times 10^1) \times 10^{-5}$$

We combine the powers of 10 by remembering that 1 place to the right plus 5 places to the left equals 4 places to the left.

$$\mathbf{2.5 \times 10^{-4}}$$

Example 2   Write 0.25×10^4 in scientific notation.

Solution   First we write 0.25 in scientific notation.

$$(2.5 \times 10^{-1}) \times 10^4$$

Since 1 place to the left plus 4 places to the right equals 3 places to the right we can write

$$\mathbf{2.5 \times 10^3}$$

With practice you will soon be able to perform these exercises mentally.

Practice   Write each number in scientific notation.

a. 0.16×10^6 b. 24×10^{-7}

c. 30×10^5 d. 0.75×10^{-8}

Problem set 89

1. The following is a list of scores Jan received in a diving competition.

7.0	6.5	6.5	7.4	7.0	6.5	6.0

(a) Which score was received most often?

(b) If the scores were arranged in order from the least to the greatest, which score would be the middle score?

(c) What is the average of all the scores?

(d) What is the difference between the highest score and the lowest score?

2. Use a ratio box to solve this problem. The team won 15 games and lost the rest. If the team's won-lost ratio was 5 to 3, how many games were played?

3. Brian swam 4 laps in 6 minutes. At that rate, how many minutes will it take Brian to swim 10 laps?

4. Write each number in scientific notation.

(a) 15×10^5 (b) 0.15×10^5

5. Draw a diagram of this statement: Two fifths of the 70 barges were loaded with coal. How many barges were loaded with coal?

6. The diameter of the tree stump was 40 cm. Find the circumference of the tree stump to the nearest centimeter.

7. Use the name of a geometric solid to describe the shape of these objects.

(a) A volleyball (b) A water pipe (c) A tepee

8. Find the circumference of each circle.

(a)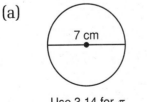

Use 3.14 for π

(b)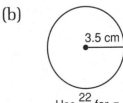

Use $\frac{22}{7}$ for π

9. Simplify:

(a) $(-4) + (-5) - (-6)$

(b) $(-2) + (-3) - (-4) - (+5)$

Refer to the figure to do Problems 10 and 11. Dimensions are in millimeters. Corners that look square are square.

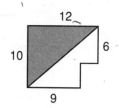

10. (a) What is the area of the hexagon?

    (b) What is the area of the shaded triangle?

11. What fraction of the hexagon is shaded?

Write equations to solve Problems 12 and 13.

12. What number is 50 percent of 200?

13. What number is 2.5 of 4.2?

14. Complete the table.

Fraction	Decimal	Percent
$\frac{3}{20}$	(a)	(b)

15. Use a unit multiplier to convert 16 pounds to ounces.

16. Evaluate: $a^2 - \sqrt{a} + ab$ if $a = 4$ and $b = 0.5$

17. Find the missing number.

In		Out
8 →	**F U N C T I O N**	→ 15
6 →		→ 11
10 →		→ 19
4 →		→ ☐

18. Divide 144 by 11 and write the answer

    (a) as a decimal with a bar over the repetend.

    (b) rounded to the nearest whole number.

Solve and check. Show your work.

19. $75 = p - 49$

20. $t + \dfrac{5}{8} = \dfrac{15}{16}$

Solve:

21. $7q = 357$

22. $\dfrac{a}{8} = \dfrac{3\frac{1}{2}}{2}$

Add, subtract, multiply, or divide, as indicated:

23.
$$\begin{array}{r} 5 \text{ ft } 7 \text{ in.} \\ + 6 \text{ ft } 8 \text{ in.} \\ \hline \end{array}$$

24. $\dfrac{350 \text{ m}}{1 \text{ sec}} \cdot \dfrac{60 \text{ sec}}{1 \text{ min}} \cdot \dfrac{1 \text{ km}}{1000 \text{ m}}$

25. $2\dfrac{1}{4} + 0.15$ (decimal)

26. $\dfrac{5}{6} + 6.5$ (fraction)

27. $6 - (0.5 \div 4)$

28. $\$7.50 \div 0.075$

29. $\left(3\dfrac{3}{4} \div 1\dfrac{2}{3}\right) \cdot 3$

30. $4\dfrac{1}{2} + \left(5\dfrac{1}{6} \div 1\dfrac{1}{3}\right)$

LESSON
90

Multiplication Rule for Equations

In Lesson 88 we learned an algebraic method for solving equations that used the addition rule. In this lesson we will discuss another rule used in algebra to solve equations. We remember that the product of any number and the number 1 is the number itself.

$$1 \cdot 4 = 4 \qquad 1 \cdot 6 = 6 \qquad 1 \cdot x = x$$

We remember that the reciprocal of a number is the number

with the top and the bottom terms interchanged.

$$\frac{4}{3} \text{ is the reciprocal of } \frac{3}{4}$$

$$\frac{3}{4} \text{ is the reciprocal of } \frac{4}{3}$$

$$\frac{1}{5} \text{ is the reciprocal of } 5$$

$$5 \text{ is the reciprocal of } \frac{1}{5}$$

If we multiply a number by its reciprocal, the answer is 1.

$$\frac{4}{3} \cdot \frac{3}{4} = \frac{12}{12} = 1 \qquad 5 \cdot \frac{1}{5} = \frac{5}{5} = 1$$

We also remember that if we divide a number by itself, the answer is also 1.

$$\frac{4}{4} = 1 \qquad \frac{\frac{3}{4}}{\frac{3}{4}} = 1 \qquad \frac{2\frac{1}{2}}{2\frac{1}{2}} = 1$$

We can solve some equations by remembering these facts and using either the multiplication rule for equations or the division rule for equations.

MULTIPLICATION RULE FOR EQUATIONS

> If both sides of an equation are multiplied by the same number (but not zero), the solution (answer) is not changed.

DIVISION RULE FOR EQUATIONS

> If both sides of an equation are divided by the same number (but not zero), the solution (answer) is not changed.

Example 1   Solve: $4x = 15$

Solution **We want *x* all by itself.** We have a choice. We can use the division rule and divide both sides of the equation by 4, or we can use the multiplication rule and multiply both sides by $\frac{1}{4}$. We will show both ways.

Using the division rule:

$$4x = 15 \qquad \text{original equation}$$

$$\frac{4x}{4} = \frac{15}{4} \qquad \text{divided both sides by 4}$$

$$1x = 3\frac{3}{4} \qquad \text{simplified both sides}$$

$$x = 3\frac{3}{4} \qquad 1x = x$$

Using the multiplication rule:

$$4x = 15 \qquad \text{original equation}$$

$$\frac{1}{4} \cdot 4x = 15 \cdot \frac{1}{4} \qquad \text{multiplied both sides by } \frac{1}{4}$$

$$1x = \frac{15}{4} \qquad \text{simplified both sides}$$

$$x = 3\frac{3}{4} \qquad \text{simplified}$$

Example 2 Solve: $\frac{2}{3}x = 150$

Solution To change $\frac{2}{3}x$ to 1*x*, we either divide both sides by $\frac{2}{3}$ or multiply both sides by $\frac{3}{2}$. With fractions it is usually easier to multiply by the reciprocal, so we will multiply both sides by $\frac{3}{2}$.

$$\frac{2}{3}x = 150 \qquad \text{original equation}$$

$$\frac{3}{2}\left(\frac{2}{3}x\right) = \frac{3}{2}(150) \qquad \text{multiplied by } \frac{3}{2}$$

$$1x = \frac{450}{2} \qquad \text{simplified both sides}$$

$$x = 225 \qquad \text{simplified}$$

Now we check our answer.

$$\frac{2}{3}x = 150 \qquad \text{original equation}$$

$$\frac{2}{3}(225) = 150 \qquad \text{substituted 225 for } x$$

$$150 = 150 \qquad \text{check}$$

Example 3   Solve: $3.5y = 280$

Solution   To change $3.5y$ to $1y$, we either divide both sides of the equation by 3.5 or multiply both sides by $\frac{1}{3.5}$. With decimal numbers it is usually easier to divide.

$$3.5y = 280 \qquad \text{original equation}$$

$$\frac{3.5y}{3.5} = \frac{280}{3.5} \qquad \text{divided both sides by 3.5}$$

$$1y = 80 \qquad \text{simplified both sides}$$

$$y = 80 \qquad 1y = y$$

Now we check our answer.

$$3.5y = 280 \qquad \text{original equation}$$

$$3.5(80) = 280 \qquad \text{substituted 80 for } y$$

$$280 = 280 \qquad \text{check}$$

Practice   Solve each equation by dividing or multiplying. Show every step. Then check each solution.

a. $8m = 416$

b. $\frac{3}{4}x = 72$

c. $\frac{5}{3}w = \frac{2}{3}$

d. $0.4n = 1.84$

Problem set 90

1. Use a ratio box to solve this problem. Four hundred fifty students attended the assembly. If the ratio of boys to girls in the assembly was 4 to 5, how many girls attended the assembly?

2. What is the average of the three numbers marked by arrows on this number line?

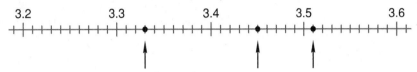

Refer to this figure to do Problems 3–5. Dimensions are in centimeters. Corners that look square are square.

3. What is the perimeter of the hexagon?

4. What is the area of the hexagon?

5. What is the area of the shaded triangle?

6. Use a unit multiplier to convert 3.5 g to milligrams.

7. Draw a diagram of this statement: In the first third of the season the Madrigals played 18 games. How many games will the Madrigals play during the whole season?

Write equations to solve Problems 8–11.

8. What number is $\frac{5}{6}$ of 72?

9. Forty-eight percent of 25 is what number?

10. What number is 0.1 of 110?

11. What is 75 percent of $40?

12. A cube has how many more edges than faces?

13. Find the circumference of each circle.

(a) 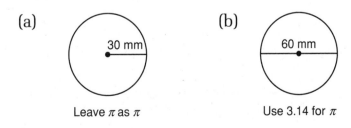 (b)

Leave π as π Use 3.14 for π

14. Complete the table.

FRACTION	DECIMAL	PERCENT
(a)	0.75	(b)

15. What number is 42 percent of 300?

16. How much is 30 percent of $5.40?

17. Sketch a picture of a triangular prism. A triangular prism has how many edges?

18. Find the missing number.

$$\begin{array}{ccc}
\text{In} & \boxed{\begin{array}{c}\text{F}\\\text{U}\\\text{N}\\\text{C}\\\text{T}\\\text{I}\\\text{O}\\\text{N}\end{array}} & \text{Out}\\
\end{array}$$

In → FUNCTION → Out

-2 → 4
4 → 10
10 → 16
16 → ☐

19. Write 37,500,000,000 in scientific notation.

20. Evaluate: $my - y^2$ if $m = 12$ and $y = 3$

Solve and check. Show each step.

21. $8x = 31.2$

22. $\frac{3}{4}y = 24$

23. $m + 3.4 = 7$

24. $p - \frac{2}{3} = \frac{3}{4}$

Add, subtract, multiply, or divide, as indicated:

25.
$$\begin{array}{r}
3 \text{ qt } 1 \text{ pt } 5 \text{ oz}\\
- 1 \text{ qt } 1 \text{ pt } 7 \text{ oz}\\
\hline
\end{array}$$

26. $\dfrac{\$300}{1 \text{ week}} \cdot \dfrac{1 \text{ week}}{5 \text{ days}} \cdot \dfrac{1 \text{ day}}{8 \text{ hr}}$

27. $7\frac{1}{2} \div \left(6\frac{2}{3} \cdot 1\frac{1}{5} \right)$

28. $5\frac{1}{2} - \left(3\frac{1}{3} - 1\frac{3}{4} \right)$

29. $3\frac{3}{5} + 0.65$ (decimal)

30. (a) $(-8) - (-7) + (-12)$ (b) $(-24) + (-18) - (+32)$

LESSON
91

Volume

Geometric solids are shapes that take up space. We use the word **volume** to describe the space occupied by a shape. To measure volume, we must use units that occupy space. The units that we use to measure volume are cubes of certain sizes. We can use sugar cubes to help us think of volume.

Example 1 This rectangular prism was constructed of sugar cubes. Its volume is how many cubes?

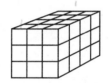

Solution To find the volume of the prism, we must calculate the number of cubes it contains. We see that there are 3 layers of cubes. Each layer contains 3 rows of cubes with 4 cubes in each row, or 12 cubes. Three layers with 12 cubes in each layer means that the volume of the prism is **36 cubes**.

Volumes are measured by using cubes of a standard size. A cube whose edges are 1 centimeter long has a volume of 1 cubic centimeter, which we abbreviate by writing 1 cm³.

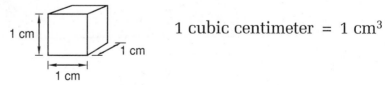

1 cubic centimeter = 1 cm³

Similarly, if each of the edges is 1 foot long, the volume is 1 cubic foot. If each of the edges is 1 meter long, the volume is 1 cubic meter.

1 cubic foot = 1 ft³   1 cubic meter = 1 m³

To calculate the volume of a solid, we can imagine constructing the solid out of sugar cubes of the same size. We would begin by constructing the base and then building up the layers to the specified height.

Example 2   Find the number of 1-cm cubes that can be placed inside a rectangular box with the dimensions shown.

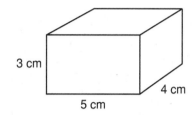

3 cm

4 cm

5 cm

Solution   The base of the box is 5 cm by 4 cm. So we can place 4 rows of 5 cubes on the base. Thus there are 20 cubes on the first layer.

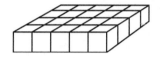

Since the box is 3 cm high, we can fit 3 layers of cubes in the box.

$$\frac{20 \text{ cubes}}{1 \text{ layer}} \times 3 \text{ layers} = 60 \text{ cubes}$$

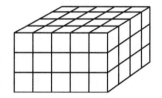

We find that **60 cubes** can be placed in the box.

Example 3   What is the volume of this cube? Dimensions are in inches.

4

4

4

Solution   The base is 4 in. by 4 in. Thus, 16 cubes can be placed on the base.

Since the big cube is 4 in. high, there are 4 layers of small cubes.

$$\frac{16 \text{ cubes}}{1 \text{ layer}} \times 4 \text{ layers} = 64 \text{ cubes}$$

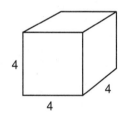

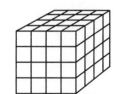

Each little cube has a volume of 1 cubic inch. Thus, the volume of the big cube is **64 cubic inches (64 in.³)**.

Practice **a.** This rectangular prism was constructed of sugar cubes. Its volume is how many sugar cubes?

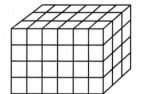

b. Find the number of 1-cm cubes that can be placed inside a box with dimensions as illustrated.

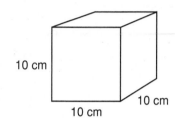

c. What is the volume of this rectangular prism? Dimensions are in feet.

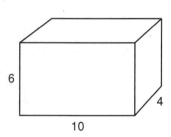

Problem set 91

1. It was 38 kilometers from the encampment to the castle. Milton galloped to the castle and cantered back. If the round trip took 4 hours, what was his average speed in kilometers per hour?

2. Use a ratio box to solve this problem. The ratio of dogs to cats in the neighborhood was 4 to 7. If there were 56 dogs in the neighborhood, how many cats were in the neighborhood?

3. Using a tape measure, Gretchen found that the circumference of the great oak was 600 cm. She estimated that its diameter was 200 cm. Was her estimate for the diameter a little too large or a little too small? Why?

4. Grapes were priced at 3 pounds for $1.29.

    (a) What was the price per pound?

    (b) How much would 10 pounds of grapes cost?

5. If the product of nine tenths and eight tenths is subtracted from the sum of seven tenths and six tenths, what is the difference?

6. Three fourths of the batter's 188 hits were singles.

   (a) How many of the batter's hits were singles?

   (b) What percent of the batter's hits were not singles?

7. Compare: $2 - 5 \bigcirc 2 + (-5)$

8. Find the number of 1-cm cubes that can be placed in this box.

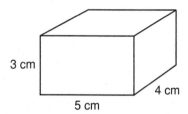

3 cm

5 cm

4 cm

9. Find the circumference of each circle.

   (a)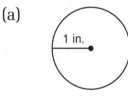

   1 in.

   Leave π as π

   (b)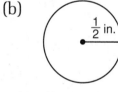

   $\frac{1}{2}$ in.

   Use 3.14 for π

10. Write each number in scientific notation.

    (a) 12×10^{-6}

    (b) 0.12×10^{-6}

11. What is the average of the three numbers marked by arrows on this number line?

0.7   0.8   0.9   1.0

12. Use a unit multiplier to convert 2000 g to kilograms.

Write equations to solve Problems 13 and 14.

13. What number is 15 percent of 2400?

14. What number is $\frac{1}{6}$ of 100?

15. Complete the table.

FRACTION	DECIMAL	PERCENT
(a)	(b)	14%

16. Simplify:

(a) $(-6) - (-4) + (+2)$

(b) $(-5) + (-2) - (-7) - (+9)$

17. Evaluate: $ab - (a - b)$ if $a = 0.4$ and $b = 0.3$

18. Round $29{,}374.6\overline{5}$ to the nearest whole number.

Solve and check. Show your work.

19. $q + 36 = 41.5$ **20.** $4.3 = x - 0.8$

Solve:

21. $5n = 24$ **22.** $\dfrac{2}{d} = \dfrac{1.2}{1.5}$

Add, subtract, multipy, or divide, as indicated:

23.   10 lb
     $\underline{-\ 6\ \text{lb}\ 7\ \text{oz}}$

24. $\dfrac{\$5.25}{1\ \text{hr}} \cdot \dfrac{8\ \text{hr}}{1\ \text{day}} \cdot \dfrac{5\ \text{days}}{1\ \text{week}}$

25. $9.2 \times 9\frac{1}{2}$ (decimal)   **26.** $11.5 - 1\frac{1}{12}$ (fraction)

27. $(0.06 \div 5) \div 0.004$ **28.** $\$15 + (0.06)(\$15)$

29. $3\frac{3}{4} \div \left(1\frac{2}{3} \cdot 3\right)$ **30.** $4\frac{1}{2} + 5\frac{1}{6} - 1\frac{1}{3}$

LESSON
92

Finding the Whole Group
When a Fraction Is Known

Drawing diagrams of fraction problems can help us understand problems such as the following.

> Three fifths of the fish in the pond were blue gill. If there were 45 blue gill in the pond, how many fish were in the pond?

The 45 blue gill are 3 of the 5 parts. We divide 45 by 3 and find that there are 15 fish in each part. Since each of the 5 parts is 15 fish, there were 75 fish in all.

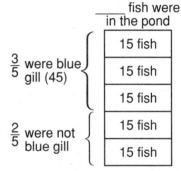

Example When Sean finished page 51, he was $\frac{3}{8}$ of the way through his book. His book had how many pages?

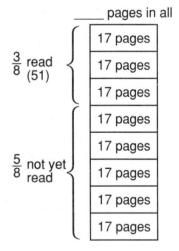

Solution Sean read 51 pages. This is 3 of 8 parts of the book. Since 51 ÷ 3 is 17, each part is 17 pages. Thus the whole book, all 8 parts, total 8 × 17, which is **136 pages**.

Practice Draw a diagram to solve this problem.

Three fifths of the students in the class are boys. If there are 15 boys in the class, how many students are there in all?

**Problem set
92**

1. George bought 5 pounds of grapes for $1.20.

    (a) What was the price per pound?

    (b) What would be the cost of 12 pounds of grapes?

2. Use a ratio box to solve this problem. The clipper sailed for 48 hours at an average speed of 6 nautical miles per hour. How far did the ship sail?

3. On his first 2 tests Nate's average score was 80 percent. On his next 3 tests Nate's average score was 90 percent. What was his average score for all 5 tests? (*Hint:* Find the total number of points scored on all tests, then divide by 5.)

4. Twenty billion is how much more than nine billion? Write the answer in scientific notation.

5. What is the sum of the first five prime numbers?

6. Use a ratio box to solve this problem. The ratio of new ones to used ones in the box was 4 to 7. In all there were 242 in the box. How many new ones were in the box?

7. Draw a diagram of this statement. Then answer the questions that follow.

   When Debbie finished page 78, she was $\frac{3}{5}$ of the way through her book.

   (a) How many pages are in her book?

   (b) How many pages does she have left to read?

8. Three fourths of 24 is what number?

9. Find the number of 1-inch cubes that can be placed in this box. Dimensions are in inches.

10. Find the circumference of each circle.

(a)

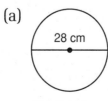

28 cm

Use 3.14 for π

(b)

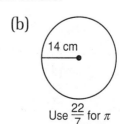

14 cm

Use $\frac{22}{7}$ for π

11. Write each number in scientific notation.

(a) 25×10^6 (b) 25×10^{-6}

12. Complete the table.

FRACTION	DECIMAL	PERCENT
(a)	0.1	(b)

13. Write an equation to solve this problem. What number is 35 percent of 80?

14. Find the missing number.

$$
\begin{array}{ccc}
\text{IN} & \boxed{\text{FUNCTION}} & \text{OUT} \\
3 \rightarrow & & \rightarrow 10 \\
0 \rightarrow & & \rightarrow \square \\
5 \rightarrow & & \rightarrow 12 \\
7 \rightarrow & & \rightarrow 14 \\
\end{array}
$$

15. A rectangular prism has how many vertices?

16. (a) Find the perimeter of the trapezoid. Dimensions are in centimeters.

(b) Find the area of the shaded isosceles triangle.

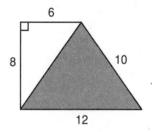

17. Compare: 0.03 of 112 ◯ $\dfrac{5}{12}$ of 84

18. Divide 256 by 24 and write the quotient as a mixed number.

19. Use a unit multiplier to convert 340 cm to meters.

20. Evaluate: $y - xy$ if $x = 0.1$ and $y = 0.01$

Solve:

21. $m + 5.75 = 26.4$

22. $4x = 4.56$

23. $\dfrac{20}{24} = \dfrac{55}{y}$

Add, subtract, multiply, or divide, as indicated:

24. $\dfrac{4^2 + \{20 - 2[6 - (5 - 2)]\}}{\sqrt{36}}$

25.
$$\begin{array}{r} 1 \text{ yd} \\ - \quad 1 \text{ ft } 1 \text{ in.} \\ \hline \end{array}$$

26. $3.5 \text{ hr} \cdot \dfrac{60 \text{ min}}{1 \text{ hr}} \cdot \dfrac{60 \text{ sec}}{1 \text{ min}}$

27. $6\dfrac{2}{3} \div \left(4\dfrac{1}{2} \cdot 2\dfrac{2}{3}\right)$

28. $7\dfrac{1}{2} - 5\dfrac{1}{6} + 1\dfrac{1}{3}$

29. $2\dfrac{2}{3} - 1.5$ (fraction)

30. (a) $(-5) + (-6) - (-7)$ (b) $(-15) - (-24) - (+8)$

LESSON 93

Implied Ratios

Many rate problems can be solved by completing a proportion. Consider the following problem.

> If 12 books weigh 20 pounds, how much would 30 books weigh?

We will illustrate two methods for solving this problem. First we will use the rate method. If 12 books weigh 20 pounds, we can write two rates.

(a) $\dfrac{12 \text{ books}}{20 \text{ pounds}}$ (b) $\dfrac{20 \text{ pounds}}{12 \text{ books}}$

To find the weight of 30 books, we could multiply 30 books by rate (b).

$$30 \, \cancel{\text{books}} \times \frac{20 \text{ pounds}}{12 \, \cancel{\text{books}}} = 50 \text{ pounds}$$

We find that 30 books would weigh 50 pounds.

Now we will solve the same problem by completing a proportion. We will record the information in a ratio box. Instead of using the words "ratio" and "actual count," we will write "case 1" and "case 2." We will use p to stand for pounds.

	CASE 1	CASE 2
Books	12	30
Pounds	20	p

From the table we write a proportion and solve it.

$$\frac{12}{20} = \frac{30}{p} \qquad \text{proportion}$$

$$12p = 20 \cdot 30 \qquad \text{cross multiplied}$$

$$\frac{\cancel{12}p}{\cancel{12}} = \frac{20 \cdot 30}{12} \qquad \text{divided by 12}$$

$$p = 50 \qquad \text{simplified}$$

We find that 30 books would weigh 50 pounds.

Example 1 If 5 pounds of grapes cost \$1.20, how much would 12 pounds of grapes cost? Use a ratio box to solve the problem.

Solution First we draw the ratio box. We use d for dollars.

	CASE 1	CASE 2
Pounds	5	12
Dollars	1.2	d

Now we write the proportion and cross multiply.

$$\frac{5}{1.2} = \frac{12}{d} \quad \longrightarrow \quad 5d = 12(1.2)$$

Now we solve by dividing both sides by 5.

$$\frac{\cancel{5}d}{\cancel{5}} = \frac{12(1.2)}{5} \quad \rightarrow \quad d = \frac{14.4}{5} = \$2.88$$

Example 2   Mrs. C can tie 25 bows in 3 minutes. How many bows can she tie in 1 hour at that rate? Work the problem (a) using rates and (b) using a ratio box.

Solution   We can use either minutes or hours but not both. **The units must be the same everywhere in a problem.** Since there are 60 minutes in 1 hour, we will use 60 minutes instead of 1 hour.

(a) $\dfrac{25 \text{ bows}}{3 \text{ \cancel{min}}} \times 60 \text{ \cancel{min}} = 500 \text{ bows}$

So Mrs. C can tie **500 bows** in 1 hour.

(b)   Bows / Minutes

	CASE 1	CASE 2
Bows	25	b
Minutes	3	60

Next we write the proportion, cross multiply, and solve by dividing by 3.

$$\frac{25}{3} = \frac{b}{60} \qquad \text{proportion}$$

$$25 \cdot 60 = 3b \qquad \text{cross multiplied}$$

$$\frac{25 \cdot 60}{3} = \frac{\cancel{3}b}{\cancel{3}} \qquad \text{divided by 3}$$

$$\mathbf{500 = \mathit{b}} \qquad \text{simplified}$$

Practice   **a.**   Use a ratio box to solve this problem. Kevin rode 30 km in 2 hours. At that rate, how long would it take him to ride 75 km?

b.   If 6 bales are needed to feed 40 head of cattle, how many bales are needed to feed 50 head of cattle? Use the rate method and then use a ratio box to solve this problem.

Problem set 93

1. Napoleon Bonaparte was born in 1769 and died in 1821. For how many years did he live?

2. In her first 4 games Jill averaged 4 points per game. In the next 6 games Jill averaged 9 points per game. What was her average number of points per game after 10 games? (*Hint:* Find the total number of points scored before dividing.)

3. Use a unit multiplier to convert 2.5 L to milliliters.

4. If the product of $\frac{1}{2}$ and $\frac{2}{5}$ is subtracted from the sum of $\frac{1}{2}$ and $\frac{2}{5}$, what is the difference?

5. Use a ratio box to solve this problem. The ratio of carnivores to herbivores in the jungle was 2 to 7. If there were 126 carnivores in the jungle, how many herbivores were there?

6. Use a ratio box to solve this problem. If 4 books weigh 9 pounds, how many pounds would 14 books weigh?

7. Write an equation to solve this problem. Two fifths of 60 is what number?

8. The diameter of a bicycle tire is 20 in. Find the distance around the tire to the nearest inch.

9. Draw a diagram of this statement. Then answer the questions that follow.

    Edmund received 150 votes. This was two thirds of the votes cast.

    (a) How many votes were cast?

    (b) How many votes were not for Edmund?

10. What is the volume of a block of ice with the dimensions shown?

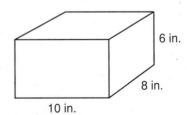

6 in.

8 in.

10 in.

11. Write each number in scientific notation.

    (a) 0.6×10^6 (b) 0.6×10^{-6}

12. What is the average of the three numbers marked by arrows on this number line?

13. Complete the table.

Fraction	Decimal	Percent
$\frac{3}{5}$	(a)	(b)

14. Write an equation to solve this problem. How much is 75 percent of $24?

15. Write the prime factorization of 540.

16. (a)   Find the area of the parallelogram.

    (b)   Find the area of the shaded triangle.

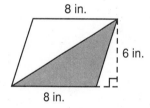

8 in.
6 in.
8 in.

17. Name each geometric solid.

    (a)   (b)   (c)

18. Find the next three numbers in this sequence:

    1, 4, 9, 16, 25, 36, ___, ___, ___

19. Compare: $\frac{2}{3}$ ◯ 0.667

20. Evaluate: $\dfrac{m}{n} - mn$ if $m = 3.6$ and $n = 0.9$

Solve:

21. $m - \dfrac{2}{3} = 1\dfrac{3}{4}$ **22.** $11t = 1760$

23. $\dfrac{w}{3} = \dfrac{\frac{2}{3}}{2}$

Add, subtract, multiply, or divide, as indicated:

24. $\dfrac{[30 - 4(5 - 2)] + 5(3^3 - 5^2)}{\sqrt{9} + \sqrt{16}}$

25. $\begin{array}{r} 2 \text{ gal } 1 \text{ qt} \\ - 1 \text{ gal } 1 \text{ qt } 1 \text{ pt} \\ \hline \end{array}$ **26.** $\dfrac{1}{2} \text{ mi} \cdot \dfrac{5280 \text{ ft}}{1 \text{ mi}} \cdot \dfrac{1 \text{ yd}}{3 \text{ ft}}$

27. $2\dfrac{2}{3} \div \left(4\dfrac{1}{2} \cdot 6\dfrac{2}{3} \right)$ **28.** $7\dfrac{1}{2} - \left(5\dfrac{1}{6} + 1\dfrac{1}{3} \right)$

29. (a) $5\dfrac{1}{4} + 1.9$ (decimal)

30. (a) $(-7) + (+5) + (-9)$ (b) $(16) + (-24) - (-18)$

**LESSON
94**

Fractional Part of a Number and Decimal Part of a Number, Part 2

In some fractional-part-of-a-number problems the fraction is unknown. In some fractional-part-of-a-number problems the total is unknown. As we discussed in Lesson 74, we can translate these problems to equations by replacing the word **of** with a multiplication sign and by replacing the word **is** with an equals sign.

Example 1   What fraction of 56 is 42?

Solution We translate this statement directly into an equation by replacing **what fraction** with W_F, replacing **of** with a multiplication symbol, and replacing **is** with an equals sign.

What fraction of 56 is 42?   question

$$W_F \quad \times \; 56 = 42 \qquad \text{equation}$$

To solve, we divide both sides by 56.

$$\frac{W_F \times 56}{56} = \frac{42}{56} \qquad \text{divided by 56}$$

$$W_F = \frac{3}{4} \qquad \text{simplified}$$

If the question had been, "What decimal part of 56 is 42?" the procedure would have been the same. As the last step we would have written $\frac{3}{4}$ as the decimal number 0.75.

$$W_D = 0.75$$

Example 2 Seventy-five is what decimal part of 20?

Solution We make a direct translation.

Seventy-five is what decimal part of 20?   question

$$75 \quad = \quad W_D \quad \times \; 20 \qquad \text{equation}$$

To solve, we divide both sides by 20.

$$\frac{75}{20} = \frac{W_D \times 20}{20} \qquad \text{divided by 20}$$

$$3.75 = W_D \qquad \text{simplified}$$

If the question had begun "What fractional part," we would have written the answer as a fraction or as a mixed number.

$$\frac{75}{20} = W_F \qquad \text{fraction}$$

$$\frac{15}{4} = W_F \qquad \text{reduced}$$

$$3\frac{3}{4} = W_F \qquad \text{mixed number}$$

Example 3   Three fourths of what number is 60?

Solution   In this problem the total is the unknown. We can still do a direct translation from the question to the equation.

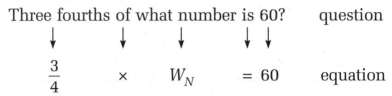

Three fourths of what number is 60?   question

$$\frac{3}{4} \quad \times \quad W_N \quad = 60 \qquad \text{equation}$$

To solve, we multiply both sides by $\frac{4}{3}$.

$$\frac{4}{3} \times \frac{3}{4} \times W_N = 60 \times \frac{4}{3} \qquad \text{multiplied by } \frac{4}{3}$$

$$W_N = 80 \qquad \text{simplified}$$

Had the question been phrased by using 0.75 instead of $\frac{3}{4}$, the procedure would have been the same.

Seventy-five hundredths of what number is 60?   question

$$0.75 \qquad \times \quad W_N \quad = 60 \qquad \text{equation}$$

To solve, we can divide both sides by 0.75.

$$\frac{0.75 \times W_N}{0.75} = \frac{60}{0.75} \qquad \text{divided by 0.75}$$

$$W_N = 80 \qquad \text{simplified}$$

Practice   **a.** What fraction of 130 is 80?

b. Seventy-five is what decimal part of 300?

c. Eighty is 0.4 of what number?

Problem set   **1.** During the first 3 days of the week, Mike read an average
94   of 28 pages per day. During the next 4 days, Mike averaged
               42 pages per day. For the whole week, Mike read an
               average of how many pages per day?

2. Twelve ounces of Brand X costs $1.14. Sixteen ounces of Brand Y costs $1.28. Brand X costs how much more per ounce than Brand Y?

3. Use a unit multiplier to convert $4\frac{1}{2}$ ft to inches.

4. The squirrel saved acorns and hazelnuts in its cache in the ratio of 7 to 3. If it had a total of 2100 nuts in its cache, how many hazelnuts had the squirrel saved? Use a ratio box to solve the problem.

5. Use a ratio box to solve this problem. If 5 pounds of apples cost $1.40, how much would 8 pounds of apples cost?

6. Draw a diagram of this statement. Then answer the questions that follow.

    Five sixths of the 300 triathletes completed the course.

    (a) How many triathletes completed the course?

    (b) What was the ratio of triathletes who completed the course to those who did not complete the course?

Write equations to solve Problems 7–10.

7. Fifteen is $\frac{3}{8}$ of what number?

8. Seventy is what decimal part of 200?

9. Two fifths of what number is 120?

10. What number is 60 percent of 180?

11. What is the volume of this cube?

3 in.

12. Find the circumference of each circle.

(a)
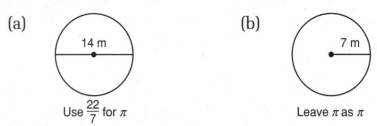
14 m

Use $\frac{22}{7}$ for π

(b)
7 m

Leave π as π

13. Complete the table.

FRACTION	DECIMAL	PERCENT
$3\frac{1}{2}$	(a)	(b)

14. What number is 20 percent of $35?

15. Find the missing number.

IN   **FUNCTION**   OUT

2 → → 24
0 → → 0
5 → → 60
3 → → ☐

16. Write four hundred twenty-five million in scientific notation.

17. Refer to the figure to answer these questions. $\overline{AE} \parallel \overline{BC}$ and $\overline{AB} \parallel \overline{EC}$.

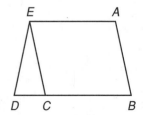

(a) What type of quadrilateral is figure *ABCE*?

(b) What type of quadrilateral is figure *ABDE*?

18. Arrange these numbers in order from least to greatest:

0.013, 0.1023, 0.0103, 0.021

19. What number is $\dfrac{7}{12}$ of 108?

20. Evaluate: $(m + n) - mn$ if $m = 1\dfrac{1}{2}$ and $n = 2\dfrac{2}{3}$

Solve:

21. $p + 3\dfrac{1}{5} = 7\dfrac{1}{2}$ **22.** $3n = 0.138$

23. $n - 0.36 = 4.8$

Add, subtract, multiply, or divide, as indicated:

24. $\sqrt{49} + \left\{5\left[3^2 - \left(2^3 - \sqrt{25}\right)\right] - 5^2\right\}$

25. $\begin{array}{r} 4\text{ hr }\quad 5\text{ min }\ 15\text{ sec} \\ -\,1\text{ hr }\ 15\text{ min }\ 30\text{ sec} \\ \hline \end{array}$ **26.** $\dfrac{24\text{ hr}}{1\text{ day}} \cdot \dfrac{60\text{ min}}{1\text{ hr}} \cdot \dfrac{60\text{ sec}}{1\text{ min}}$

27. $3\dfrac{1}{8} + \left(2\dfrac{1}{2}\right)\left(1\dfrac{3}{4}\right)$ **28.** $5\dfrac{5}{6} \div 1\dfrac{2}{3} \div 1\dfrac{3}{4}$

29. $8\dfrac{1}{3} + 7.5$ (fraction)

30. (a) $(-9) + (-11) - (+14)$ (b) $(26) + (-43) - (-36)$

LESSON 95

Multiplying and Dividing Signed Numbers

We can develop the rules for the multiplication and division of signed numbers if we remember that multiplication is a shorthand notation for repeated addition. We remember that 2 times 3 means $3 + 3$ and that 2 times -3 means $(-3) + (-3)$, so

$$2(3) = 6 \quad \text{and} \quad 2(-3) = -6$$

We remember that division undoes multiplication, so the following must be true.

If $2(3) = 6$ then $\dfrac{6}{2} = 3$ and $\dfrac{6}{3} = 2$

and if $2(-3) = -6$ then $\dfrac{-6}{2} = -3$ and $\dfrac{-6}{-3} = +2$

We use these examples to illustrate the fact that when we multiply or divide two positive numbers the answer is a positive number. Also, when we multiply or divide two numbers whose signs are different, the answer is a negative number. But what happens if we multiply two negative numbers? Since 2 times -3 equals -6

$$2(-3) = -6$$

then the *opposite of* 2 times -3 should equal the *opposite of* -6, which is 6.

$$(-2)(-3) = +6$$

And since division undoes multiplication, these division examples must also be true.

$$\dfrac{+6}{-2} = -3 \qquad \text{and} \qquad \dfrac{+6}{-3} = -2$$

This gives us the rules for the multiplication and division of signed numbers.

Rules for Multiplication and Division

1. If the two numbers that are multiplied or divided have the same sign, the answer is a positive number.

2. If the two numbers that are multiplied or divided have different signs, the answer is a negative number.

Here are some examples.

Multiplication	Division
$(+6)(+2) = +12$	$\dfrac{+6}{+2} = +3$

$$(-6)(-2) = +12 \qquad \frac{-6}{-2} = +3$$

$$(-6)(+2) = -12 \qquad \frac{-6}{+2} = -3$$

$$(+6)(-2) = -12 \qquad \frac{+6}{-2} = -3$$

Example Divide or multiply:

(a) $\dfrac{-12}{+4}$ (b) $\dfrac{-12}{-3}$ (c) $(6)(-3)$ (d) $(-6)(-4)$

Solution We divide or multiply as indicated. If both signs are the same, the answer is positive. If one sign is positive and the other is negative, the answer is negative.

(a) **−3**   (b) **+4**   (c) **−18**   (d) **+24**

Practice Divide or multiply:

a. $(-7)(3)$ **b.** $(+4)(-8)$ **c.** $(8)(+5)$

d. $(-6)(-4)$ **e.** $\dfrac{25}{-5}$ **f.** $\dfrac{-27}{-3}$

g. $\dfrac{-28}{4}$ **h.** $\dfrac{+30}{6}$

Problem set 95

1. Use a ratio box to solve this problem. If Mrs. C can wrap 12 packages in 5 minutes, how many packages can she wrap in 1 hour?

2. Lydia walked for 30 minutes a day for 5 days. The next 3 days she walked for an average of 46 minutes per day. What was the average amount of time she spent walking during those 8 days?

3. If the sum of 0.2 and 0.5 is divided by the product of 0.2 and 0.5, what is the quotient?

4. Use a unit multiplier to convert 23 cm to millimeters.

5. Use a ratio box to solve this problem. The ratio of paperback books to hardbound books in the school library was 3 to 11. If there were 9240 hardbound books in the library, how many books were there in all?

6. Write each number in scientific notation.

   (a) 24×10^{-5} (b) 24×10^{7}

7. Draw a diagram of this statement. Then answer the questions that follow.

       The 30 true-false questions amounted to $\frac{1}{4}$ of the test's questions.

   (a)   How many questions were on the test?

   (b)   How many of the questions were not true-false?

Write equations to solve Problems 8–11.

8. Forty-five is $\dfrac{5}{9}$ of what number?

9. Twenty-four is 0.4 of what number?

10. What number is 80 percent of 760?

11. What decimal part of 30 is 21?

12. Divide or multiply:

   (a) $\dfrac{-36}{9}$ (b) $\dfrac{-36}{-6}$

   (c) $9(-3)$ (d) $(+8)(+7)$

13. Find the number of 1-ft cubes that will fit inside a closet with dimensions as shown.

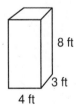

8 ft
3 ft
4 ft

14. Find the circumference of each circle.

(a)

21 m

Use 3.14 for π

(b)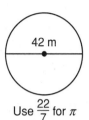

42 m

Use $\frac{22}{7}$ for π

15. Complete the table.

FRACTION	DECIMAL	PERCENT
(a)	2.5	(b)

Use angle measurements to classify each triangle. Then find the area of each triangle. Dimensions are in centimeters.

16.

8

10

6

17.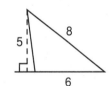

5

8

6

18.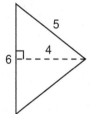

5

4

6

19. Name each three-dimensional figure.

(a)

(b)

(c)

20. Compare: $\frac{2}{3}$ of 96 $\bigcirc$ $\frac{5}{6}$ of 84

21. Evaluate: $ab - (a - b)$ if $a = \frac{5}{6}$ and $b = \frac{3}{4}$

Solve and check. Show your work.

22. $a + \frac{3}{4} = 1\frac{1}{8}$

23. $b - 1.6 = 0.16$

24. $20w = 5.6$

Add, subtract, multiply, or divide, as indicated:

25.   2 yd 1 ft 7 in.
        + 1 yd 2 ft 8 in.

26. 0.5 m $\cdot$ $\dfrac{100 \text{ cm}}{1 \text{ m}}$ $\cdot$ $\dfrac{10 \text{ mm}}{1 \text{ cm}}$

27. $12\dfrac{1}{2} \cdot 4\dfrac{1}{5} \cdot 2\dfrac{2}{3}$

28. $7\dfrac{1}{2} \div \left(6\dfrac{2}{3} \cdot 1\dfrac{1}{5}\right)$

29. $2.25 \times 1\dfrac{1}{3}$ (fraction)

30. (a) $(-8) + (-7) - (-15)$ (b) $(-15) + (+11) - (+24)$

LESSON 96

Area of a Complex Figure • Area of a Trapezoid

Area of a complex figure   We have practiced finding the areas of figures that can be divided into two or more rectangles. In this lesson we will begin finding the areas of figures that include triangular regions as well.

Example 1   Find the area of this figure. Corners that look square are square. Dimensions are in millimeters.

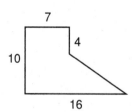

Solution   We draw a dotted line that divides the figure into a rectangle and a triangle.

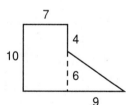

Area of rectangle = 7 × 10 =   70 mm²

Area of triangle   = $\dfrac{6 \times 9}{2}$ = + 27 mm²

―――――――――――――――――――――――――

Total area   =   **97 mm²**

Example 2   Find the area of this figure. Corners that look square are square. Dimensions are in centimeters.

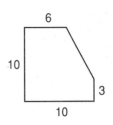

Solution   There are many ways to divide this figure.

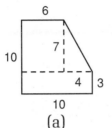
(a)

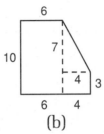
(b)

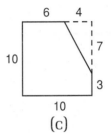
(c)

We decide to use (c). We will find the area of the big rectangle and subtract from it the area of the triangle.

$$\text{Area of rectangle} = 10 \times 10 = \quad 100 \text{ cm}^2$$

$$\text{Area of triangle} \quad = \quad \frac{4 \times 7}{2} = -\ 14 \text{ cm}^2$$

$$\text{Area of figure} \qquad\qquad = \quad \mathbf{86 \text{ cm}^2}$$

Area of a trapezoid   We remember that a quadrilateral with just one pair of parallel sides is a trapezoid. One way to find the area of a trapezoid is to divide the trapezoid into two triangular regions and find the combined area of the triangles.

Example 3   Find the area of this trapezoid. Dimensions are in centimeters.

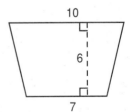

Solution   We can divide the trapezoid into two triangles by drawing either diagonal. We show both ways.

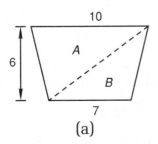
(a)

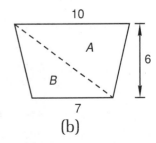
(b)

Both figures have an upper triangle (A) and a lower triangle (B). The height of all four triangles is 6 cm.

$$\text{Area of triangle } A = \frac{10 \times 6}{2} = \quad 30 \text{ cm}^2$$

$$\text{Area of triangle } B = \frac{7 \times 6}{2} = + 21 \text{ cm}^2$$

$$\text{Total area} \qquad\qquad = \quad \textbf{51 cm}^2$$

Practice Find the area of each figure. Dimensions are in centimeters. Corners that look square are square.

a. 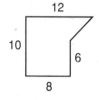   b. 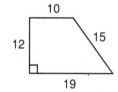   c.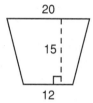

Problem set 96

1. Pablo ran an 8-lap race. For the first 5 laps he averaged 72 seconds per lap. For the rest of the race he averaged 80 seconds per lap. What was his average lap time for the whole race?

2. If 30 ounces of cereal cost $2.49, what is the cost per ounce?

3. One thousand five hundred meters is how many kilometers?

4. The sum of $\frac{1}{2}$ and $\frac{3}{5}$ is how much greater than the product of $\frac{1}{2}$ and $\frac{3}{5}$?

5. Use a ratio box to solve this problem. The ratio of Marci's age to Chelsea's age is 3 to 2. If Marci is 60 years old, she is how many years older than Chelsea?

6. Write each number in scientific notation.

   (a) 12.5×10^4 (b) 12.5×10^{-4}

7. Use a ratio box to solve this problem. Martha rode 40 miles in 3 hours. At this rate, how long would it take Martha to ride 100 miles?

8. Draw a diagram of this statement. Then answer the questions that follow.

   Two fifths of the library's 21,000 books were checked out during the school year.

   (a) How many books were checked out?

   (b) How many books were not checked out?

Write equations to solve Problems 9 – 12.

9. Sixty is $\dfrac{5}{12}$ of what number?

10. Seventy percent of what number is 35?

11. Thirty-five is what fraction of 80?

12. Fifty-six is what decimal part of 70?

13. Simplify:

   (a) $\dfrac{-120}{4}$

   (b) $(-12)(11)$

   (c) $\dfrac{-120}{-5}$

   (d) $12(+20)$

14. Find the volume of this rectangular prism. Dimensions are in centimeters.

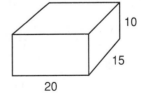

15. The diameter of the plate was 11 inches. Find its circumference to the nearest half inch.

16. Find the area of this trapezoid. Dimensions are in inches.

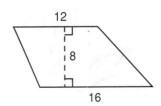

17. A corner was trimmed from a square sheet of paper to leave the paper in this shape. Dimensions are in centimeters.

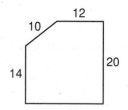

(a) Find the perimeter of the figure.

(b) Find the area of the figure.

18. Complete the table.

FRACTION	DECIMAL	PERCENT
(a)	(b)	125%

19. What is 20 percent of $12.50?

20. Evaluate: $x^3 - xy - \dfrac{x}{y}$ if $x = 2$ and $y = 0.5$

Solve and check. Show each step.

21. $\dfrac{5}{8}x = 40$ **22.** $1.2w = 26.4$

23. $y + 3.6 = 8.47$

Add, subtract, multiply, or divide, as indicated:

24. $9^2 - [3^3 - (9 \cdot 3 - \sqrt{9})]$

25.   2 hr 48 min 20 sec   **26.** $100 \text{ yd} \cdot \dfrac{3 \text{ ft}}{1 \text{ yd}} \cdot \dfrac{12 \text{ in.}}{1 \text{ ft}}$
        $-$ 1 hr 23 min 48 sec

27. $5\dfrac{1}{3} \cdot \left(3 \div 1\dfrac{1}{3}\right)$ **28.** $3\dfrac{1}{5} + 2\dfrac{1}{2} - 1\dfrac{1}{4}$

29. $1\dfrac{3}{5} + 0.47$ (decimal)

30. (a) $(-26) + (-15) - (-40)$

　　　(b) $(-5) + (-4) - (-3) - (+2)$

LESSON 97

Inverting the Divisor

Hidden ones
Every number has a coefficient of 1. Every number has a divisor of 1. Every number has an exponent of 1.

$$4 = 1 \cdot 4$$

$$4 = \frac{4}{1}$$

$$4 = 4^1$$

In this lesson we will be concerned with divisors of 1. Even fractions have a divisor of 1.

$$\frac{\frac{4}{3}}{1} = \frac{4}{3} \qquad \frac{\frac{9}{10}}{1} = \frac{9}{10}$$

Since Lesson 29 we have been dividing fractions by inverting the divisor and multiplying. Now we will show the reason for this rule.

We remember that we can multiply a number by a number that equals 1 without changing the value of the number.

$$\frac{2}{5} \times \frac{3}{3} = \frac{6}{15}$$

Because 3 over 3 equals 1, we have multiplied $\frac{2}{5}$ by 1. This lets us see why $\frac{2}{5}$ and $\frac{6}{15}$ have the same value. We can use this fact to simplify complex fractions.

Example 1 Simplify: $\dfrac{\frac{3}{5}}{\frac{2}{3}}$

Solution
The product of a number and its reciprocal is 1. The product of $\frac{2}{3}$ and $\frac{3}{2}$ is 1. If we multiply the denominator by $\frac{3}{2}$, we have to multiply the numerator by $\frac{3}{2}$.

$$\frac{\frac{3}{5}}{\frac{2}{3}} \times \frac{\frac{3}{2}}{\frac{3}{2}} = \frac{\frac{9}{10}}{1} = \frac{9}{10}$$

We could have inverted the divisor and multiplied as we show here.

$$\frac{\dfrac{3}{5}}{\dfrac{2}{3}} = \frac{3}{5} \cdot \frac{3}{2} = \frac{9}{10}$$

If instead we multiply above and below by $\frac{3}{2}$, we can simplify the expression without using a rote rule.

Example 2   Simplify: $\dfrac{25\frac{2}{3}}{100}$

Solution   **First we write both numerator and denominator as fractions.**

$$\frac{\dfrac{77}{3}}{\dfrac{100}{1}}$$

Now we multiply above and below by $\dfrac{1}{100}$.

$$\frac{\dfrac{77}{3}}{\dfrac{100}{1}} \cdot \frac{\dfrac{1}{100}}{\dfrac{1}{100}} = \frac{\dfrac{77}{300}}{1} = \mathbf{\frac{77}{300}}$$

Example 3   Simplify: $\dfrac{15}{7\frac{1}{3}}$

Solution   **We begin by writing both numerator and denominator as fractions.**

$$\frac{\dfrac{15}{1}}{\dfrac{22}{3}}$$

Now we multiply above and below by $\dfrac{3}{22}$.

$$\frac{\dfrac{15}{1}}{\dfrac{22}{3}} \cdot \frac{\dfrac{3}{22}}{\dfrac{3}{22}} = \frac{\dfrac{45}{22}}{1} = \mathbf{2\frac{1}{22}}$$

Example 4   Change $83\frac{1}{3}$ percent to a fraction.

Solution   A percent is a fraction that has a denominator of 100. Thus $83\frac{1}{3}\%$ is

$$\frac{83\frac{1}{3}}{100}$$

Next we write both numerator and denominator as fractions.

$$\frac{\frac{250}{3}}{\frac{100}{1}}$$

Now we multiply above and below by $\frac{1}{100}$.

$$\frac{\frac{250}{3} \cdot \frac{1}{100}}{\frac{100}{1} \cdot \frac{1}{100}} = \frac{\frac{250}{300}}{\frac{1}{1}} = \frac{5}{6}$$

Practice   Simplify:

a. $\dfrac{37\frac{1}{2}}{100}$

b. $\dfrac{12}{\frac{5}{6}}$

Change each percent to a fraction.

c. $66\frac{2}{3}\%$

d. $8\frac{1}{3}\%$

Problem set 97

1. Nestor finished a 42-kilometer bicycle race in 1 hour 45 minutes. What was his average speed in kilometers per hour?

2. Kim's scores in the diving competition were 7.9, 8.3, 8.1, 7.8, 8.4, 8.1, and 8.2. The highest and lowest scores were not counted. What was the average of the remaining scores?

3. Use a ratio box to solve this problem. The ratio of good guys to bad guys in the movie was 2 to 5. If there were 35 guys in the movie, how many of them were good?

4. Use unit multipliers to convert 2 kilometers to centimeters.

5. Change $16\frac{2}{3}$ percent to a fraction.

6. Use a ratio box to solve this problem. Thirty is to 80 as 24 is to what number?

7. One sixth of the rock's mass was quartz. If the weight of the rock was 144 grams, what was the weight of the quartz in the rock?

8. If $a = 8$, what does $\sqrt{2a}$ equal?

9. Simplify:

    (a) $\dfrac{-60}{-12}$ (b) $(-8)(6)$

    (c) $\dfrac{40}{-8}$ (d) $(-5)(-15)$

10. What is the circumference of this circle?

30 cm

Leave π as π

11. The figure shows a pyramid with a square base. Copy the figure and find the number of its (a) faces, (b) edges, and (c) vertices.

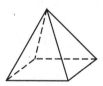

Write equations to solve Problems 12 – 16.

12. What is 10 percent of $37.50?

13. What number is $\dfrac{5}{8}$ of 72?

14. Twenty-five is what fraction of 60?

15. Sixty is what decimal part of 80?

16. Twenty percent of 30 is what number?

17. Complete the table.

Fraction	Decimal	Percent
$\frac{5}{6}$	(a)	(b)

18. A square sheet of paper with an area of 81 in.² has a corner cut off, forming a pentagon as shown.

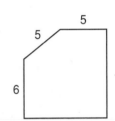

(a) What is the perimeter of the pentagon?

(b) What is the area of the pentagon?

19. What kind of angle is made by the hands of the clock at 3 o'clock?

20. Multiply two thousand by three thousand and write the product in scientific notation.

Solve each equation. Show each step.

21. $x - 25 = 96$

22. $\frac{2}{3}m = 12$

23. $2.5p = 6.25$

24. $10 = f + 3\frac{1}{3}$

Add, subtract, multiply, or divide, as indicated:

25. $\sqrt{13^2 - 5^2}$

26. 1 ton − 400 lb

27. $3\frac{3}{4} \times 4\frac{1}{6} \times 0.16$ (fraction)

28. $3\frac{1}{8} + 6.7 + 8\frac{1}{4}$ (decimal)

29. $(-3) + (-5) - (-3) - (+5)$

30. $(-73) + (-24) - (-50)$

LESSON 98

More on Percent

We remember that 40 percent means 40 hundredths.

$$40\% \quad \text{means} \quad \frac{40}{100}$$

If a problem asks what percent, we will write $\frac{W_P}{100}$. We will write our answer with a percent sign.

Example 1 What percent of 40 is 25?

Solution We can translate the question to an equation and solve.

$$\text{What percent of 40 is 25?} \quad \text{question}$$

$$\frac{W_P}{100} \quad \times\ 40\ =\ 25 \quad \text{equation}$$

To solve, we first divide both sides by 40.

$$\frac{\dfrac{W_P}{100} \cdot \cancel{40}}{\cancel{40}} = \frac{25}{40} \quad \text{divided by 40}$$

$$\frac{W_P}{100} = \frac{25}{40} \quad \text{simplified}$$

Then we multiply both sides by $\dfrac{100}{1}$.

$$\frac{\cancel{100}}{1} \cdot \frac{W_P}{\cancel{100}} = \frac{25}{40} \cdot \frac{100}{1} \quad \text{multiplied by } \frac{100}{1}$$

$$W_P = \frac{2500}{40} \quad \text{simplified}$$

$$W_P = 62.5\% \qquad \text{simplified}$$

Since W_P stands for "what percent," the answer to the question is **62.5%.**

Example 2   Fifteen percent of what number is 600?

Solution   We translate the question to an equation and solve.

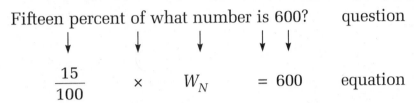

To solve, we multiply both sides by 100 over 15.

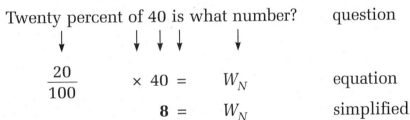

$$W_N = 4000 \qquad \text{simplified}$$

Example 3   Twenty percent of 40 is what number?

Solution   We write the equation and solve.

Twenty percent of 40 is what number?   question

$$\frac{20}{100} \qquad \times 40 = \qquad W_N \qquad \text{equation}$$
$$8 = \qquad W_N \qquad \text{simplified}$$

Practice   **a.** Twenty-four is what percent of 40?

b. What percent of 6 is 2?

c. Fifteen percent of what number is 45?

Problem set   **1.** Use a ratio box to solve this problem. Tammy saved
98   nickels and pennies in a jar. The ratio of nickels to
          pennies was 2 to 5. If there were 70 nickels in the jar,
          how many coins were there in all?

Refer to the line graph to answer questions 2–4.

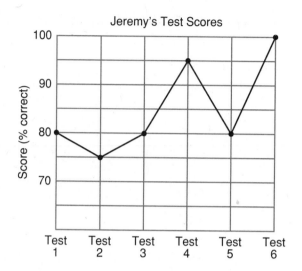

Jeremy's Test Scores

2. If there were 50 questions on Test 1, how many questions did Jeremy answer correctly?

3. What was Jeremy's average score?

4. (a) Which score did Jeremy make most often?

   (b) What was the difference between his highest score and his lowest score?

5. Name the shape of each object.

   (a) A marble

   (b) A length of pipe

   (c) A box of tissue

6. Use a ratio box to solve this problem. One hundred inches equal 254 centimeters. How many centimeters equal 250 inches?

7. Draw a diagram of this statement. Then answer the questions that follow.

   Three fifths of those present agreed, but 12 disagreed.

   (a) What fraction of those present disagreed?

(b)   How many were present?

(c)   How many of those present agreed?

Write equations to solve Problems 8 – 11.

8. Forty is $\frac{4}{25}$ of what number?

9. Twenty-four percent of 10,000 is what number?

10. Twelve percent of what number is 240?

11. Twenty is what percent of 25?

12. Simplify:

(a) $25(-5)$ (b) $-15(-5)$

(c) $\frac{-250}{-5}$ (d) $\frac{-225}{15}$

13. Complete the table.

Fraction	Decimal	Percent
(a)	0.2	(b)

14. What is 4.5 percent of $20?

15. Simplify: (a) $\dfrac{14\frac{2}{7}}{100}$ (b) $\dfrac{60}{\frac{2}{3}}$

16. Find the area of this figure. Dimensions are in feet. Corners that look square are square.

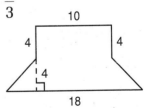

17. Sketch a picture of a cube with edges 2 cm long. What is the volume of the cube?

18. Write twelve billion in scientific notation.

19. Find the circumference of each circle.

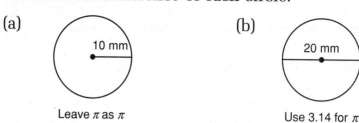

(a) 10 mm

Leave π as π

(b) 20 mm

Use 3.14 for π

Solve each equation. Show each step.

20. $3x = 26.7$

21. $y - 3\frac{1}{3} = 7$

22. $\frac{2}{3}x = 48$

23. Find the missing number.

IN	FUNCTION	OUT
3 →		→ ☐
0 →		→ −2
−1 →		→ −3
4 →		→ 2

Add, subtract, multiply, or divide, as indicated:

24. $5^2 - \{2^3 + 3[4^2 - (4)(\sqrt{9})]\}$

25. 4 gal 3 qt 1 pt
     + 1 gal 2 qt 1 pt

26. $1 \text{ ft}^2 \cdot \dfrac{12 \text{ in.}}{1 \text{ ft}} \cdot \dfrac{12 \text{ in.}}{1 \text{ ft}}$

27. $5\frac{1}{3} \div \left(1\frac{1}{3} \div 3\right)$

28. $3\frac{1}{5} - 2\frac{1}{2} + 1\frac{1}{4}$

29. $3\frac{1}{3} \div 2.5$ (fraction)

30. (a) $(-3) + (-4) - (+5)$ (b) $(-6) - (-16) - (+30)$

LESSON
99

Graphing Inequalities

The symbols ≥ and ≤

We have used the symbols >, <, and = to compare two numbers. In this lesson we will introduce the symbols ≥ and ≤. We will also practice graphing on the number line.

The symbols ≥ and ≤ combine the greater than/less than sign with the equals sign. We read the first end of the symbol and do not read the other end. Thus, reading from left to right, the symbol

$$\geq$$

is read, "greater than or equal to," and, reading from left to right, the symbol

$$\leq$$

is read, "less than or equal to."

Graphing on the number line

To graph a number on the number line, we draw a dot at the point that represents the number. Thus, when we graph 4 on the number line, it looks like this:

This time we will graph **all the numbers that are greater than or equal to 4**. We might think the graph should look like this:

It is true that all the dots mark points that represent numbers that are greater than or equal to 4. However, we did not graph **all** the numbers that are greater than 4. For instance, we did not graph 10, 11, 12, and so on. Also, we did not graph $4\frac{1}{2}$, $5\frac{1}{3}$, $6\frac{3}{4}$, and so on. If we graph all of these numbers, the dots are so close together that we end up with a solid line that goes on and on. Thus a graph of all the numbers greater than or equal to 4 looks like this:

The large dot marks the 4. The shaded line marks the numbers greater than 4. The arrowhead shows that this shaded line goes on without end.

Graphing inequalities

Expressions such as the following are called **inequalities**.

$$\text{(a)} \quad x \le 4 \qquad \text{(b)} \quad x > 4$$

We read (a) as "x is less than or equal to 4." We read (b) as "x is greater than 4."

We can graph inequalities on the number line by graphing all the numbers that make the inequality a true inequality.

Example 1   Graph on a number line: $x \le 4$

Solution   We are told to graph all numbers that are less than or equal to 4. We draw a dot at the point that represents 4, and then we shade all the points to the left of the dot. The arrowhead shows that the shading continues without end.

Example 2   Graph on a number line: $x > 4$

Solution   We are told to graph all numbers greater than 4 **but not including 4**. We do not start the graph at 5 because we also need to graph numbers like $4\frac{1}{2}$ and 4.001. To show that the graph does not include 4, **we draw an empty circle at 4** and then shade the number line to the right of the circle.

Practice   **a.**   On a number line, graph all the numbers less than 2.

b.   On a number line, graph all the numbers greater than or equal to 1.

Graph each inequality on a number line.

c. $x \le -1$ **d.** $x > -1$

**Problem set
99**

1. Use a ratio box to solve this problem. If 4 cartons are needed to feed 30 hungry children, how many cartons are needed to feed 75 hungry children?

2. Gabriel's average score after 4 tests was 88. What score must Gabriel average on the next 2 tests to have a 6-test average of 90?

3. If the sum of $\frac{2}{3}$ and $\frac{3}{4}$ is divided by the product of $\frac{2}{3}$ and $\frac{3}{4}$, what is the quotient?

4. Use a ratio box to solve this problem. The ratio of monocotyledons to dicotyledons in the nursery was 3 to 4. If there were 84 dicotyledons in the nursery, how many monocotyledons were there?

5. The diameter of a nickel is 21 mm. Find the circumference of a nickel to the nearest millimeter.

6. Graph each inequality on a separate number line.

    (a) $x > 2$ (b) $x \leq 1$

7. Use a unit multiplier to convert 1.5 kg to grams.

8. Five sixths of 30 people who participated in the taste test preferred the taste of Brand X. The rest preferred Brand Y. How many more people preferred Brand X than preferred Brand Y?

Write equations to solve Problems 9 – 12.

9. Forty-two is seven tenths of what number?

10. Sixty percent of what number is 600?

11. Forty percent of 50 is what number?

12. Forty is what percent of 50?

13. Write 1.5×10^{-3} in standard form.

14. Simplify:

(a) $\dfrac{-45}{9}$

(b) $\dfrac{-450}{15}$

(c) $15(-20)$

(d) $-15(-12)$

15. Complete the table.

Fraction	Decimal	Percent
(a)	(b)	50%

16. Simplify: $\dfrac{83\frac{1}{3}}{100}$

17. Find the area of this trapezoid. Dimensions are in millimeters.

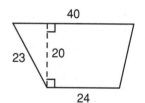

18. A box of tissues is 24 cm long, 12 cm wide, and 10 cm high. Sketch a picture of the box and find its volume.

19. Find the missing number.

In	FUNCTION	Out
2 →		→ 9
3 →		→ 13
4 →		→ □
5 →		→ 21

20. How much is 30 percent of $18.50?

Solve each equation. Show each step.

21. $m + 8.7 = 10.25$

22. $\dfrac{4}{3}w = 36$

23. $0.7y = 48.3$

Add, subtract, multiply, or divide, as indicated:

24. $\{4^2 + 10[2^3 - (3)(\sqrt{4})]\} - \sqrt{36}$

25. 1 yd − 1 ft 3 in.

26. $1 \text{ m}^2 \cdot \dfrac{100 \text{ cm}}{1 \text{ m}} \cdot \dfrac{100 \text{ cm}}{1 \text{ m}}$

27. $7\dfrac{1}{2} \cdot 3 \cdot \dfrac{5}{9}$ **28.** $3\dfrac{1}{5} - \left(2\dfrac{1}{2} - 1\dfrac{1}{4}\right)$

29. $7.2 - 1\dfrac{3}{5}$ (decimal)

30. (a) $(-10) - (-8) - (+6)$

      (b) $(+10) + (-20) - (-30)$

LESSON
100

Insufficient Information · Quantitative Comparisons

Insufficient information   Sometimes we encounter problems for which there is insufficient (not enough) information to determine the answer. The following problem provides insufficient information to answer the question.

> A 10-pound bag of potatoes costs $1.49. What is the average price of each potato?

Since we do not know the number of potatoes in the bag, we do not have enough information to find the average price of each potato.

We will practice recognizing problems with insufficient information as we answer quantitative comparison problems like the examples in this lesson.

Quantitative comparisons   We have practiced comparing numbers using the symbols $>$, $<$, and $=$. In this lesson we will begin considering comparison problems in which insufficient information has been provided to determine the comparison.

Example 1   The numbers x and y are whole numbers. Compare:

$$x \bigcirc y$$

Solution   We are told that x and y are whole numbers, but we are not given information that will let us determine which is greater, or if x and y are equal. Since we do not have enough information to determine the comparison, as our answer we write **insufficient information**.

Example 2   The number x is positive and y is negative. Compare:

$$x \bigcirc y$$

Solution   We are not given enough information to determine what each number is. However, we are given enough information to determine the comparison. Any positive number is greater than any negative number. Thus, the answer is

$$x > y$$

Example 3 $a - b = 0$ Compare:

$$a \bigcirc b$$

Solution   The equation does not provide enough information to determine the value of either number. However, since their difference is zero, the two numbers must be equal.

$$a = b$$

Practice   Answer each comparison by writing $>$, $<$, $=$, or insufficient information.

a. $x - y = 1$ Compare: $x \bigcirc y$

b. $\dfrac{m}{n} = 1$ Compare: $m \bigcirc n$

c. $a \cdot b = 1$ Compare: $a \bigcirc b$

d. x is not positive. y is not negative.

Compare: $x \bigcirc y$

Problem set 100

1. The average number of students in 4 classrooms was 33.5. If the students were regrouped into 5 classrooms, what would be the average number of students in each room?

2. Nelda drove 315 kilometers and used 35 liters of gasoline. Her car averaged how many kilometers per liter of gas?

3. Use a ratio box to solve this problem. The ratio of winners to losers was 7 to 5. If the total number of winners and losers was 1260, how many more winners were there than losers?

4. Write each number in scientific notation.
   (a) 37.5×10^{-6} (b) 37.5×10^{6}

5. Compare: $x \bigcirc y$ if $\dfrac{y}{x} = 2$

6. Graph each inequality on a separate number line.
   (a) $x < 1$ (b) $x \geq -1$

7. Use a ratio box to solve this problem. Four inches of snow fell in 3 hours. At that rate, how long would it take for 1 foot of snow to fall?

8. Draw a diagram of this statement. Then answer the questions that follow.

   Twelve students earned A's. This was $\dfrac{3}{8}$ of the students in the class.

   (a)   How many students did not earn A's?

   (b)   What percent of the students did not earn A's?

Write equations to solve Problems 9 – 12.

9. Thirty-five is $\dfrac{7}{11}$ of what number?

10. What percent of 20 is 17?

11. What number is 5 percent of 360?

12. Three hundred sixty is 75 percent of what number?

13. Simplify:

(a) $\dfrac{144}{-8}$

(b) $\dfrac{-144}{+6}$

(c) $-12(12)$

(d) $-16(-9)$

14. Complete the table.

FRACTION	DECIMAL	PERCENT
$\dfrac{1}{25}$	(a)	(b)

15. What is 60 percent of $8.40?

16. Simplify: $\dfrac{62\frac{1}{2}}{100}$

17. A square sheet of paper with an area of 100 in.^{2} has a corner cut off as shown in the figure. Dimensions are in inches.

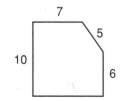

(a) What is the perimeter of the shape?

(b) What is the area of the shape?

18. In the figure, each small cube is 1 cubic centimeter. What is the volume of this rectangular prism?

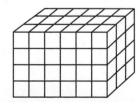

19. Find the circumference of each circle.

(a)

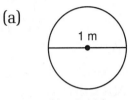

Use 3.14 for π

(b)

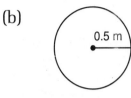

Leave π as π

20. Identify each angle as acute, right, or obtuse.

(a)

(b)

(c)

Solve each equation. Show every step.

21. $1.2x = 2.88$

22. $3\dfrac{1}{3} = x + \dfrac{5}{6}$

23. $\dfrac{3}{2}w = \dfrac{9}{10}$

Add, subtract, multiply, or divide, as indicated:

24. $\dfrac{\sqrt{100} + 5[3^3 - 2(3^2 + 3)]}{5}$

25. $\begin{array}{r} 3 \text{ hr } 15 \text{ min } 24 \text{ sec} \\ - 2 \text{ hr } 45 \text{ min } 30 \text{ sec} \end{array}$

26. $1 \text{ yd}^2 \cdot \dfrac{3 \text{ ft}}{1 \text{ yd}} \cdot \dfrac{3 \text{ ft}}{1 \text{ yd}}$

27. $7\dfrac{1}{2} \cdot \left(3 \div \dfrac{5}{9}\right)$

28. $4\dfrac{5}{6} + 3\dfrac{1}{3} + 7\dfrac{1}{4}$

29. $3\dfrac{3}{4} \div 1.5$ (decimal)

30. $(-10) - (+20) - (-30)$

LESSON 101

Measuring Angles with a Protractor

In Lesson 18 we discussed angles and classified angles as acute, right, or obtuse. In this lesson we will begin measuring angles.

Angles are commonly measured in units called **degrees**. The abbreviation for degrees is a small circle written above and to the right of the number. One full rotation, a full circle, measures 360 degrees.

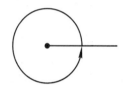 A full circle measures 360°.

A half circle measures half of 360°, which is 180°.

 A half circle measures 180°.

One fourth of a full rotation is a right angle. A right angle measures one fourth of 360°, which is 90°.

 A right angle measures 90°.

Thus, the measure of an acute angle is less than 90°, and the measure of an obtuse angle is greater than 90° but less than 180°. An angle that measures 180° is a straight angle. The chart below summarizes the types of angles and their measures.

TYPE OF ANGLE	MEASURE
Acute angle	Greater than 0° but less than 90°
Right angle	Exactly 90°
Obtuse angle	Greater than 90° but less than 180°
Straight angle	Exactly 180°

A **protractor** may be used to measure angles. The protractor is placed on the angle to be measured so that the vertex is under the dot or circle or cross-mark of the protractor, and one side of the angle is under the line at either end of the scale of the protractor.

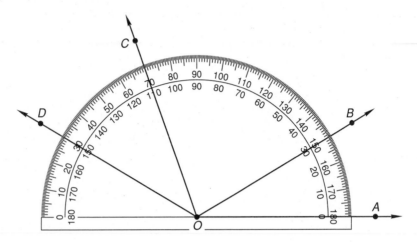

The measures of the three angles shown are as follows.

$\angle AOB = 30°$ $\angle AOC = 110°$ $\angle AOD = 150°$

Notice that there are two scales on a protractor, one starting from the left side, the other from the right. One way to check whether you are reading from the correct scale is to consider whether you are measuring an acute angle or an obtuse angle.

Example Find the measure of each angle.

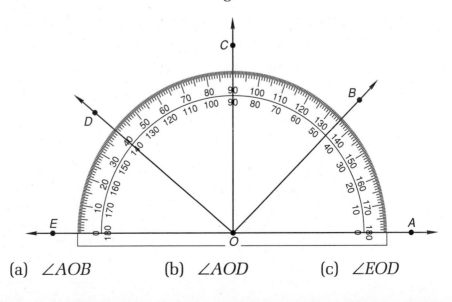

(a) $\angle AOB$ (b) $\angle AOD$ (c) $\angle EOD$

Solution   (a) ∠*AOB* is an acute angle. We use the scale with the numbers less than 90. Ray *OB* passes through the mark halfway between 40 and 50. Thus the measure of ∠*AOB* is **45°**.

(b) ∠*AOD* is an obtuse angle. We read from the scale with numbers greater than 90. The measure of ∠*AOD* is **140°**.

(c) ∠*EOD* is an acute angle. The measure of ∠*EOD* is **40°**.

Practice   Find the measure of each angle.

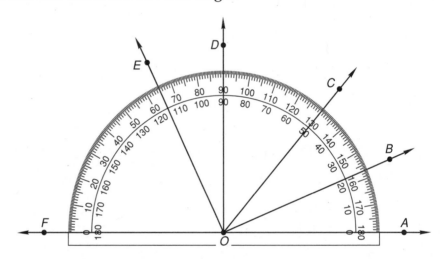

a. ∠*AOD*   **b.** ∠*AOC*   **c.** ∠*AOE*

d. ∠*FOE*   **e.** ∠*FOC*   **f.** ∠*AOB*

Problem set 101

1. Tina mowed lawns for 4 hours and earned $5.25 per hour. Then she washed windows for 3 hours and earned $3.50 per hour. What was Tina's average hourly pay for the 7-hour period?

2. Evaluate: $x + (x^2 - xy) - y$ if $x = 4$ and $y = 3$

3. Compare: $a \bigcirc b$ if $ab = 2$

4. Use a ratio box to solve this problem. When Nelson cleaned his room, he found that the ratio of clean clothes to dirty clothes was 2 to 3. If 30 articles of clothing were discovered, how many were clean?

5. The diameter of a half dollar is 3 cm. Find the circumference of a half dollar to the nearest millimeter.

6. Use a unit multiplier to convert $1\frac{1}{2}$ quarts to pints.

7. Graph each inequality on a separate number line.

    (a) $x > -2$ (b) $x \leq 0$

8. Use a ratio box to solve this problem. In 25 minutes, 400 customers entered the attraction. At this rate, how many customers would enter the attraction in 1 hour?

9. Draw a diagram of this statement. Then answer the questions that follow.

    Nathan found that it was 18 inches from his knee joint to his hip joint. This was $\frac{1}{4}$ of his total height.

    (a) What was Nathan's total height in inches?

    (b) What was Nathan's total height in feet?

Write equations to solve Problems 10 – 13.

10. Six hundred is $\frac{5}{9}$ of what number?

11. Two hundred eighty is what percent of 400?

12. What number is 4 percent of 400?

13. Sixty is 60 percent of what number?

14. Simplify:

    (a) $\dfrac{600}{-15}$ (b) $\dfrac{-600}{-12}$

    (c) $20(-30)$ (d) $+15(40)$

15. What is 80 percent of $12.50?

16. Complete the table.

FRACTION	DECIMAL	PERCENT
(a)	0.3	(b)

17. Express in scientific notation.

(a) 30×10^6 (b) 30×10^{-6}

18. Find the area of this trapezoid. Dimensions are in meters.

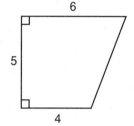

19. If each edge of a cube is 5 inches, what is the volume of the cube?

20. Find the measure of each angle.

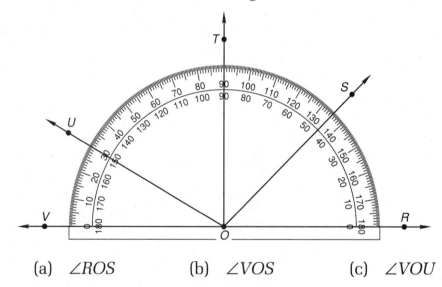

(a) $\angle ROS$ (b) $\angle VOS$ (c) $\angle VOU$

Solve each equation. Show each step.

21. $17a = 408$ **22.** $\dfrac{3}{8}m = 48$

23. $1.4 = x - 0.41$

Add, subtract, multiply, or divide, as indicated:

24. $\dfrac{2^3 + 4 \cdot 5 - 2 \cdot 3^2}{\sqrt{25} \cdot \sqrt{4}}$

25. 10 lb   6 oz
      − 7 lb 11 oz

26. $1 \text{ cm}^2 \cdot \dfrac{10 \text{ mm}}{1 \text{ cm}} \cdot \dfrac{10 \text{ mm}}{1 \text{ cm}}$

27. $7\dfrac{1}{2} \div \left(3 \cdot \dfrac{5}{9}\right)$

28. $4\dfrac{5}{6} + 3\dfrac{1}{3} - 7\dfrac{1}{4}$

29. $7\dfrac{1}{7} \times 1.4$ (fraction)

30. $(-3) - (+7) + (-10)$

LESSON
102

Using Proportions to Solve Percent Problems

A percent is a ratio. Thus, percent problems can be solved using the same method we use to solve ratio problems. Consider the following problem and explanation.

> Thirty percent of the class passed the test. If 21 students did not pass the test, how many students were in the class?

The problem is about 2 parts of a whole class. We recognize this as a part-part-whole problem. One part of the class passed the test; the other part of the class did not pass the test. The whole class is 100 percent. The part that passed was 30 percent. Thus, the part that did not pass must have been 70 percent. We will record these numbers in a ratio box just as we do with ratio problems.

	PERCENT	ACTUAL COUNT
Passed	30	
Did not pass	70	
Whole class	100	

As we read the problem, we find an actual count as well. There were 21 students who did not pass the test. We record

21 in the appropriate place of the table and use letters in the remaining places of the table.

	PERCENT	ACTUAL COUNT
Passed	30	P
Did not pass	70	21
Whole class	100	W

We will use this table to help us write a proportion so that we can solve the problem. **We will use the numbers in two of the three rows to write a proportion. This time we will use the numbers in the second row because we know both numbers. Since the problem asks for the total number of students in the class, we will also use the third row.** Then we solve the proportion.

	PERCENT	ACTUAL COUNT
Passed	30	P
Did not pass	70	21
Whole class	100	W

$$\frac{70}{100} = \frac{21}{W}$$

$$70W = 2100$$

$$W = 30$$

By solving the proportion, we find that there were 30 students in the whole class.

Example 1   Forty percent of the leprechauns had never seen the pot of gold. If 480 leprechauns had seen the pot of gold, how many of the leprechauns had not seen it?

Solution   We may solve this problem just as we solve a ratio problem. We use the percents to fill the ratio column of the table. All the leprechauns was 100 percent. The part that had never seen the pot of gold was 40 percent. Therefore, the part that had seen the pot of gold was 60 percent. The number 480 was the actual count of the leprechauns who had seen the gold. We write these numbers in the table.

	PERCENT	ACTUAL COUNT
Had not seen	40	N
Had seen	60	480
Total	100	T

Now we use the table to write a proportion. Since we know both numbers in the second row, we will also use that row in the proportion. Since the problem asks us to find the actual count of leprechauns who had not seen the pot of gold, we will also use the first row in the proportion.

	PERCENT	ACTUAL COUNT
Had not seen	40	N
Had seen	60	480
Total	100	T

$$\frac{40}{60} = \frac{N}{480}$$

$$60N = 19{,}200$$

$$N = 320$$

We find that **320 leprechauns** had not seen the pot of gold.

Example 2 Twenty-seven of the 45 elves who worked in the toy factory had to work the night shift. What percent of the elves had to work the night shift?

Solution We make a ratio box and write in the numbers. The total number of elves was 45, so 18 elves worked the day shift.

	PERCENT	ACTUAL COUNT
Night shift	P_N	27
Day shift	P_D	18
Total	100	45

We use P_N to stand for the percent who worked the night shift. We use this row and the total row to write the proportion.

	PERCENT	ACTUAL COUNT
Night shift	P_N	27
Day shift	P_D	18
Total	100	45

$$\frac{P_N}{100} = \frac{27}{45}$$

$$45P_N = 2700$$

$$P_N = 60$$

We find that **60 percent** of the elves worked the night shift.

Practice Use a ratio box to solve each problem.

a. Twenty-one of the 70 acres were planted in alfalfa. What percent of the acres was not planted in alfalfa?

b. Lori figures she still has 60 percent of the book to read. If she has read 120 pages, how many pages does she still have to read?

Problem set 102   Use the information given to answer questions 1–3.

On his first 15 tests, Paul earned these scores: 70, 85, 80, 85, 90, 80, 85, 80, 90, 95, 85, 90, 100, 85, 90.

1. What was Paul's average test score?

2. If Paul's scores were arranged in order from lowest to highest, what would be the middle score? This number is called the **median**.

3. (a) Which score did Paul earn most often? This score is called the **mode**.

(b) What was the difference between Paul's highest score and his lowest score? This number is called the **range**.

4. Danny is 6' 1" (6 ft 1 in.) tall. His sister is 5' $6\frac{1}{2}$" tall. Danny is how many inches taller than his sister?

5. Use a ratio box to solve this problem. Michelle bought 5 pencils for 75¢. At this rate, how much would she pay for a dozen pencils?

6. Graph each inequality on a separate number line.

(a) $x < 4$ (b) $x \geq -2$

7. Draw a diagram of this statement. Then answer the questions that follow.

Gilbert answered 48 questions correctly. This was $\frac{4}{5}$ of the questions on the test.

(a) How many questions were on the test?

(b) What was the ratio of Gilbert's correct answers to his incorrect answers?

8. If point B is located halfway between points A and C, what is the length of segment AB?

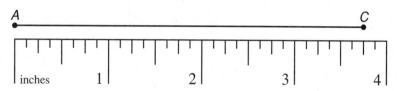

9. Use a ratio box to solve this problem. The ratio of gleeps to blobbles was 9 to 5. If the total number of gleeps and blobbles was 2800, then how many gleeps were there?

10. If $x = 9$, what does $x^2 + \sqrt{x}$ equal?

11. Compare: $m \bigcirc n$ if $\dfrac{m}{n} = 0.5$

12. Complete the table.

Fraction	Decimal	Percent
$2\frac{1}{4}$	(a)	(b)

Write equations to solve Problems 13 and 14.

13. What percent of 40 is 12?

14. Fifty percent of what number is 0.4?

15. Simplify: $\dfrac{16\frac{2}{3}}{100}$

Use a ratio box to solve Problems 16 and 17.

16. Nathan correctly answered 21 of the 25 questions. What percent of the questions did he answer correctly?

17. Twenty percent of the 4000 acres were left fallow. How many acres were not left fallow?

18. Find the measure of each angle.

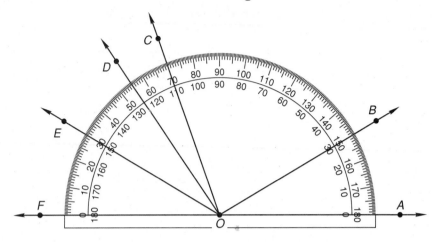

(a) ∠*AOB*   (b) ∠*FOB*   (c) ∠*AOD*   (d) ∠*FOC*

19. Which angle shown in Problem 18 measures 55°?

20. What is the area of this trapezoid? Dimensions are in inches.

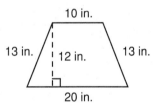

21. Find the missing number.

$$
\begin{array}{ccc}
\text{In} & \boxed{\begin{array}{c}\text{F}\\\text{U}\\\text{N}\\\text{C}\\\text{T}\\\text{I}\\\text{O}\\\text{N}\end{array}} & \text{Out}\\
4 \rightarrow & & \rightarrow \square \\
-1 \rightarrow & & \rightarrow -5 \\
6 \rightarrow & & \rightarrow 2 \\
2 \rightarrow & & \rightarrow -2
\end{array}
$$

22. Write each number in scientific notation.

   (a) 56×10^7 (b) 56×10^{-7}

Solve each equation. Show each step.

23. $5x = 16.5$

24. $3\frac{1}{2} + a = 5\frac{3}{8}$

Add, subtract, multiply, or divide, as indicated:

25. $3^2 + 5[6 - (10 - 2^3)]$

26.   1 day   8 hr 15 min   **27.** $2\frac{2}{3} \times 4.5 \div 6$ (fraction)
      + 2 day 15 hr 45 min

28. $8\frac{3}{4} - (5 - 3.4)$ (decimal)

29. (a) $(-12)(-9)$ (b) $\dfrac{-100}{5}$

30. (a) $(-3) + (-4) - (-5)$ (b) $(-18) - (+20) + (-7)$

LESSON
103

Area of a Circle

We can find the areas of some polygons by multiplying the two perpendicular dimensions.

- We find the area of a rectangle by multiplying the length times the width.

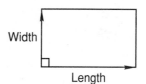

- We find the area of a parallelogram by multiplying the base times the height.

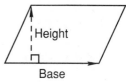

- We find the area of a triangle by multiplying the base times the height, then dividing by 2.

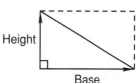

To find the area of a circle, we again begin by multiplying two perpendicular dimensions. We multiply the radius times the radius. This gives us the area of a square built on the radius.

If the radius of the circle is 3, the area of the square is 3^2, which is 9. If the radius of the circle is r, the area of the square is r^2. We see that the area of the circle is less than the area of four of these squares.

However, the area of three of these squares does not quite equal the area of the circle.

The number of squares whose area exactly equals the area of the circle is a number between 3 and 4. The exact number is π. Thus, to find the area of a circle, we first find the area of a square built on the radius. Then we multiply that area by π. This is summarized by the equation

$$\text{Area of circle} = \pi r^2$$

Example Find the area of each circle.

(a)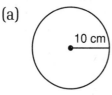

Use 3.14 for π

(b)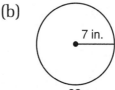

Use $\frac{22}{7}$ for π

(c)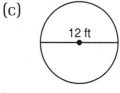

Leave π as π

Solution (a) The area of a square built on the radius is 100 cm². We multiply this by π.

$$A = \pi r^2$$

$$A \approx (3.14)(100 \text{ cm}^2)$$

$$A \approx \textbf{314 cm}^2$$

(b) The area of a square built on the radius is 49 in.² We multiply this by π.

$$A = \pi r^2$$

$$A \approx \frac{22}{\cancel{7}} \cdot \cancel{49}^{\,7} \text{ in.}^2$$

$$A \approx \textbf{154 in.}^2$$

(c) Since the diameter is 12 ft, the radius is 6 ft. The area of a square built on the radius is 36 ft². We multiply this by π.

$$A = \pi r^2$$

$$A = \pi \cdot 36 \, \text{ft}^2$$

$$A = 36\pi \, \textbf{ft}^2$$

Notice the answer is written as 36 times π followed by the unit.

Practice Find the area of each circle.

a.

Use 3.14 for π

b.

Leave π as π

c.

Use $\frac{22}{7}$ for π

Problem set 103

1. Find the volume of this rectangular prism. Dimensions sions are in feet.

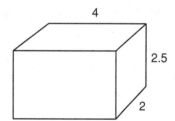

2. The heights of the 5 basketball starters were 6' 3", 6' 5", 5' 11", 6' 2", and 6' 1". Find the average height of the 5 starters. (*Hint*: Change all measures to inches before dividing.)

3. Use a ratio box to solve this problem. The student/teacher ratio at the high school was 20 to 1. If there were 48 high school teachers, then how many students were there?

4. Use a unit multiplier to convert 66 inches to feet.

5. Graph each inequality on a separate number line.

    (a) $x < -2$

    (b) $x \geq 0$

6. Use a ratio box to solve this problem. Don's heart beats 225 times in 3 minutes. At that rate, how many times will his heart beat in 5 minutes?

7. Draw a diagram of this statement. Then answer the questions that follow.

    > Two fifths of the students in the class were boys. There were 15 girls in the class.

    (a) How many students were in the class?

    (b) What was the ratio of girls to boys in the class?

8. Evaluate: $y + xy - x$ if $x = 3$ and $y = 4$

9. What percent of this circle is shaded?

10. Compare: $a \bigcirc b$ if $a - b$ is negative

11. Find the circumference of each circle.

     (a)

     7 cm

     Use 3.14 for π

     (b)

     14 cm

     Use $\frac{22}{7}$ for π

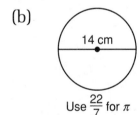

12. Find the area of each circle in Problem 11.

13. Complete the table.

Fraction	Decimal	Percent
(a)	1.6	(b)

14. Write an equation to solve this problem. What is 8 percent of $25?

15. Express in scientific notation.

     (a) 12×10^5 (b) 12×10^{-5}

16. Use a ratio box to solve this problem. Sixty-four percent of the 175 students correctly described the process of photosynthesis. How many students correctly described this process?

17. Use a ratio box to solve this problem. Ginger still has 40 percent of her book to read. If she has read 180 pages, how many pages does she still have to read?

18. Find the area of this figure. Dimensions are in inches. Corners that look square are square.

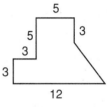

19.

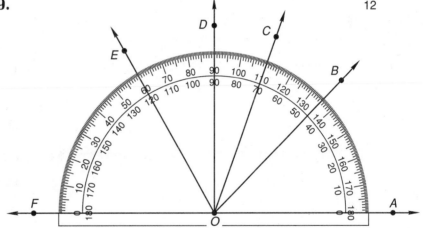

(a)   Find the measure of ∠AOE.

(b)   Find the measure of ∠FOC.

(c)   Find two angles that each measure one half of a right angle.

20. Write the prime factorization of 816.

21. Find the next two numbers in this sequence.

27, 9, 3, ___, ___

22. Write one hundred million in scientific notation.

Solve each equation. Show each step.

23. $\frac{3}{4}x = 36$

24. $3.2 + a = 3.46$

Add, subtract, multiply, or divide, as indicated:

25. $\dfrac{\sqrt{3^2 + 4^2}}{5}$

26. $\begin{array}{r} 4 \text{ gal } 2 \text{ qt} \\ - \ 1 \text{ gal } 3 \text{ qt } 1 \text{ pt} \\ \hline \end{array}$

27. $3\dfrac{1}{2} \div (7 \div 0.2)$ (decimal)

28. $4.5 + 2\dfrac{2}{3} - 3$ (fraction)

29. (a) $\dfrac{(-3)(-4)}{(-2)}$ (b) $(-2)(+3)(-4)$

30. (a) $(-3) + (-4) - (-2)$ (b) $(-20) + (+30) - (-40)$

LESSON 104   Multiplying Powers of 10 • Multiplying Numbers in Scientific Notation

Multiplying powers of 10   Here we show two powers of 10.

$$10^3 \qquad 10^4$$

We remember that

$$10^3 \text{ means } 10 \cdot 10 \cdot 10$$

and

$$10^4 \text{ means } 10 \cdot 10 \cdot 10 \cdot 10$$

We can multiply powers of 10.

$$10^3 \qquad \cdot \qquad 10^4$$

$$\underbrace{10 \cdot 10 \cdot 10} \cdot \underbrace{10 \cdot 10 \cdot 10 \cdot 10}$$

We see that 10^3 times 10^4 means that 7 tens are multiplied. We can write this as 10^7.

$$10^3 \cdot 10^4 = 10^7$$

As we focus our attention on the exponents, we see that

$$3 + 4 = 7$$

This example illustrates an important rule of mathematics.

> **When we multiply powers of 10, we add the exponents.**

Multiplying numbers in scientific notation   To multiply numbers that are written in scientific notation, we multiply the decimal numbers to find the decimal number part of the product. Then we multiply the powers of 10 to find the power-of-10 part of the product. We remember that when we multiply powers of 10, we add the exponents.

Example 1   Multiply: $(1.2 \times 10^5)(3 \times 10^7)$

Solution   We multiply 1.2 by 3 and get 3.6. Then we multiply 10^5 by 10^7 and get 10^{12}. The product is

$$\mathbf{3.6 \times 10^{12}}$$

Example 2   Multiply: $(4 \times 10^6)(3 \times 10^5)$

Solution   We multiply 4 by 3 and get 12. Then we multiply 10^6 by 10^5 and get 10^{11}. The product is

$$12 \times 10^{11}$$

We simplify this expression by writing

$$(1.2 \times 10^1) \times 10^{11} = \mathbf{1.2 \times 10^{12}}$$

Example 3   Multiply: $(2 \times 10^{-5})(3 \times 10^{-7})$

Solution   We multiply 2 by 3 and get 6. To multiply 10^{-5} by 10^{-7} we add the exponents and get 10^{-12}. Thus, the product is

$$\mathbf{6 \times 10^{-12}}$$

Example 4   Multiply: $(5 \times 10^3)(7 \times 10^{-8})$

Solution   We multiply 5 by 7 and get 35. We multiply 10^3 by 10^{-8} and get 10^{-5}. The product is

$$35 \times 10^{-5} \qquad\qquad \text{product}$$

$$(3.5 \times 10^1) \times 10^{-5}$$

$$\mathbf{3.5 \times 10^{-4}} \qquad\qquad \text{simplified}$$

Practice   Multiply and write each product in scientific notation.

a. $(4.2 \times 10^6)(1.4 \times 10^3)$

b. $(5 \times 10^5)(3 \times 10^7)$

c. $(4 \times 10^{-3})(2.1 \times 10^{-7})$

d. $(6 \times 10^{-2})(7 \times 10^{-5})$

Problem set 104

1. The 16-ounce box cost $1.12. The 24-ounce box cost $1.32. The smaller box cost how much more per ounce than the larger box?

2. Use a ratio box to solve this problem. The ratio of good apples to bad apples in the basket was 5 to 2. If there were 70 apples in the basket, how many of them were good?

3. Jan's average score after 15 tests was 82. Her average score on the next 5 tests was 90. What was her average score for all 20 tests?

4. Jackson earns $6 per hour at a part-time job. How much does he earn if he works for 2 hours 30 minutes?

5. Convert 24 shillings to pence. (1 shilling = 12 pence)

6. Graph $x \le -1$ on a number line.

7. Use a ratio box to solve this problem. Five is to 12 as 20 is to what number?

8. If $a = 3$, then what does $4a + 5$ equal?

9. Four fifths of the football team's 30 points were scored on pass plays. How many points did the team score on pass plays?

10. Compare: $x \bigcirc y$ if $x + y = 10$

11. Find the circumference of each circle.

(a)

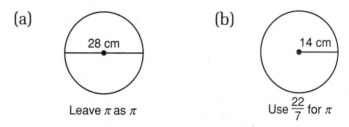

28 cm

Leave π as π

(b)

14 cm

Use $\dfrac{22}{7}$ for π

12. Find the area of each circle in Problem 11.

13. What is the volume of a cube with edges 10 cm long?

14. Complete the table.

FRACTION	DECIMAL	PERCENT
(a)	(b)	250%

15. Write an equation to solve this problem. What is 6 percent of $8.50?

Use a ratio box to solve Problems 16 and 17.

16. Judy found that there were 12 minutes of commercials during every hour of prime time programming. Commercials were shown for what percent of each hour?

17. Thirty percent of the boats that traveled up the river on Monday were steam-powered. If 42 of the boats that traveled up the river were not steam-powered, then how many boats were there in all?

18. Simplify: $\dfrac{33\frac{1}{3}}{100}$

19. Find the area of this trapezoid.

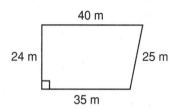

20. (a) Sketch a right angle. A right angle measures how many degrees?

(b) Sketch an acute angle that is half of a right angle. An angle that is half of a right angle measures how many degrees?

21. Find the missing number in each diagram.

IN	F U N C T I O N	OUT
□ →		→ 21
5 →		→ 11
7 →		→ 15
2 →		→ 5

22. Multiply and write each product in scientific notation.

(a) $(3 \times 10^4)(6 \times 10^5)$ (b) $(1.2 \times 10^{-3})(4 \times 10^{-6})$

Solve each equation. Show every step.

23. $b - 1\frac{2}{3} = 4\frac{1}{2}$ **24.** $0.4y = 1.44$

Add, subtract, multiply, or divide, as indicated:

25. $5^2 + 3[4^2 - 2(5 - 2)]$ **26.**   3 yd 2 ft 7 in.
                                               − 1 yd 2 ft 8 in.

27. $0.6 \times 3\frac{1}{3} \div 2$ (fraction)

28. $5.63 - \left(4 - 1\dfrac{2}{5}\right)$ (decimal)

29. (a) $\dfrac{(-4)(-6)}{(-2)(-3)}$　　　　　(b) $(-3)(-4)(-5)$

30. (a) $(-3) + (-4) - (-5)$　　(b) $(-15) - (+14) + (+10)$

LESSON
105

Mean, Median, Mode, and Range

The words **mean**, **median**, **mode**, and **range** are used in the study of statistics. An example is the best way to explain the meanings of these words.

On his first 5 tests Paul earned these scores.

100　　80　　70　　70　　90

If we arrange these scores in order from least to greatest, we get

1. The scores ranged from a low of 70 to a high of 100. The difference between 70 and 100 is 30. We say that the **range** of this group of numbers is 30.
2. In English the word "mode" means the customary fashion or style. The number 70 occurs more than any other number, and it is called the **mode** of this group of numbers. It is the "fashionable number."

3. The middle of a divided highway is called the "median" of the highway. The **median** of a group of numbers is the middle number when the numbers are arranged in order. The median of this group of numbers is 80.

4. The **mean** of a group of numbers is the average of the group. The mean of these numbers is 82.

$$\text{Mean} = \text{average} = \frac{70 + 70 + 80 + 90 + 100}{5}$$

$$= \frac{410}{5} = 82$$

The nonsense phrase "mean old average" can be used as a mnemonic device to associate the words "mean" and "average."

The median is not always one of the numbers listed. If there are an **even number** of numbers, the median is the number half way between the middle two numbers. Thus the median is the average of the middle two numbers. For example, the middle of the following list of numbers is between 70 and 80. The median is 75, which is the number halfway between 70 and 80.

$$65 \qquad 65 \qquad [70 \qquad 80] \qquad 80 \qquad 100$$

There are two modes, 65 and 80.

Example   Amanda is in the Strikers bowling league. Her bowling scores for the first 10 games were 95, 103, 96, 110, 103, 109, 110, 103, 116, and 105.

(a) What was her mean score?

(b) What was her median score?

(c) What was the mode of her scores?

(d) What was the range of her scores?

Solution   (a) To find the "mean old average," we add the 10 numbers and divide by 10. The sum is 1050.

$$\frac{1050}{10} = 105$$

(b) To find the median, we arrange the numbers in order and select the middle number. Since there are two middle numbers, we find their average.

95   96   103   103   [103   105]   109   110   110   116

↑

104

(c) The mode is the number that appears most often, which is **103**.

(d) The range is the difference between the largest number and the smallest number.

$$116 - 95 = 21$$

Practice   Find the following given this list of measurements:

64 cm, 72 cm, 68 cm, 72 cm, 59 cm, 67 cm, 74 cm

a. Mode   **b.** Mean   **c.** Median   **d.** Range

Problem set 105   Refer to this graph to answer questions 1–3.

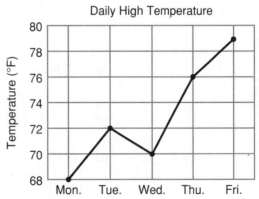

Daily High Temperature

1. What was the range in the daily high temperature from Monday to Friday?

2. Which day had the greatest increase in temperature from the previous day?

3. Wednesday's high temperature was how much lower than the average high temperature for these 5 days?

4. Frank's scores on 10 tests were as follows:

$$90, 90, 100, 95, 95, 85, 100, 100, 80, 100$$

For this set of scores find the (a) mean, (b) median, (c) mode, and (d) range.

5. Use a ratio box to solve this problem. The ratio of rowboats to sailboats in the bay was 3 to 7. If the total number of rowboats and sailboats in the bay was 210, how many sailboats were in the bay?

6. Graph $x < 3$ on a number line.

7. Write a proportion to solve this problem. If 4 cost $1.40, how much would 10 cost?

8. Five eighths of the members supported the treaty, whereas 36 opposed the treaty. How many members supported the treaty?

9. Evaluate: $a - (ab - a^2)$ if $a = 3$ and $b = 4$

10. Compare: $f \bigcirc g$ if $\dfrac{f}{g} = 1$

11. (a) Find the circumference of this circle.

(b) Find the area of this circle.

6 in.

Use 3.14 for π

12. Use unit multipliers to convert 4.8 meters to centimeters.

13. A rectangular prism has how many faces?

14. Complete the table.

Fraction	Decimal	Percent
$1\frac{4}{5}$	(a)	(b)

15. Write an equation to solve this problem. What is 30 percent of $18?

16. Simplify: $\dfrac{12\frac{1}{2}}{100}$

Use a ratio box to solve Problems 17 and 18.

17. When the door was left open, 36 pigeons flew the coop. If this was 40 percent of all the pigeons, how many pigeons were there originally?

18. Sixty percent of the gnomes were 3 feet tall or less. If there were 300 gnomes in all, how many were more than 3 feet tall?

19. A square sheet of paper with a perimeter of 48 in. has a corner cut off, forming a pentagon as shown.

    (a) What is the perimeter of the pentagon?

    (b) What is the area of the pentagon?

20.

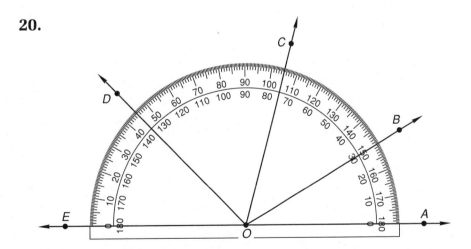

Find the measure of each angle.

  (a) ∠AOC   (b) ∠EOC   (c) ∠AOD

21. Find the next two numbers in this sequence:

$$-8, -6, -4, \underline{}, \underline{}$$

22. Multiply and write the product in scientific notation.

$$(1.5 \times 10^{-3})(3 \times 10^6)$$

Solve each equation. Show each step.

23. $b - 4.75 = 5.2$ **24.** $\dfrac{2}{3}y = 36$

Add, subtract, multiply, or divide, as indicated:

25. $\sqrt{5^2 - 4^2} + 2^3$ **26.** $1 \text{ m} - 45 \text{ mm}$

27. $0.9 \div 2\dfrac{1}{4} \cdot 24$ (decimal)

28. $7.8 - \left(5\dfrac{1}{3} + 0.2\right)$ (fraction)

29. (a) $\dfrac{(-8)(+6)}{(-3)(+4)}$ (b) $(+3)(-5)(+2)$

30. $(+30) - (-50) - (+20)$

LESSON 106

Order of Operations with Signed Numbers • Functions, Part 2

Order of operations with signed numbers

To simplify expressions that involve several operations, we perform the operations in a prescribed order. We have practiced simplifying expressions with whole numbers. In this lesson we will begin simplifying expressions that contain both whole numbers and negative numbers.

Example 1 Simplify: $(-2) + (-2)(-2) - \dfrac{(-2)}{(+2)}$

Solution First we multiply and divide in order from left to right.

$$(-2) + (-2)(-2) - \frac{(-2)}{(+2)}$$

$$(-2) + \quad (+4) \quad - (-1)$$

Then we add and subtract in order from left to right.

$$(-2) + (+4) - (-1)$$

$$(+2) - (-1)$$

$$+3$$

Example 2 Simplify: $(-2) - [(-3) - (-4)(-5)]$

Solution We simplify within brackets first. Within the brackets we follow the order of operations, multiplying and dividing before adding and subtracting.

$$(-2) - [(-3) - (-4)(-5)]$$

$$(-2) - [(-3) - (+20)]$$

$$(-2) - (-23)$$

$$+21$$

Functions We remember that a function is a relationship between two sets of numbers. We have practiced finding missing numbers in functions when some number pairs have been given. For instance, the missing numbers in the functions below are 14 and 7.

We have found the missing numbers by first finding the rule of the function. The rule of the function on the left is, "Multiply the IN number by 2 to find the OUT number." The rule of the function on the right is, "Subtract 7 from the IN number to find the OUT number."

Often the rule of a function is expressed as an equation with x standing for the IN number and y standing for the OUT number. If we write the rule of the function on the left as an equation, we get

$$y = 2x$$

If we write the rule of the function on the right as an equation, we get

$$y = x - 7$$

Beginning with this lesson we will practice finding missing numbers in functions when the rule is given as an equation.

Example 3 Find the missing numbers in this function. $y = 2x + 1$

x	y
4	☐
7	☐
0	☐

Solution The letter y stands for the OUT number. The letter x stands for the IN number. We are given three IN numbers and are asked to find the OUT number for each by using the rule of the function. The expression $2x + 1$ shows us what to do to find y, the OUT number. It shows us we should multiply the x number by 2 and then add 1. The first x number is 4. We multiply by 2 and add 1.

$$y = 2(4) + 1 \qquad \text{substituted}$$

$$y = 8 + 1 \qquad \text{multiplied}$$

$$y = 9 \qquad \text{added}$$

We find that the y number is 9 when x is 4. The next x number is 7. We multiply by 2 and add 1.

$$y = 2(7) + 1 \qquad \text{substituted}$$

$$y = 14 + 1 \qquad \text{multiplied}$$

$$y = 15 \qquad \text{added}$$

The third x number is 0. We multiply by 2 and add 1.

$$y = 2(0) + 1 \qquad \text{substituted}$$
$$y = 0 + 1 \qquad \text{multiplied}$$
$$y = 1 \qquad \text{added}$$

The missing numbers are **9, 15,** and **1.**

Practice Simplify:

a. $(-3) + (-3)(-3) - \dfrac{(-3)}{(+3)}$

b. $(-3) - [(-4) - (-5)(-6)]$

c. $(-2)[(-3) - (-4)(-5)]$

Find the missing numbers in each function.

d. $y = 3x - 1$

x	y
3	☐
1	☐
0	☐

e. $y = \dfrac{1}{2}x$

x	y
6	☐
0	☐
☐	4

f. $y = 8 - x$

x	y
7	☐
1	☐
4	☐

Problem set 106

1. Use a ratio box to solve this problem. The team's ratio of games won to games played was 3 to 4. If the team played 24 games, how many games did the team fail to win?

2. Find the (a) mean, (b) median, (c) mode, and (d) range of the following group of scores:

70, 80, 90, 80, 70, 90, 75, 95, 100, 90

3. Use a ratio box to solve the problem. Mary was chagrined to find that the ratio of dandelions to marigolds in the garden was 11 to 4. If there were 44 marigolds in the garden, how many dandelions were there?

4. Use unit multipliers to convert 0.98 liter to milliliters.

5. Graph $x > 0$ on a number line.

6. Use a ratio box to solve this problem. If sound travels 2 miles in 10 seconds, how far does sound travel in 1 minute?

7. Draw a diagram of this statement. Then answer the questions that follow.

> Thirty-five thousand dollars was raised in the charity drive. This was seven tenths of the goal.

(a) The goal of the charity drive was to raise how much money?

(b) The drive fell short of the goal by what percent?

8. Compare: $2a \bigcirc a^2$ if a is a whole number

9. What is the circumference of a circle whose radius is 4 meters?

10. What is the area of a circle whose radius is 4 meters?

11. What fraction of this circle is shaded?

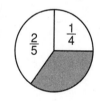

12. A certain rectangular prism is 5 inches long, 4 inches wide, and 3 inches high. Sketch the figure and find its volume.

13. Complete the table.

Fraction	Decimal	Percent
$\frac{1}{40}$	(a)	(b)

14. Write an equation to solve this problem. What is 5 percent of $1.20?

15. Ten is 20 percent of what number?

16. Simplify: $\dfrac{8\frac{1}{3}}{100}$

Use a ratio box to solve Problems 17 and 18.

17. Max was delighted when he found that he had correctly answered 38 of the 40 questions. What percent of the questions had he answered correctly?

18. Before the clowns arrived, only 35 percent of the children wore happy faces. If 91 children did not wear happy faces, how many children were there in all?

19. (a) Name this shape.

(b) Find its perimeter.

(c) Find its area.

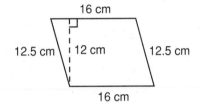

20.

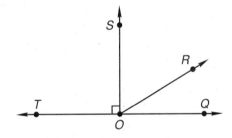

(a) Find $m\angle TOS$. (The m means "the measure of.")

(b) Find $m\angle QOT$.

(c) $\angle QOR$ is one third of a right angle. Find $m\angle QOR$.

(d) Find $m\angle TOR$.

21. Find the missing numbers in this function.

$y = 2x - 1$

x	y
5	☐
3	☐
1	☐

22. Multiply and write the product in scientific notation.

$$(5 \times 10^{-3})(6 \times 10^{8})$$

Solve each equation. Show each step.

23. $13.2 = 1.2w$ **24.** $c + \dfrac{5}{6} = 1\dfrac{1}{4}$

Add, subtract, multiply, or divide, as indicated:

25. $3\{20 - [6^2 - 3(10 - 4)]\}$ **26.**

$$\begin{array}{r} 3 \text{ hr } 15 \text{ min } 25 \text{ sec} \\ - 2 \text{ hr } 45 \text{ min } 30 \text{ sec} \\ \hline \end{array}$$

27. $0.6 \times 5\dfrac{1}{3} \div 4$ (fraction)

28. $6\dfrac{3}{5} + 4.9 + 12.25$ (decimal)

29. $(-2) + (-2)(+2) - \dfrac{(-2)}{(-2)}$

30. $(-3) - [(-2) - (+4)(-5)]$

LESSON 107

Number Families

In mathematics we give special names to certain sets of numbers. Some of these sets or families of numbers are the counting numbers, the whole numbers, the integers, and the rational numbers. In this lesson we will describe each of these number families and discuss how they are related.

- **The Counting Numbers.** Counting numbers are the numbers we say when we count. The first counting number is 1, the next is 2, then 3, and so on.

    Counting numbers: 1, 2, 3, 4, 5, . . .

The three dots mean that the list goes on and on.

- **The Whole Numbers.** The whole number family includes all of the counting number family and has one more member, which is zero.

    Whole numbers: 0, 1, 2, 3, 4, 5, . . .

If we use a dot to mark each of the whole numbers on the number line, the graph looks like this.

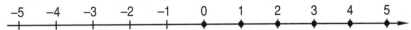

Notice that there are no dots on the negative side of the number line, because no whole number is a negative number. Also notice that there are no dots between the whole numbers because numbers between whole numbers are not "whole." We put an arrowhead on the right end of the number line to indicate that the whole numbers continue without end.

- **The Integers**. The integer family includes all of the whole numbers. The integer family also includes the opposites (negatives) of the positive whole numbers. The list of integers goes on and on in both directions as indicated by the dots.

Integers: . . . , $-4, -3, -2, -1, 0, 1, 2, 3, 4, . . .$

A graph of the integers looks like this.

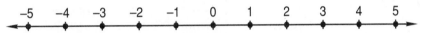

Thus the set of integers includes the numbers $-2, -1, 0, 1,$ 2, etc…, but does not include $\frac{1}{2}, \frac{5}{3}$, and other fractions. Note the arrowheads on both ends of the number line to indicate that the integers continue without end in both directions.

- **The Rational Numbers**. The rational number family includes all numbers that can be written as a fraction of two integers. Thus all fractions name a rational number. Here are some examples of rational numbers.

$$\frac{1}{2} \quad \frac{5}{3} \quad \frac{-3}{2} \quad \frac{-4}{1} \quad \frac{0}{2} \quad \frac{3}{1}$$

Notice that the family of rational numbers includes all the integers, because every integer can be written as a fraction whose denominator is the number 1. For example, we can write -4 as a fraction by writing

$$\frac{-4}{1}$$

The set of rational numbers also includes all the positive and negative mixed numbers, because these numbers can be written as fractions. For example, we can write $4\frac{1}{5}$ as

$$\frac{21}{5}$$

Sometimes rational numbers are written in decimal form, in which case the decimal will either terminate

$$\frac{1}{8} = 0.125$$

or it will repeat.

$$\frac{5}{6} = 0.8333 \ldots$$

The following diagram may be helpful in visualizing the relationships between these families of numbers. The diagram shows that the set of rational numbers includes all the other number families described in this lesson.

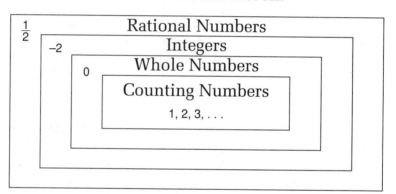

Example 1   Graph the integers that are less than 4.

Solution   We sketch a number line and draw a dot at every integer that is less than 4. Since the set of integers includes whole numbers, we draw dots at 3, 2, 1, and 0. Since the set of integers also includes the negatives of the positive whole numbers, we continue to draw dots at $-1, -2, -3$, and so on. We draw an arrowhead to indicate that the graph of integers which are less than 4 continues without end.

Example 2   Answer true or false.

(a)   All whole numbers are integers.

(b)   All rational numbers are integers.

Solution (a) **True.** Every whole number is included in the family of integers.

(b) **False.** Although every integer is a rational number, it is not true that every rational number is an integer. There are some rational numbers, such as $\frac{1}{2}$ and $\frac{5}{3}$, that are not integers.

Practice a. Graph the integers that are greater than -4.

b. Graph the whole numbers that are less than 4.

Answer true or false.

c. Every integer is a whole number.

d. Every integer is a rational number.

Problem set 1. Heavenly Scent was priced at $28.50 for 3 ounces, while
107 Eau de Rue cost only $4.96 for 8 ounces. Heavenly Scent cost how much more per ounce than Eau de Rue?

2. Use a ratio box to solve this problem. The ratio of rookies to veterans in the camp was 2 to 7. Altogether there were 252 in the camp. How many of them were rookies?

3. The seven linemen weighed 197 lb, 213 lb, 246 lb, 205 lb, 238 lb, 213 lb, and 207 lb. Find the (a) mode, (b) median, (c) mean, and (d) range of this group of measures.

4. Convert 12 bushels to pecks. (1 bushel = 4 pecks)

5. The Martins drove the car from 7 a.m. to 4 p.m. and traveled 468 miles. Their average speed was how many miles per hour?

6. Graph the integers that are less than or equal to 3.

7. Use a ratio box to solve this problem. Nine is to 6 as what number is to 30?

8. Nine tenths of the school's 1800 students attended the homecoming game.

    (a) How many of the school's students attended the homecoming game?

    (b) What percent of the school's students did not attend the homecoming game?

9. Evaluate: $b^2 - 4ac$ if $a = 1$, $b = 5$, and $c = 6$

10. Compare: $a^2 \bigcirc a$ if a is positive

11. (a) Find the circumference of this circle.

    (b) Find the area of this circle.

12 in.

Leave π as π

12. Simplify: $\dfrac{60}{2\frac{1}{2}}$

13. The figure shown is a triangular prism. Copy the shape on your paper and find the number of its (a) faces, (b) edges, and (c) vertices.

14. Complete the table.

Fraction	Decimal	Percent
(a)	0.9	(b)

15. Write an equation to solve this problem. What is 80 percent of $6.50?

Use a ratio box to solve Problems 16 and 17.

16. The sale price of $24 was 60 percent of the regular price. What was the regular price?

17. Forty-eight corn seeds sprouted. This was 75 percent of the seeds that were planted. How many of the planted seeds did not sprout?

18. Thirty is 40 percent of what number?

19. (a) Classify this quadrilateral.

(b) Find its perimeter.

(c) Find its area.

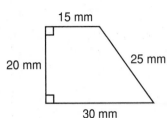

20.

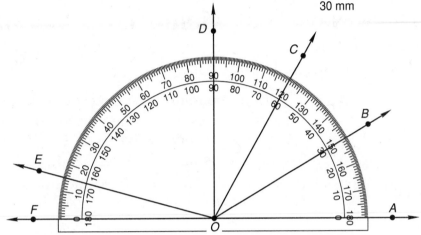

Find the measure of these angles.

(a) ∠*COF*   (b) ∠*AOE*

21. Find the missing numbers in this function.

$y = 3x + 1$

x	y
4	☐
7	☐
0	☐

22. Multiply and write this product in scientific notation.

$$(1.2 \times 10^5)(1.2 \times 10^{-8})$$

Solve each equation. Show each step.

23. $56 = \dfrac{7}{8}w$ **24.** $4.8 + c = 7.34$

Add, subtract, multiply, or divide, as indicated:

25. $\sqrt{10^2 - 6^2} - \sqrt{10^2 - 8^2}$ **26.** $\begin{array}{r} 5 \text{ lb} \ \ 9 \text{ oz} \\ + \ 4 \text{ lb} \ \ 7 \text{ oz} \\ \hline \end{array}$

27. $12 - \left(5\dfrac{1}{6} - 1.75\right)$ (fraction)

28. $1.4 \div 3\frac{1}{2} \times 10^3$ (decimal)

29. $(-4)(-5) - (-4)(+3)$ **30.** $(-2)[(-3) - (-4)(+5)]$

LESSON
108

Memorizing Common Fraction-Percent Equivalents

Certain percents are encountered so frequently that it is worth the effort to memorize some of the more common fraction-percent equivalents. The list below includes percent equivalents for halves, thirds, fourths, fifths, sixths, eighths, tenths, and twelfths. Also included are $\frac{1}{20}$, $\frac{1}{25}$, $\frac{1}{50}$, and $\frac{1}{100}$.

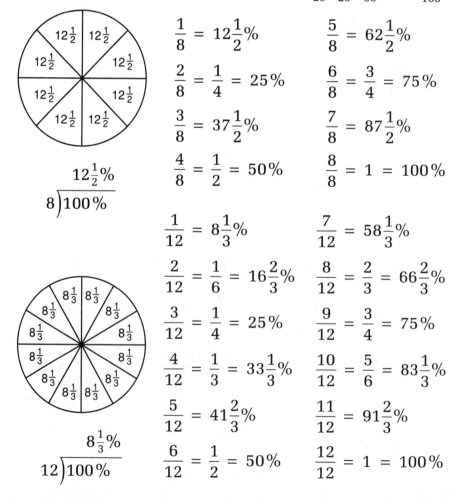

$$\frac{1}{8} = 12\frac{1}{2}\%$$ $$\frac{5}{8} = 62\frac{1}{2}\%$$

$$\frac{2}{8} = \frac{1}{4} = 25\%$$ $$\frac{6}{8} = \frac{3}{4} = 75\%$$

$$\frac{3}{8} = 37\frac{1}{2}\%$$ $$\frac{7}{8} = 87\frac{1}{2}\%$$

$$\frac{4}{8} = \frac{1}{2} = 50\%$$ $$\frac{8}{8} = 1 = 100\%$$

$$\frac{12\frac{1}{2}\%}{8)100\%}$$

$$\frac{1}{12} = 8\frac{1}{3}\%$$ $$\frac{7}{12} = 58\frac{1}{3}\%$$

$$\frac{2}{12} = \frac{1}{6} = 16\frac{2}{3}\%$$ $$\frac{8}{12} = \frac{2}{3} = 66\frac{2}{3}\%$$

$$\frac{3}{12} = \frac{1}{4} = 25\%$$ $$\frac{9}{12} = \frac{3}{4} = 75\%$$

$$\frac{4}{12} = \frac{1}{3} = 33\frac{1}{3}\%$$ $$\frac{10}{12} = \frac{5}{6} = 83\frac{1}{3}\%$$

$$\frac{5}{12} = 41\frac{2}{3}\%$$ $$\frac{11}{12} = 91\frac{2}{3}\%$$

$$\frac{8\frac{1}{3}\%}{12)100\%}$$

$$\frac{6}{12} = \frac{1}{2} = 50\%$$ $$\frac{12}{12} = 1 = 100\%$$

The upper circle shows the division of 100 percent into eighths. The lower circle shows the division of 100 percent into twelfths. Use of these circles may make the following fraction-percent equivalents easier to memorize.

FRACTION-PERCENT EQUIVALENTS

FRACTION	PERCENT	FRACTION	PERCENT	FRACTION	PERCENT
$\frac{1}{2}$	50%	$\frac{1}{6}$	$16\frac{2}{3}\%$	$\frac{9}{10}$	90%
$\frac{1}{3}$	$33\frac{1}{3}\%$	$\frac{5}{6}$	$83\frac{1}{3}\%$	$\frac{1}{12}$	$8\frac{1}{3}\%$
$\frac{2}{3}$	$66\frac{2}{3}\%$	$\frac{1}{8}$	$12\frac{1}{2}\%$	$\frac{5}{12}$	$41\frac{2}{3}\%$
$\frac{1}{4}$	25%	$\frac{3}{8}$	$37\frac{1}{2}\%$	$\frac{7}{12}$	$58\frac{1}{3}\%$
$\frac{3}{4}$	75%	$\frac{5}{8}$	$62\frac{1}{2}\%$	$\frac{11}{12}$	$91\frac{2}{3}\%$
$\frac{1}{5}$	20%	$\frac{7}{8}$	$87\frac{1}{2}\%$	$\frac{1}{20}$	5%
$\frac{2}{5}$	40%	$\frac{1}{10}$	10%	$\frac{1}{25}$	4%
$\frac{3}{5}$	60%	$\frac{3}{10}$	30%	$\frac{1}{50}$	2%
$\frac{4}{5}$	80%	$\frac{7}{10}$	70%	$\frac{1}{100}$	1%

Practice Study the lists and tables in the lesson. Practice remembering percents for fractions and practice remembering fractions for percents. Timed written tests give valuable practice for helping you to remember these commonly encountered percents.

Problem set 108

1. How far will the jet travel in 2 hours 30 minutes if its average speed is 450 miles per hour?

2. Use a unit multiplier to convert 3 yd to feet.

3. Use a ratio box to solve this problem. If 240 of the 420 students in the auditorium were girls, what was the ratio of boys to girls in the auditorium?

4. Geoff and his two brothers are very tall. Geoff's height is 18' 3". The heights of his two brothers are 17' 10" and 17' 11". What is the average height of Geoff and his brother giraffes?

5. The Martins' car traveled 468 miles on 18 gallons of gas. Their car averaged how many miles per gallon?

6. Graph the whole numbers that are less than or equal to 5.

7. Use a ratio box to solve this problem. The road was steep. Every 100 yards the elevation increased 36 feet. How many feet did the elevation increase in 1500 yards?

8. Draw a diagram of this statement. Then answer the questions that follow.

   Fifty-six antelope were seen in the clearing. This was $\frac{7}{8}$ of the herd.

   (a) How many antelope were in the herd?

   (b) What percent of the herd was not seen in the clearing?

9. If $x = 4$ and $y = 3x - 1$, y equals what number?

10. Find the circumference of this circle.

11. Find the area of this circle.

70 mm

Use $\frac{22}{7}$ for π

12. The shape shown was built of 1-inch cubes. What is the volume of the shape?

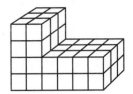

13. Complete the table.

FRACTION	DECIMAL	PERCENT
(a)	(b)	$12\frac{1}{2}\%$

14. Write an equation to solve this problem. What number is 25 percent of 4?

Use a ratio box to solve Problems 15 and 16.

15. The sale price of $24 was 80 percent of the regular price. What was the regular price?

16. David had finished 60 percent of the race, but he still had 2000 meters to run. How long was his race?

17. Seventy is what percent of 80?

18. Find the area of this figure. Dimensions are in centimeters.

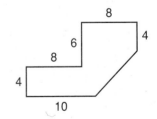

19. In the figure, $\angle AOE$ is a straight angle and $m\angle AOB = m\angle BOC = m\angle COD$.

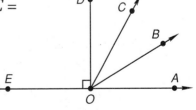

    (a) Find $m\angle AOB$.

    (b) Find $m\angle AOC$.

    (c) Find $m\angle EOC$.

20. Simplify: $\dfrac{66\frac{2}{3}}{100}$

21. Find the missing numbers in this function.

$y = \dfrac{24}{x}$

x	y
3	☐
4	☐
12	☐

22. Multiply and write the product in scientific notation.

$$(4 \times 10^{-5})(2.1 \times 10^{-7})$$

Solve each equation. Show each step.

23. $d - 8.47 = 9.1$ **24.** $0.25m = 3.6$

Add, subtract, multiply, or divide, as indicated:

25. $\dfrac{3 + 5.2 - 1}{4 - 3 + 2}$ **26.** $1 \text{ kg} - 75 \text{ g}$

27. $6\frac{2}{3} \div 0.02 \times 12$ (fraction)

28. $3.7 + 2\frac{5}{8} + 15$ (decimal)

29. $\dfrac{(-3) + (-3)(+4)}{(+3) + (-4)}$ **30.** $(-5) - (-2)[(-3) - (+4)]$

LESSON 109

Multiple Unit Multipliers • Conversion of Units of Area

Multiple unit multipliers
We may multiply a number by 1 repeatedly without changing the number.

$$5 \cdot 1 = 5$$
$$5 \cdot 1 \cdot 1 = 5$$
$$5 \cdot 1 \cdot 1 \cdot 1 = 5$$

Since unit multipliers are forms of 1, we may also multiply a measure by several unit multipliers without changing the measure.

$$10 \cancel{yd} \cdot \frac{3 \cancel{ft}}{1 \cancel{yd}} \cdot \frac{12 \text{ in.}}{1 \cancel{ft}} = 360 \text{ in.}$$

Ten yards is the same distance as 360 inches. We used one unit multiplier to change from yards to feet and a second unit multiplier to change from feet to inches. Of course, we did not need to use two unit multipliers. Instead we could have changed from yards to inches using the unit multiplier

$$\frac{36 \text{ in.}}{1 \text{ yd}}$$

However, sometimes it prevents mistakes if we use more than one unit multiplier to perform this conversion.

Example 1 Use two unit multipliers to convert 5 hours to seconds.

Solution We are changing units from hours to seconds.

$$\text{hours} \rightarrow \text{seconds}$$

We will perform the conversion in two steps. We will change from hours to minutes with one unit multiplier and from minutes to seconds with a second unit multiplier.

$$\text{hours} \rightarrow \text{minutes} \rightarrow \text{seconds}$$

$$5 \cancel{hr} \cdot \frac{60 \cancel{min}}{1 \cancel{hr}} \cdot \frac{60 \text{ sec}}{1 \cancel{min}} = \textbf{18,000 sec}$$

Conversion of units of area To convert from one unit of area to another, it is helpful to use two unit multipliers.

Consider this rectangle, which has an area of 6 ft².

3 ft

2 ft

$$3 \text{ ft} \cdot 2 \text{ ft} = 6 \text{ ft}^2$$

Recall that the expression 6 ft² means 6 ft · ft. Thus, to convert 6 ft² to in.², we convert from

$$ft \cdot ft \quad to \quad in. \cdot in.$$

To cancel feet twice, we use two unit multipliers.

$$6 \text{ ft}^2 \quad \longrightarrow \quad 6 \cancel{ft} \cdot \cancel{ft} \cdot \frac{12 \text{ in.}}{1 \cancel{ft}} \cdot \frac{12 \text{ in.}}{1 \cancel{ft}} = 864 \text{ in.} \cdot \text{in.}$$

$$\longrightarrow \quad \mathbf{864 \text{ in.}^2}$$

Example 2   Convert 5 yd² to square feet.

Solution   Since yd² means yd · yd, we use two unit multipliers to cancel yards twice.

$$5 \text{ yd}^2 \quad \longrightarrow \quad 5 \cancel{yd} \cdot \cancel{yd} \cdot \frac{3 \text{ ft}}{1 \cancel{yd}} \cdot \frac{3 \text{ ft}}{1 \cancel{yd}} = 45 \text{ ft} \cdot \text{ft}$$

$$\longrightarrow \quad \mathbf{45 \text{ ft}^2}$$

Example 3   Convert 1.2 m² to square centimeters.

Solution $$1.2 \cancel{m^2} \cdot \frac{100 \text{ cm}}{1 \cancel{m}} \cdot \frac{100 \text{ cm}}{1 \cancel{m}} = \mathbf{12,000 \text{ cm}^2}$$

Practice   Use two unit multipliers to perform each conversion.

a. 5 yd to inches   **b.** $1\frac{1}{2}$ hr to seconds

c. 15 yd² to square feet   **d.** 20 cm² to square millimeters

Problem set 109

1. Jackson earns $6 per hour at a part-time job. How much does he earn working 3 hours 15 minutes?

2. Mikki was not happy with her test average after 6 tests. On the next 4 tests, Mikki's average score was 93, which raised her average score for all 10 tests to 84. What was Mikki's average score on the first 6 tests?

3. Use two unit multipliers to convert 6 ft² to square inches.

4. Use a ratio box to solve this problem. The ratio of woodwinds to brass instruments in the orchestra was 3 to 2. If there were 15 woodwinds, how many brass instruments were there?

5. Graph the counting numbers that are less than 4.

6. Use a ratio box to solve this problem. Artichokes were on sale 8 for $2. At that price, how much would 3 dozen artichokes cost?

7. Draw a diagram of this statement. Then answer the questions that follow.

> When Sandra walked through the house, she saw that 18 lights were on and only $\frac{1}{3}$ of the lights were off.

  (a) How many lights were off?

  (b) What percent of the lights were on?

8. Evaluate: $a - [b - (a - b)]$ if $a = 5$ and $b = 3$

9. Compare: $x \bigcirc y$ if x and y are negative and $\dfrac{x}{y} = 2$

10. A horse was tied to a stake by a rope that was 30 feet long so that the horse could move about in a circle.

  (a) What is the circumference of the circle?

  (b) What is the area of the circle?

11. What percent of this circle is shaded?

12. Sketch a cube with edges 3 cm long. What is the volume of the cube?

13. Write an equation to solve this problem. What number is 75 percent of 100?

14. Complete the table.

Fraction	Decimal	Percent
(a)	0.125	(b)

15. Simplify: $\dfrac{60}{1\frac{1}{4}}$

Use a ratio box to solve Problems 16 and 17.

16. The regular price was $24. The sale price was $18. The sale price was what percent of the regular price?

17. The auditorium seated 375, but this was enough for only 30 percent of those who wanted a seat. How many wanted a seat but could not get one?

18. Write an equation to solve this problem. Twenty-four is 25 percent of what number?

19. (a) Classify this quadrilateral.

(b) Find the perimeter of the quadrilateral.

(c) Find the area of the quadrilateral.

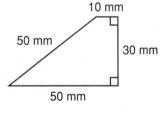

20. Find the missing numbers in this function.

$y = x - 5$

x	y
10	☐
7	☐
5	☐

21. Multiply and write this product in scientific notation.

$$(9 \times 10^{-6})(4 \times 10^{-8})$$

Solve each equation. Show each step.

22. $8\dfrac{5}{6} = d - 5\dfrac{1}{2}$

23. $\dfrac{5}{6}m = 90$

24.

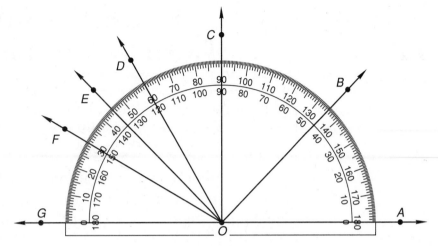

Find the measure of these angles.

(a) $\angle GOF$ (b) $\angle AOE$

Add, subtract, multiply, or divide, as indicated:

25. $6\{5 \cdot 4 - 3[6 - (3 - 1)]\}$

26.   1 yd 1 ft $3\frac{1}{2}$ in.

     $-$ 2 ft $6\frac{1}{2}$ in.

27. $7\frac{1}{2} \div 12\frac{1}{2} \div 10^2$ (decimal)

28. $9.5 - \left(5 - \dfrac{5}{9}\right)$ (fraction)

29. $\dfrac{(-3)(-4) - (-3)}{(-3) - (+4)(+3)}$

30. $(+5) + (-2)[(+3) - (-4)]$

LESSON 110

Sum of the Angle Measures of a Triangle • Straight Angles

Project

> Materials needed: Paper (preferably construction paper)
>   Straightedge
>   Scissors
>
> Using a pencil and straightedge, draw a triangle on a sheet of paper. Then cut the triangle from the paper. Next label the angles 1, 2, and 3 as shown below.
>
>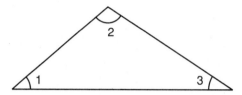
>
> After labeling the angles, cut each angle from the triangle as shown by the dashed lines in the illustration. Then arrange the pieces to meet at one point as illustrated. You will find that the angles form a semi-circle.
>
>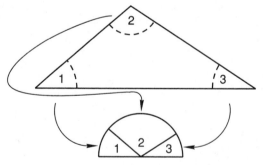
>
> Compare your results with the results of others in the class. The triangles all have different sizes and shapes. We note that in every example the angles form half of a circle.

Sum of the angle measures of a triangle

The project in this lesson illustrates that the sum of the measures of the angles of a triangle equals 180° (half of a circle). If we know the measures of two angles of a triangle, we can calculate the measure of the third angle.

Example 1   What is the measure of ∠B in this triangle?

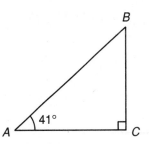

Solution   The sum of the angle measures is 180° for any triangle. We see that the measure of ∠A is 41°. The square at ∠C shows that the measure of ∠C is 90°. Thus the sum of the measures of ∠A and ∠C is 131°. The measure of ∠B plus the measures of the other two angles must total 180°. We find the measure of ∠B by subtracting 131° from 180°.

$$m\angle A + m\angle C = 131° \qquad \begin{array}{r} 180° \\ -\,131° \\ \hline 49° \end{array}$$

The measure of ∠B is **49°**.

Straight angles   The angle below forms a straight line. This angle is called a **straight angle** and its measure is 180°.

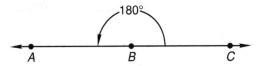

When two or more angles together form a straight angle, then the sum of their measures is 180°.

Example 2   What is the measure of ∠AOB?

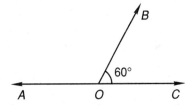

Solution   Since ∠AOB and ∠BOC combine to form a straight angle, their measures must total 180°. We can find the measure of ∠AOB by subtracting 60° from 180°.

$$\text{Measure of } \angle AOB = 180° - 60°$$
$$= 120°$$

The measure of ∠AOB is **120°**.

Example 3   What is the measure of ∠x?

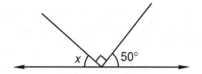

Solution   We see that three angles combine to form a straight angle. We are given the measures of two of the angles (50° and 90°). The sum of the measures of these angles is 140°, so the measure of ∠x is

$$180° - 140° = \mathbf{40°}$$

Example 4   Find the measure of ∠A.

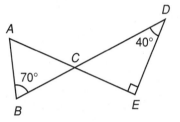

Solution   Finding the measure of ∠A requires several steps. We begin with what we know and gradually work our way to angle A.

First we copy the figure, naming three of the four angles at C with numbers to make it easier to discuss them.

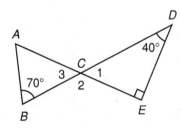

(a) Angle 1 + 90° + 40° = 180°, so angle 1 = 50°.

(b) Angle 1 + angle 2 = 180°, so angle 2 = 130°.

(c) Angle 2 + angle 3 = 180°, so angle 3 = 50°.

(d) Angle 3 + 70° + angle A = 180°, so angle A = **60°**.

Practice   Use this figure to find the measure of each angle.

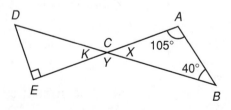

a. ∠X   **b.** ∠Y

c. ∠K   **d.** ∠D

**Problem set
110**

1. Use a ratio box to solve this problem. Jason's remote control car traveled 440 feet in 10 seconds. At that rate, how long would it take the car to travel a mile?

2. In the forest there were lions, tigers, and bears. The ratio of lions to tigers was 3 to 2. The ratio of tigers to bears was 3 to 4. If there were 18 lions, how many bears were there? Use a ratio box to find how many tigers there were. Then use another ratio box to find the number of bears.

3. Bill measured the shoe box and found that it was 30 cm long, 15 cm wide, and 12 cm high. What was the volume of the shoe box?

4. A baseball player's batting average is found by dividing the number of hits by the number of at-bats and writing the result as a decimal number rounded to the nearest thousandth. If Erika had 24 hits in 61 at-bats, what was her batting average?

5. Use two unit multipliers to convert 18 square feet to square yards.

6. Graph the integers greater than -4.

7. Draw a diagram of this statement. Then answer the questions that follow.

   Jimmy bought the shirt for \$12. This was $\frac{3}{4}$ of the regular price.

   (a) What was the regular price of the shirt?

   (b) Jimmy bought the shirt for what percent of the regular price?

8. Use this figure to find the measure of each angle.

   (a) $\angle a$ (b) $\angle b$

   (c) $\angle c$ (d) $\angle d$

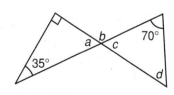

9. What is the circumference of this circle?

21 in.

Use $\frac{22}{7}$ for π

10. Simplify: $\dfrac{91\frac{2}{3}}{100}$

11. Evaluate: $\dfrac{ab + a}{a + b}$ if $a = 10$ and $b = 5$

12. Compare: $a^2 \bigcirc a$ if $a = 0.5$

13. Complete the table.

FRACTION	DECIMAL	PERCENT
$\frac{7}{8}$	(a)	(b)

14. Write an equation to solve this problem. What number is 100 percent of 50?

Use a ratio box to solve Problems 15 and 16.

15. Forty-five percent of the 3000 fast-food customers ordered a hamburger. How many of the customers ordered a hamburger?

16. The sale price of $24 was 75 percent of the regular price. The sale price was how many dollars less than the regular price?

17. Write an equation to solve this problem. Twenty is what percent of 200?

18. Find the area of this trapezoid.

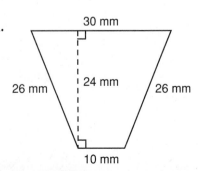

30 mm

26 mm   24 mm   26 mm

10 mm

19. A full circle contains 360°. How many degrees is $\frac{1}{12}$ of a full circle?

20. Find the missing numbers in this function.

$y = \frac{1}{3}x$

x	y
12	☐
9	☐
☐	6

21. Multiply and write the product in scientific notation.

$(1.25 \times 10^{-3})(8 \times 10^{-5})$

22. The lengths of two sides of an isosceles triangle are 4 cm and 10 cm. Sketch the triangle and find its perimeter.

Solve each equation. Show each step.

23. $\frac{4}{9}p = 72$

24. $12.3 = 4.56 + f$

Add, subtract, multiply, or divide, as indicated:

25. $\dfrac{9 \cdot 8 - 7 \cdot 6}{6 \cdot 5}$

26. $1 \text{ hr} - 15 \text{ min } 45 \text{ sec}$

27. $3.6 \times \dfrac{1}{20} \times 10^2$ (decimal)

28. $13\frac{1}{3} - \left(4.75 + \frac{3}{4}\right)$ (fraction)

29. $\dfrac{(+3) + (-4)(-6)}{(-3) + (-4) - (-6)}$

30. $(-5) - (+6)(-2) + (-2)(-3)(-1)$

LESSON
111

Equations with Mixed Numbers

We have been solving equations like this one

$$\frac{4}{5}x = 7$$

by multiplying both sides of the equation by the reciprocal of the coefficient of x. Here the coefficient of x is $\frac{4}{5}$, so we multiply both sides by the reciprocal of $\frac{4}{5}$, which is $\frac{5}{4}$.

$$\frac{\cancel{5}}{\cancel{4}} \cdot \frac{\cancel{4}}{\cancel{5}}x = \frac{5}{4} \cdot 7 \qquad \text{multiplied by } \frac{5}{4}$$

$$x = \frac{35}{4} \qquad \text{simplified}$$

$$x = 8\frac{3}{4} \qquad \text{mixed number}$$

When solving an equation that has a mixed number, we convert the mixed number to an improper fraction as the first step. Then we multiply both sides by the reciprocal of the improper fraction.

Example 1   Solve: $3\frac{1}{3}x = 5$

Solution   First we write $3\frac{1}{3}$ as an improper fraction.

$$\frac{10}{3}x = 5$$

Then we multiply both sides of the equation by $\frac{3}{10}$, which is the reciprocal of $\frac{10}{3}$.

$$\frac{\cancel{3}}{\cancel{10}} \cdot \frac{\cancel{10}}{\cancel{3}}x = \frac{3}{10} \cdot 5$$

$$x = \frac{3}{2}$$

In arithmetic, we usually convert an improper fraction such as $\frac{3}{2}$ to a mixed number. In algebra, we usually leave improper fractions in fraction form, which we will do beginning with this lesson.

Example 2   Solve: $2\frac{1}{2}y = 1\frac{7}{8}$

Solution   Since we will be multiplying on both sides to find y, we first convert both mixed numbers to improper fractions.

$$\frac{5}{2}y = \frac{15}{8}$$

Then we multiply both sides by $\frac{2}{5}$, which is the reciprocal of $\frac{5}{2}$.

$$\frac{2}{5} \cdot \frac{5}{2}y = \frac{2}{5} \cdot \frac{15}{8}$$

$$y = \frac{3}{4}$$

Practice   Solve:

a. $1\frac{1}{8}x = 36$

b. $3\frac{1}{2}a = 490$

c. $2\frac{3}{4}w = 6\frac{3}{5}$

d. $2\frac{2}{3}y = 1\frac{4}{5}$

Problem set 111

1. The sum of 0.8 and 0.9 is how much greater than the product of 0.8 and 0.9? Use words to write the answer.

2. For this set of scores find the (a) mean, (b) median, (c) mode, and (d) range.

   8, 6, 9, 10, 8, 7, 9, 10, 8, 10, 9, 8

3. The 24-ounce container was priced at $1.20. This container costs how much more per ounce than the 32-ounce container priced at $1.44?

4. Twenty-two of the ninety 2-digit numbers are prime numbers. What is the ratio of the number of prime numbers to the number of composite 2-digit numbers?

Use a ratio box to solve Problems 5–7.

5. Twenty-seven is to 36 as 36 is to what number?

6. The sale price of $36 was 90 percent of the regular price. What was the regular price?

7. Seventy-five percent of the citizens voted for Graham. If there were 800 citizens, how many of them did not vote for Graham?

8. Twenty-four is what percent of 30?

9. Use two unit multipliers to convert 100 yd to inches.

10. Draw a diagram of this statement. Then answer the questions that follow.

> Three hundred doctors recommended Brand X. This was $\frac{2}{5}$ of the doctors surveyed.

(a) How many doctors were surveyed?

(b) How many doctors surveyed did not recommend Brand X?

11. If $x = 4.5$ and $y = 2x + 1$, y equals what number?

12. Compare: $a \bigcirc ab$ if $a < 0$ and $b > 1$

13. If the perimeter of a square is 1 foot, what is the area of the square in square inches?

14. Complete the table.

Fraction	Decimal	Percent
(a)	1.75	(b)

15. Write an equation to solve this problem. What is 6 percent of $325?

16. Multiply and write the product in scientific notation.

$$(6 \times 10^4)(8 \times 10^{-7})$$

17. What is the volume of a cereal box with dimensions as shown? Dimensions are in inches.

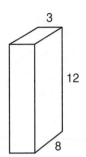

18. Find the circumference of this circle.

19. Find the area of this circle.

Use 3.14 for π

20.

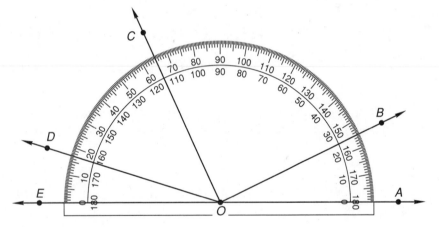

    (a)   Find $m\angle AOB$.

    (b)   Find $m\angle EOD$.

    (c)   Find $m\angle COA$.

21. What is $\dfrac{2}{3}$ of 20?

22. Sixteen is 40 percent of what number?

23. Graph $x \leq 4$.

Solve:

24. $2\dfrac{2}{3}w = 24$ **25.** $x + 3.5 = 4.28$

Add, subtract, multiply, or divide, as indicated:

26. $10^4 - (10^3 - 10^2)$ **27.** 1 ton $-$ 100 pounds

28. $3\frac{1}{5} - \left(2\frac{1}{2} - 1.2\right)$ (decimal)

29. $(-3) + (-4)(-5) - (-6)$

30. $[(-3) - (+2)][(+2) - (-3)]$

LESSON 112   Evaluations with Signed Numbers • Signed Numbers without Parentheses

Evaluation with signed numbers

We have practiced evaluating expressions such as

$$x - xy - y$$

by using positive numbers in place of x and y. In this lesson we will practice evaluating such expressions by using negative numbers as well. When evaluating expressions that contain signed numbers, it is helpful to replace each letter with parentheses as the first step. Doing this will help prevent making mistakes in signs.

Example 1 Evaluate: $x - xy - y$ if $x = -2$ and $y = -3$

Solution We write parentheses for each variable.

$$(\ ) - (\ )(\ ) - (\ )   \text{parentheses}$$

Now we write the proper number within the parentheses.

$$(-2) - (-2)(-3) - (-3)   \text{insert numbers}$$

We multiply first.

$$(-2) - (+6) - (-3)   \text{multiplied}$$

Then we add algebraically from left to right.

$$(-8) - (-3) \qquad \text{added} -2 \text{ and } -6$$

$$\mathbf{-5} \qquad \text{added} -8 \text{ and } +3$$

Signed numbers without parentheses

Signed numbers are often written without parentheses. To simplify an expression such as

$$-3 + 4 - 5 - 2$$

we may mentally insert parentheses and then add algebraically.

$$-3 \qquad +4 \qquad -5 \qquad -2 \qquad \text{(we see)}$$

$$(-3) + (+4) + (-5) + (-2) = -6 \qquad \text{(we think)}$$

Example 2 Simplify: $+3 - 4 + 2 - 1$

Solution We may think of this expression as

$$(\text{pos } 3) \text{ plus } (\text{neg } 4) \text{ plus } (\text{pos } 2) \text{ plus } (\text{neg } 1)$$

$$(+3) \quad + \quad (-4) \quad + \quad (+2) \quad + \quad (-1)$$

Then we add algebraically from left to right.

$$(+3) + (-4) + (+2) + (-1) = \mathbf{0}$$

Example 3 Simplify: $-2 + 3(-2) - 2(+4)$

Solution We may think of this expression as

$$(\text{neg } 2) \text{ plus } (\text{pos } 3) \text{ times } (\text{neg } 2) \text{ plus } (\text{neg } 2) \text{ times } (\text{pos } 4)$$

$$(-2) \quad + \quad (+3)(-2) \quad + \quad (-2)(+4)$$

We multiply first.

$$(-2) + (-6) + (-8)$$

Then we add algebraically from left to right.

$$(-2) + (-6) + (-8) = \mathbf{-16}$$

Practice   Evaluate each expression. Write parentheses as the first step.

    a. $x + xy - y$ if $x = 3$ and $y = -2$

    b. $-m + n - mn$ if $m = -2$ and $n = -5$

Simplify:

    c. $-3 + 4 - 5 - 2$ **d.** $-2 + 3(-4) - 5(-2)$

    e. $-3(-2) - 5(2) + 3(-4)$

Problem set 112

1. For his first 6 tests, Ted's average score was 86. For his next 4 tests, his average score was 94. What was his average score for his first 10 tests?

2. Ten billion is how much more than nine hundred eighty million? Use words to write the answer.

3. The Martins completed the 130-mile trip in $2\frac{1}{2}$ hours. What was their average speed in miles per hour?

Use a ratio box to solve Problems 4–7.

4. The ratio of laborers to supervisors at the job site was 3 to 5. Of the 120 individuals at the job site, how many were laborers?

5. Vera bought 3 notebooks for $8.55. At this rate, how much would 5 notebooks cost?

6. The sale price was 90 percent of the regular price. If the regular price was $36, what was the sale price?

7. Forty people came to the party. This was 80 percent of those who were invited. How many were invited?

8. Write an equation to solve this problem. Twenty is 40 percent of what number?

9. Use two unit multipliers to convert 3600 in.^{2} to square feet.

10. Draw a diagram of this statement. Then answer the questions that follow.

>   Three fourths of the questions on the test were multiple-choice. There were 60 multiple-choice questions.

(a) How many questions were on the test?

(b) What percent of the questions on the test were not multiple-choice?

11. Evaluate: $x - y - xy$ if $x = -3$ and $y = -2$

12. Compare: $m \bigcirc n$ if m is an integer and n is a whole number

13. (a) Classify this quadrilateral.

(b) Find the perimeter of this figure.

(c) Find the area of this figure.

15 mm

13 mm

12 mm

20 mm

14. Complete the table.

Fraction	Decimal	Percent
$1\frac{2}{3}$	(a)	(b)

15. What number is $33\frac{1}{3}$ percent of 30? (*Hint*: Change $33\frac{1}{3}$ percent to a fraction.)

16. Multiply and write the product in scientific notation.

$$(2.4 \times 10^{-4})(5 \times 10^{-7})$$

17. A pyramid with a square base has how many

(a) faces?

(b) edges?

(c) vertices?

18. (a) Find the circumference of this circle.

(b) Find the area of this circle.

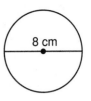

8 cm

Use 3.14 for π

19. Find the missing numbers in this function.

$y = 2x - 5$

x	y
2	☐
3	☐
5	☐

20. Use the information in the figure to answer these questions.

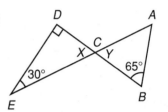

(a) What is $m\angle X$?

(b) What is $m\angle Y$?

(c) What is $m\angle A$?

21. Write an equation to solve this problem. Twenty is what percent of 30?

22. Graph $x \geq -3$.

23. Segment AB is how many millimeters longer than segment BC?

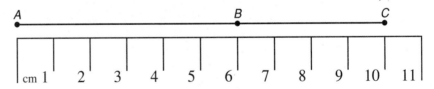

Solve:

24. $5 = y - 4.75$

25. $3\frac{1}{3}y = 7\frac{1}{2}$

Add, subtract, multiply, or divide, as indicated:

26. $\sqrt{3^2 + 2^4}$

27. $\dfrac{32\text{ ft}}{1\text{ sec}} \cdot \dfrac{60\text{ sec}}{1\text{ min}}$

28. $5\dfrac{1}{3} + 2.5 + \dfrac{1}{6}$ (fraction)

29. $\dfrac{2\frac{3}{4} + 3.5}{2\frac{1}{2}}$ (decimal)

30. (a) $\dfrac{(-3) - (-4)(+5)}{(-2)}$ (b) $-3 + 4 - 5 + 6 - 7$

LESSON
113

Sales Tax

To find the amount of **sales tax** on a purchase, we multiply the full price of the purchase by the tax rate.

Example 1 A bicycle is on sale for $119.95. The tax rate is 6 percent.

(a) What is the tax on the bicycle?

(b) What is the total price including tax?

Solution (a) To find the tax, we change 6 percent to the decimal 0.06 and multiply $119.95 by 0.06. We round the result to the nearest cent.

$$
\begin{array}{r}
119.95 \\
\times \quad .06 \\
\hline
7.1970
\end{array}
\quad \longrightarrow \quad \textbf{\$7.20}
$$

(b) To find the total price, including tax, we add the tax to the price.

$$
\begin{array}{rl}
119.95 & \text{price} \\
+ \quad 7.20 & \text{tax} \\
\hline
\textbf{\$127.15} & \text{total}
\end{array}
$$

Example 2 Find the total price, including tax, of an $18.95 book, a $1.89 pen, and a $2.29 pad of paper. The tax rate is 5 percent.

Solution   We begin by finding the combined price of the items.

$$
\begin{array}{rl}
\$18.95 & \text{book} \\
1.89 & \text{pen} \\
+\quad 2.29 & \text{paper} \\
\hline
\$23.13 &
\end{array}
$$

Next we multiply the combined price by 0.05 (5 percent) and round the product to the nearest cent.

$$
\begin{array}{r}
\$23.13 \\
\times \quad 0.05 \\
\hline
\$1.1565 \quad \longrightarrow \quad \$1.16 \text{ tax}
\end{array}
$$

Then we add the tax to the price to find the total.

$$
\begin{array}{r}
\$23.13 \\
+ \quad 1.16 \\
\hline
\mathbf{\$24.29} \quad \text{total}
\end{array}
$$

Practice   **a.** Find the sales tax on a $36.89 radio if the tax rate is 7 percent.

b. Find the total price of the radio, including tax.

c. Find the total price, including 6 percent tax, for a $6.95 dinner, a 95¢ beverage, and a $2.45 dessert.

Problem set 113   **1.** The following marks are Darren's 100-meter dash times, in seconds, during track season. Find the (a) median, (b) mode, and (c) range of these times.

12.3, 11.8, 11.9, 11.7, 12.0, 11.9, 12.1, 11.6, 11.8

2. How much money does Jackson earn working for 3 hours 45 minutes at $6 per hour?

Use a ratio box to solve Problems 3–6.

3. The recipe called for 3 cups of flour and 2 eggs to make 6 servings. If 15 cups of flour were used to make more servings, how many eggs should be used?

4. Lester can type 48 words per minute. At that rate, how many words can he type in 90 seconds?

5. Ten students scored 100 percent. This was 40 percent of the class. How many students were in the class?

6. The dress was on sale for 60 percent of the regular price. If the regular price was $24, what was the sale price?

7. Write an equation to solve this problem. What percent of 60 is 24?

8. Use two unit multipliers to convert 3 gal to pints.

9. Draw a diagram of this statement. Then answer the questions that follow.

> The Trotters won $\frac{5}{6}$ of their games. They won 20 games and lost the rest.

(a) How many games did they play?

(b) What was the Trotters' ratio of games won to games lost?

10. If $x = -2$ and $y = 2x + 1$, y equals what number?

11. Compare: $w \bigcirc m$ if w is 0.5 and m is the reciprocal of w

12. The figure at right is a hexagon with dimensions given in centimeters. Corners that look square are square. Find the area of the hexagon.

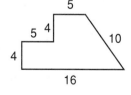

13. A television was on sale for $325. The tax rate was 6 percent.

(a) What was the tax on the television?

(b) What was the total price, including tax?

14. Multiply and write the product in scientific notation.

$$(8 \times 10^8)(4 \times 10^{-2})$$

15. Complete the table.

Fraction	Decimal	Percent
(a)	0.02	(b)

16. What is 65 percent of $24?

17. Sketch this cube. Then find its volume. Dimensions are in inches.

18. Find the circumference of this circle.

19. Find the area of this circle.

Leave π as π

20. Find the missing numbers in this function.

$$y = 3x + 1$$

21. Divide 1.23 by 9 and write the quotient

(a) with a bar over the repetend.

(b) rounded to three decimal places.

22. The ratio of the measure of angle B to the measure of angle C is 2 to 3. What is the measure of angle B?

23. Graph the counting numbers greater than -2. This will be the graph of the positive integers.

24. Draw a pair of parallel lines. Draw another pair of parallel lines that are perpendicular to the first pair. What kind of quadrilateral is formed?

Solve:

25. $3\frac{1}{7}d = 88$

26. $n + 1.61 = 10.6$

Add, subtract, multiply, or divide, as indicated:

27. $5^2 + \left(3^3 - \sqrt{81}\right)$

28.
$$\begin{array}{r} 1 \text{ hr } 15 \text{ min } 30 \text{ sec} \\ - \qquad 48 \text{ min } 45 \text{ sec} \\ \hline \end{array}$$

29. $\left(4\frac{4}{9}\right)(2.7)\left(1\frac{1}{3}\right)$

30. $(-2)(-3) - (-4)(-5)$

LESSON 114

Percents Greater than 100, Part 1

The methods we have used to work with percents less than 100 may also be used to work with percents greater than 100. In this lesson we will practice writing equations to solve problems that include percents greater than 100.

Example 1 Fifty is what percent of 40?

Solution We translate directly to an equation.

Fifty is what percent of 40?   question

$$50 \quad = \quad \frac{W_P}{100} \quad \times 40 \qquad \text{equation}$$

We solve by multiplying both sides by $\dfrac{100}{40}$.

$$\frac{100}{40} \cdot 50 = \frac{W_P \cdot \cancel{40}}{\cancel{100}} \cdot \frac{\cancel{100}}{\cancel{40}} \qquad \text{multiplied by } \frac{100}{40}$$

$$125 = W_P \qquad \text{simplified}$$

Since 125 is a percent, we can write

$$\mathbf{125\% = W_P} \qquad \text{solution}$$

This answer makes sense because 50 is greater than 40, so 50 is more than 100 percent of 40.

Example 2   What is 106 percent of $12.40?

Solution   We translate directly. We remember that 106 percent means 106 over 100.

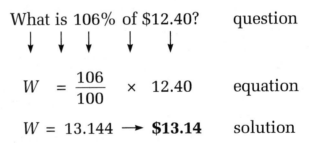

$$W = \frac{106}{100} \times 12.40 \qquad \text{equation}$$

$$W = 13.144 \longrightarrow \mathbf{\$13.14} \qquad \text{solution}$$

Example 3   Sixty is 150 percent of what number?

Solution   We translate directly.

Sixty is 150% of what number?   question

↓   ↓   ↓   ↓   ↓

$$60 = \frac{150}{100} \times W_N \qquad \text{equation}$$

We solve by multiplying both sides by $\dfrac{100}{150}$.

$$\frac{100}{150} \cdot 60 = \frac{\cancel{150}}{\cancel{100}} \cdot W_N \cdot \frac{\cancel{100}}{\cancel{150}} \qquad \text{multiplied by } \frac{100}{150}$$

$$\mathbf{40 = W_N} \qquad \text{solution}$$

Practice   **a.** What is 150 percent of 56?

b. Sixty is 250 percent of what number?

Problem set   **1.** The mean of these numbers is how much greater than the
114   median?

$$3, 12, 7, 5, 18, 6, 9, 28$$

2. What is the quotient when the sum of $\frac{5}{6}$ and $\frac{3}{4}$ is divided
by the product of $\frac{5}{6}$ and $\frac{3}{4}$?

3. The lengths of two sides of an isosceles triangle are 5 in.
and 1 ft. Sketch the triangle and find its perimeter in
inches.

Use a ratio box to solve Problems 4–7.

4. The ratio of youths to adults at the convocation was 3 to
7. If 4500 attended the convocation, how many adults
were present?

5. Every time the knight went over 2, he went up 1. If the
knight went over 8, how far did he go up?

6. Eighty percent of those who were invited came to the
party. If 40 people were invited to the party, how many
did not come?

7. The dress was on sale for 60 percent of the regular price.
If the sale price was $24, what was the regular price?

8. Write an equation to solve this problem. One hundred
forty percent of what number is 70?

9. Use two unit multipliers to convert 1,000,000 cm² to square
meters.

10. Draw a diagram of this statement. Then answer the questions that follow.

> Exit polls showed that 7 out of 10 voters cast their ballot for the incumbent. The incumbent received 1400 votes.

(a) How many voters cast their ballots?

(b) What percent of the voters did not vote for the incumbent?

11. Evaluate: $x + xy - xy$ if $x = 3$ and $y = -2$

12. Compare: $a \bigcirc a - a$ if $a < 0$

13. If the perimeter of a square is 1 meter, what is the area of the square in square centimeters?

14. Find the total price, including tax, of a $12.95 bat, a $7.85 baseball, and a $49.50 glove. The tax rate is 7 percent.

15. Multiply and write the product in scientific notation.

$$(3.5 \times 10^5)(3 \times 10^6)$$

16. Complete the table.

Fraction	Decimal	Percent
(a)	(b)	$33\frac{1}{3}\%$

Write an equation to solve Problems 17 and 18.

17. What is 125 percent of 84?

18. What is 106 percent of $180?

19. What is the volume of this rectangular prism? Dimensions are in feet.

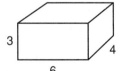

20. (a) Find the circumference of this circle.

(b) Find the area of this circle.

7 m

Use $\frac{22}{7}$ for π

21. Find the missing numbers in this function.

$y = 3x$

x	y
2	☐
−1	☐
☐	15

22. Polygon $ZWXY$ is a rectangle. What is the measure of each angle?

(a) $\angle a$ (b) $\angle b$ (c) $\angle c$

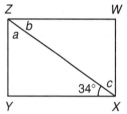

23. Graph $x \geq -2$.

24. What numbers correspond to the points marked A and B on this number line?

Solve:

25. $p + 7 = 50.2$

26. $1\frac{3}{4}f = 5\frac{1}{4}$

Add, subtract, multiply, or divide, as indicated:

27. $\sqrt{100 - 64} - \left(\sqrt{100} - \sqrt{64}\right)$

28.   8 lb 7 oz
      + 2 lb 9 oz

29. $\left(4\frac{1}{2}\right)(0.2)(10^2)$

30. $\frac{(-4)(+3)}{(-2)} - (-1)$

LESSON
115

Percents Greater than 100, Part 2

The percent problems that we have considered until now have used a percent to describe part of a whole. In this lesson we will consider percent problems that use a percent to describe an amount of change. The change may be an increase or a decrease.

Increase: $\dfrac{\text{Original}}{\text{number}} + \dfrac{\text{amount of}}{\text{change}} = \dfrac{\text{new}}{\text{number}}$

Decrease: $\dfrac{\text{Original}}{\text{number}} - \dfrac{\text{amount of}}{\text{change}} = \dfrac{\text{new}}{\text{number}}$

We may use a ratio box to help us with increase-decrease problems just as we have with ratio problems and part-part-whole problems. However, there are some differences in the way we set up the ratio box. When we make a table for a part-part-whole problem, the bottom number in the percent column is 100 percent.

	PERCENT	ACTUAL COUNT
Part		
Part		
Whole	100	

When we set up a ratio box for an increase-decrease problem, we also have three rows. The three rows represent the original number, the amount of change, and the new number. We will use the words **original-change-new** on the left side. Most increase-decrease problems consider the original amount to be 100 percent. So the top number in the percent column will be 100 percent.

	PERCENT	ACTUAL COUNT
Original	100	
Change		
New		

If the change is an increase, we add it to the original amount to get the new amount. If the change is a decrease, we subtract it from the original amount to get the new amount.

Example 1 The county's population increased 15 percent from 1980 to 1990. If the population in 1980 was 120,000, what was the population in 1990?

Solution First we identify the type of problem. The percent describes an amount of change. This is an increase problem. We make a table and write the words "original," "change," and "new" down the side. Since the change was an increase, we write a plus sign in front of change. In the percent column we write 100 percent for the original (1980 population), 15 percent for the change, and add to get 115 percent for the new (1990 population).

	PERCENT	ACTUAL COUNT
Original	100	120,000
+ Change	15	C
New	115	N

$$\frac{100}{115} = \frac{120,000}{N}$$

In the actual count column we write 120,000 for "original," and use letters for "change" and "new." We are asked for the new number. Since we know both numbers in the first row, we use the first and third rows to write the proportion.

$$\frac{100}{115} = \frac{120,000}{N}$$

$$100N = 13,800,000$$

$$N = 138,000$$

The county's population in 1990 was **138,000**.

Example 2 The price was reduced 30 percent. If the sale price was $24.50, what was the original price?

Solution First we identify the problem. This is a decrease problem. We make a table and write original, change, and new down the side with a minus sign in front of change. In the percent column we write 100 percent for original, 30 percent for change, and 70 percent for new. The sale price is the new actual count. We are asked to find the original price.

	PERCENT	ACTUAL COUNT
Original	100	R
– Change	30	C
New	70	24.50

$$\frac{100}{70} = \frac{R}{24.50}$$

$$70R = 2450$$

$$R = 35$$

The original price before it was reduced was **$35**.

Practice Use a ratio box to solve each problem.

a. The regular price was $24.50, but the item was on sale for 30 percent off. What was the sale price?

b. The number of students taking algebra increased 20 percent in one year. If 60 students are taking algebra this year, how many took algebra last year?

c. Bikes were on sale for 20 percent off. Tom bought one for $120. How much money did he save by buying the bike at the sale price instead of at the regular price?

Problem set 115

1. The product of the first three prime numbers is how much less than the sum of the next three prime numbers?

2. After 5 tests Amanda's average score was 88. What score must she average on the next 2 tests to have a 7-test average of 90?

3. Jenna finished a 2-mile race in 15 minutes. What was her average speed in miles per hour?

Use a ratio box to solve Problems 4–7.

4. Forty-five of the 80 students in the club were girls. What was the ratio of boys to girls in the club?

5. Two dozen sparklers cost $3.60. At that rate, how much would 60 sparklers cost?

6. The county's population increased 20 percent from 1980 to 1990. If the county's population in 1980 was 340,000, what was the county's population in 1990?

7. Because of unexpected cold weather, the cost of tomatoes increased 50 percent in one month. If the cost after the increase was 96¢ per pound, what was the cost before the increase?

8. Write an equation to solve this problem. Sixty is what percent of 75?

9. Use two unit multipliers to convert 100 cm² to square millimeters.

10. Draw a diagram of this statement. Then answer the questions that follow.

> Five eighths of the trees in the grove were deciduous. There were 160 deciduous trees in the grove.

(a) How many trees were in the grove?

(b) How many of the trees in the grove were not deciduous?

11. If $x = -5$ and $y = 3x - 1$, then y equals what number?

12. Compare: 30% of 20 ◯ 20% of 30

13. Find the area of this figure.

14. The price of the stereo was $179.50. The tax rate was 6 percent.

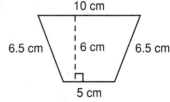

(a) What was the tax on the stereo?

(b) What was the total price, including tax?

15. Multiply and write the product in scientific notation.

$$(8 \times 10^{-5})(3 \times 10^{12})$$

16. Complete the table.

Fraction	Decimal	Percent
$2\frac{1}{3}$	(a)	(b)

17. Write an equation to solve this problem. What is 6.5 percent of $14,500?

18. What number is 250 percent of 60?

19. A triangular prism has how many

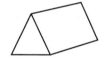

(a)   triangular faces?

(b)   rectangular faces?

20. John measured the diameter of his bicycle tire and found that it was 24 inches. What is the distance around the tire to the nearest inch? (Use 3.14 for π.)

21. Find the missing numbers in this function.

$$y = 7x + 1$$

x	y
6	☐
3	☐
−2	☐

22. The ratio of the measures of the two angles was 4 to 5. If the sum of their measures was 180°, what was the measure of the smaller angle?

23. Graph the whole numbers greater than −5.

24. Draw a pair of parallel lines. Draw a second pair of parallel lines that intersect but are not perpendicular to the first pair. What kind of quadrilateral is formed?

Solve:

25. $3\frac{1}{7}x = 66$

26. $w - 0.15 = 4.9$

Add, subtract, multiply, or divide, as indicated:

27. $(2 \cdot 3)^2 - 2(3^2)$

28. $1 \text{ L} - 100 \text{ mL}$

29. $5 - \left(3\frac{1}{3} - 1.5\right)$

30. $\dfrac{(-8)(-6)(-5)}{(-4)(-3)(-2)}$

LESSON 116

Solving Two-Step Equations

We have solved one-step equations by adding or subtracting. We have also solved one-step equations by multiplying or dividing. In this lesson we will begin solving two-step equations. When we solve equations such as $3x + 2 = 7$, we work to get the equation in this form:

$$1x + 0 = \text{a particular number}$$

Then we can write

$$x = \text{a particular number}$$

Thus, to solve $3x + 2 = 7$, we take one step to change 2 to 0 and a second step to change 3 to 1. The order of the steps is important. **We use the addition rule first.** Then we use the multiplication rule.

To change $+2$ to 0, we add -2 to (or subtract 2 from) both sides of the equation.

$$
\begin{array}{rll}
3x + 2 = 7 & \quad \text{equation} \\
\underline{-2 \quad -2} & \quad \text{add } -2 \\
3x + 0 = 5 & \quad \text{added} \\
3x = 5 & \quad 3x + 0 = 3x
\end{array}
$$

Now we change $3x$ to $1x$ by dividing both sides of the equation by 3 or by multiplying both sides by $\frac{1}{3}$.

<table>
<tr><td align="center">DIVIDE BY 3</td><td align="center">MULTIPLY BY $\frac{1}{3}$</td></tr>
<tr><td align="center">$\frac{3x}{3} = \frac{5}{3}$</td><td align="center">$\frac{1}{3} \cdot 3x = \frac{1}{3} \cdot 5$</td></tr>
<tr><td align="center">$1x = \frac{5}{3}$</td><td align="center">$1x = \frac{5}{3}$</td></tr>
<tr><td align="center">$x = \frac{5}{3}$</td><td align="center">$x = \frac{5}{3}$</td></tr>
</table>

Example 1   Solve: $5x - 9 = 36$

Solution   First we change -9 to 0 by adding 9 to both sides.

$$
\begin{array}{ll}
5x - 9 = 36 & \text{equation} \\
\underline{\quad +9 \quad +9} & \text{add 9} \\
5x + 0 = 45 & \text{added} \\
5x = 45 & 5x + 0 = 5x
\end{array}
$$

Now we will divide both sides by 5.

$$
\begin{array}{ll}
\frac{5x}{5} = \frac{45}{5} & \text{divided by 5} \\
1x = 9 & \text{simplified} \\
\mathbf{x = 9} & 1x = x
\end{array}
$$

Example 2   Solve: $\frac{3}{4}x + 6 = 18$

Solution   First we add -6 to or subtract 6 from both sides.

$$
\begin{array}{ll}
\frac{3}{4}x + 6 = 18 & \text{equation} \\
\underline{\quad\quad -6 \quad -6} & \text{add } -6 \\
\frac{3}{4}x + 0 = 12 & \text{added} \\
\frac{3}{4}x = 12 & \frac{3}{4}x + 0 = \frac{3}{4}x
\end{array}
$$

Then we multiply both sides by $\frac{4}{3}$.

$$\frac{4}{3} \cdot \frac{3}{4}x = \frac{4}{3} \cdot 12 \qquad \text{multiplied by } \frac{4}{3}$$

$$1x = \frac{48}{3} \qquad \text{multiplied}$$

$$x = 16 \qquad \text{simplified}$$

Example 3 Solve: $0.2x - 1.4 = 3$

Solution First we add +1.4 to both sides.

$$
\begin{array}{ll}
0.2x - 1.4 = 3 & \text{simplified} \\
\underline{+1.4 \quad +1.4} & \text{add 1.4} \\
0.2x + 0 = 4.4 & \text{added} \\
0.2x = 4.4 & \text{simplified}
\end{array}
$$

Then we divide by 0.2.

$$\frac{0.2x}{0.2} = \frac{4.4}{0.2} \qquad \text{divided by 0.2}$$

$$1x = 22 \qquad \text{divided}$$

$$x = 22 \qquad 1x = x$$

Practice Solve. Show each step.

a. $3x - 2 = 7$ **b.** $2x + 3 = 15$

c. $\frac{2}{3}x - 5 = 7$ **d.** $\frac{5}{6}x + 7 = 37$

e. $0.4x + 1.4 = 3$ **f.** $1.2x - 0.7 = 8.9$

Problem set 116

1. From Don's house to the lake is 30 km. If he completed the round trip on his bike in 2 hours 30 minutes, what was his average speed in kilometers per hour?

2. Find (a) the mean and (b) the range for this set of numbers.

$$3, 9, 7, 5, 10, 4, 5, 8, 5, 4, 8, 40$$

Use a ratio box to solve Problems 3–6.

3. The ratio of red marbles to blue marbles in the bag was 7 to 5. If there was a total of 600 red and blue marbles in the bag, how many marbles were blue?

4. The machine could punch out 500 plastic pterodactyls in 20 minutes. At that rate, how many could it punch out in $1\frac{1}{2}$ hours?

5. The price was reduced by 25 percent. If the regular price was $24, what was the sale price?

6. The price was reduced by 25 percent. If the sale price was $24, what was the regular price?

7. Write an equation to solve this problem. Seventy-five is what percent of 60?

8. Use two unit multipliers to convert 7 days to minutes.

9. Draw a diagram of this statement. Then answer the questions that follow.

> Five ninths of the 45 cars pulled by the locomotive were not cattle cars.

(a) How many cattle cars were pulled by the locomotive?

(b) What percent of the cars pulled by the locomotive were not cattle cars?

10. Compare: $\frac{1}{3}$ ◯ 33%

11. Evaluate: $ab - a - b$ if $a = -3$ and $b = -1$

12. Find the total price, including 5 percent tax, for a $7.95 dinner, a 90¢ beverage, and a $2.35 dessert.

13. A corner was cut from a square sheet of paper, resulting in this pentagon. Dimensions are in inches.

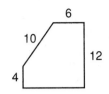

(a) What is the perimeter of this pentagon?

(b) What is the area of this pentagon?

14. Complete the table.

FRACTION	DECIMAL	PERCENT
(a)	0.08	(b)

15. Write an equation to solve this problem. What number is 120 percent of 360?

16. Multiply and write the product in scientific notation.

$$(8 \times 10^{-3})(6 \times 10^{7})$$

17. Each edge of the picture cube was 10 cm. What was the volume of the cube?

18. What is the area of this circle?

19. What is the circumference of this circle?

Use 3.14 for π

20. Find the missing numbers in this function.

$y = 2x + 3$

x	y
1	☐
0	☐
−2	☐

21. Write an equation to solve this problem. Sixty is $\frac{3}{8}$ of what number?

22. Graph the integers that are greater than −2 and less than 2.

23. Find the measure of each angle.

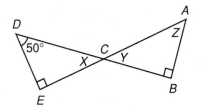

(a) $\angle X$

(b) $\angle Y$

(c) $\angle Z$

24. What is the sum of the numbers labeled A and B on this number line?

Solve:

25. $3x + 2 = 9$ **26.** $\dfrac{2}{3}w + 4 = 14$

27. $0.2y - 1 = 7$

Add, subtract, multiply, or divide, as indicated:

28. $3\left(2^3 + \sqrt{16}\right) - \sqrt{36}$ **29.** $5\dfrac{1}{4} + 3\dfrac{5}{6} + 2.5$

30. (a) $\dfrac{(-9)(+6)(-5)}{(-4) - (-1)}$ (b) $-3(4) + 2(3) - 1$

LESSON 117

Simple Probability

The singular of dice is die. A die has six faces. If we roll a die, the number of dots on the top face will be 1, 2, 3, 4, 5, or 6. Each face is equally likely to occur, and each roll of the die is called an **outcome**. **Probability** is the likelihood that a particular event will occur. Probability is written as a ratio.

$$\text{Probability} = \frac{\text{number of favorable outcomes}}{\text{number of possible outcomes}}$$

We will consider three examples.

Example 1   A single die is rolled. What is the probability that it will come up

(a) a 4?

(b) a number greater than 4?

(c) a number greater than 6?

(d) a number less than 7?

Solution   There are 6 different faces on a die, so there are 6 equally likely outcomes. Thus, 6 will be the bottom number of each of these ratios.

(a)   There is only one way to roll a 4 with one die, so the probability of rolling a 4 is $\frac{1}{6}$.

(b)   The numbers greater than 4 on a die are 5 and 6, so there are 2 ways to roll a number greater than 4. Thus the probability is $\frac{2}{6}$, which we reduce to $\frac{1}{3}$.

(c)   On a die, there are no numbers greater than 6, so there is no way to roll a number greater than 6. Thus the probability is $\frac{0}{6}$, which is **0. An event that cannot happen has a probability of zero.**

(d)   On a die, there are 6 numbers less than 7, so there are 6 ways to roll a number less than 7. Thus the probability is $\frac{6}{6}$, which is **1. An event that is certain to happen has a probability of 1.**

Example 2   If this spinner is spun, what is the probability of the spinner

(a)   stopping on 3?

(b)   not stopping on 3?

Solution   There are 5 equally likely outcomes, so 5 is the bottom number of each ratio.

(a)   There is one way for the spinner to stop on 3, so the probability is $\frac{1}{5}$.

(b)   There are 4 ways for the spinner to not stop on 3, so the probability is $\frac{4}{5}$.

Notice that the probability of an event happening plus the probability of an event not happening is 1.

Example 3   What is the probability of this spinner stopping on 1?

Solution   There are 3 possible outcomes, but the outcomes are not equally likely. Since region 1 occupies half the area, the probability of the spinner stopping in region 1 is $\frac{1}{2}$. Regions 2 and 3 each occupy $\frac{1}{4}$ of the area, so the probability of the spinner stopping on 2 is $\frac{1}{4}$ and on 3 is $\frac{1}{4}$.

Practice   **a.**   What is the probability of rolling a number less than 4 with one roll of a die?

b.   What is the probability of this spinner stopping on 3?

c.   What is the probability of this spinner stopping on 5?

d.   What is the probability of this spinner stopping on a number less than 6?

e.   What is the probability of this spinner stopping on C?

f.   What is the probability of this spinner not stopping on B?

Problem set 117

1.   Twenty-one billion is how much more than nine billion, eight hundred million? Write the answer in scientific notation.

2.   The train traveled at an average speed of 48 miles per hour for the first 2 hours and at 60 miles per hour for the next 4 hours. What was the train's average speed for the 6-hour trip? (Average equals total miles divided by total time.)

3. A 10-pound box of detergent costs $8.40. A 15-pound box costs $10.50. Which costs the most per pound? How much more?

4. In a rectangular prism, what is the ratio of faces to edges?

Use a ratio box to solve Problems 5–8.

5. The team's won-lost ratio was 3 to 2. If the team won 12 games and did not tie any games, then how many games did the team play?

6. Twenty-four is to 36 as 42 is to what number?

7. What number is 20 percent less than 360?

8. During his slump, Matt's batting average dropped by 20 percent to .260. What was Matt's batting average before his slump?

9. Use two unit multipliers to perform each conversion.

   (a) 12 ft^2 to square inches

   (b) 1 km to millimeters

10. Draw a diagram of this statement. Then answer the questions that follow.

      The duke conscripted two fifths of the male serfs in his dominion. He conscripted 120 serfs in all.

   (a) How many male serfs were in the duke's dominion?

   (b) How many male serfs in his dominion were not conscripted?

11. If a die is tossed once, what is the probability that the number rolled is

   (a) a prime number?

   (b) a number greater than 6?

12. If $y = 4x - 3$ and $x = -2$, then y equals what number?

13. The perimeter of a certain square is 4 yards. Find the area of the square in square feet.

14. The price of the new car was $14,500. The tax rate was 6.5 percent.

   (a)   What was the sales tax on the car?

   (b)   What was the total price including tax?

15. Complete the table.

Fraction	Decimal	Percent
(a)	(b)	$66\frac{2}{3}\%$

16. What is 200 percent of $7.50?

17. Multiply and write the product in scientific notation.

$$(2 \times 10^8)(8 \times 10^2)$$

18. Robbie stores his 1-inch blocks in a box with inside dimensions as shown. How many blocks will fit in this box?

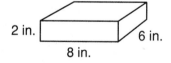

19. The length of each side of the square equals the diameter of the circle. The area of the square is how much greater than the area of the circle?

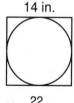

Use $\frac{22}{7}$ for π

20. Divide 7.2 by 0.11 and write the quotient with a bar over the repetend.

21. Find the missing numbers in this function.

$y = 5x - 12$

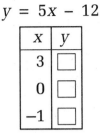

x	y
3	☐
0	☐
−1	☐

22. Graph all of the numbers that are greater than −2 and are also less than 2.

23. In the figure, the measure of $\angle AOC$ is half the measure of $\angle AOD$. The measure of $\angle AOB$ is one third the measure of $\angle AOD$.

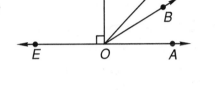

(a) Find $m\angle AOB$.

(b) Find $m\angle EOC$.

24. The length of segment BC is how much less than the length of segment AB?

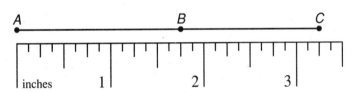

Solve:

25. $3x - 2 = 9$

26. $6\frac{2}{3}m = 1\frac{1}{9}$

27. $1.2p + 4 = 28$

Add, subtract, multiply, or divide, as indicated:

28. 1 yd 6 in. − 1 ft 7 in.

29. $4\frac{7}{8} - 2.5 - \frac{1}{4}$

30. (a) $\dfrac{(-8) - (-6) - (4)}{-3}$

(b) $-5(-4) - 3(-2) - 1$

LESSON 118

Volume of a Right Solid

A **right solid** is a geometric solid whose sides are perpendicular to the base. The volume of a right solid is the area of the base times the height. This rectangular solid is a right solid. It is 5 m long and 2 m deep, so the area of the base is 10 m².

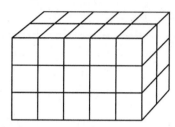

One cube will fit on each square meter of the base and the cubes are stacked 3 high, so

$$\text{Volume} = \text{area of base} \times \text{height}$$

$$= 10 \text{ m}^2 \times 3 \text{ m}$$

$$= 30 \text{ m}^3$$

The volume of any right solid is the area of the base times the height. If the base of a right solid is a circle, the solid is called a **right circular cylinder.** If the base of the solid is a polygon, the solid is called a **prism.**

Right square prism

Right triangular prism

Right circular cylinder

Example 1 Find the volume of this right triangular prism. Dimensions are in centimeters. We show two views of the prism.

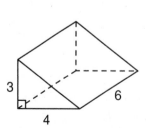

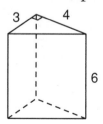

Solution The area of the base is the area of the triangle.

$$\text{Area of base} = \frac{(4 \text{ cm})(3 \text{ cm})}{2} = 6 \text{ cm}^2$$

The volume equals the area of the base times the height.

$$\text{Volume} = (6 \text{ cm}^2)(6 \text{ cm}) = \mathbf{36 \text{ cm}^3}$$

Example 2 The diameter of this right circular cylinder is 20 cm. Its height is 25 cm. What is its volume? Leave π as π.

Solution First we find the area of the base. The diameter of the circular base is 20 cm, so the radius is 10 cm.

$$\text{Area} = \pi r^2 = \pi(10 \text{ cm})^2 = 100\pi \text{ cm}^2$$

The volume equals the area of the base times its height.

$$\text{Volume} = (100\pi \text{ cm}^2)(25 \text{ cm}) = \mathbf{2500\pi \text{ cm}^3}$$

Practice Find the volume of each right solid shown. Dimensions are in centimeters.

a.

b.

c.

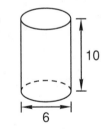

Leave π as π

d.

Base

Prism

e.

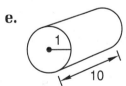

Leave π as π

Problem set 118

1. The taxi ride cost $1.40 plus 35¢ for each tenth of a mile. What was the average cost per mile for a 4-mile taxi ride?

2. The product of the first four prime numbers is how much greater than the sum of the next four prime numbers?

3. What is the average of these fractions?

$$\frac{1}{4}, \frac{1}{6}, \frac{1}{12}$$

4. If Jackson is paid $6 per hour, how much will he earn in 4 hours 20 minutes?

5. What is the ratio of the shaded area to the unshaded area of this rectangle?

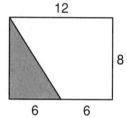

Use a ratio box to solve Problems 6–8.

6. If 600 pounds of sand cost $7.20, what would be the cost of 1 ton of sand at the same price per pound?

7. What is 30 percent more than $3.90?

8. The cost of production rose 30 percent. If the new cost is $3.90 per unit, what was the old cost per unit?

9. Use two unit multipliers to convert 1000 mm² to square centimeters.

10. Draw a diagram of this statement. Then answer the questions that follow.

> Three fifths of the Lilliputians believed in giants. The other 60 Lilliputians did not believe in giants.

(a) How many Lilliputians were there?

(b) How many Lilliputians believed in giants?

11. Compare: $a \bigcirc b$ if a is a counting number and b is an integer

12. Evaluate: $m(m + n)$ if $m = -2$ and $n = -3$

13. If a die is tossed once, what is the probability that the number rolled is

(a)   an even number?

(b)   a number less than 7?

14. Find the volume of this triangular prism. Dimensions are in millimeters.

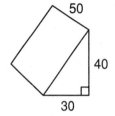

15. The diameter of a soup can was 6 cm. Its height was 10 cm. What was the volume of the soup can?

16. Find the total cost, including 6 percent tax, of 3 tacos at $1.25 each, 2 soft drinks at 95¢ each, and a shake at $1.30.

17. Complete the table.

FRACTION	DECIMAL	PERCENT
$2\frac{3}{4}$	(a)	(b)

18. What is 6.5 percent of $36?

19. Multiply and write the product in scientific notation.

$$(8 \times 10^{-6})(4 \times 10^{4})$$

20. Find the missing numbers in this function.

$y = \frac{1}{2}x + 1$

x	y
6	☐
4	☐
−2	☐

21. Divide 1000 by 48 and write the quotient as a mixed number.

22. Find the measures of the following angles.

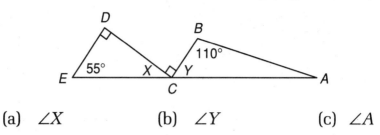

(a) ∠X   (b) ∠Y   (c) ∠A

23. Graph all the negative numbers that are greater than −2.

24. What is the average of the numbers labeled A and B on this number line?

Solve:

25. $5w + 11 = 51$ **26.** $\dfrac{4}{3}x - 2 = 14$

27. $0.9m + 1.2 = 3$

Add, subtract, multiply, or divide, as indicated:

28. $\sqrt{1^3 + 2^3} + (1 + 2)^3$ **29.** $5 - 2\dfrac{2}{3}\left(1\dfrac{3}{4}\right)$

30. (a) $\dfrac{(-10) + (-8) - (-6)}{(-2)(+3)}$

(b) $-8 + 3(-2) - 6$

LESSON
119

Rectangular Coordinates

By drawing two perpendicular number lines and extending the marks, we can create a grid or graph over an entire plane. Then we can identify any point on the coordinate plane with two numbers.

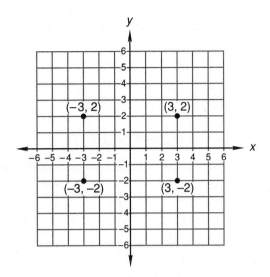

The horizontal number line is called the **x-axis**. The vertical number line is called the **y-axis**. The point at which the *x*-axis and the *y*-axis intersect is called the **origin**. The two numbers that indicate the location of a point are the **coordinates** of the point. The coordinates are written as a pair of numbers in parentheses, such as (3, 2). The first number shows the horizontal (↔) direction and distance from the origin. The second number shows the vertical ($\updownarrow$) direction and distance from the origin. The sign of the number indicates the direction. Positive coordinates are to the right or up. Negative coordinates are to the left or down.

The two axes divide the plane into four regions called **quadrants**, which are numbered counterclockwise beginning with the upper right as first, second, third, and fourth. The signs of the coordinates of each quadrant are shown below.

	y	
II Second Quadrant (−, +)		I First Quadrant (+, +)
	→ *x*	
III Third Quadrant (−, −)		IV Fourth Quadrant (+, −)

To graph a point on a coordinate plane, we draw a dot at the point indicated by the coordinates.

Example 1 Graph the following points on a coordinate plane.

(a) (3, 4)   (b) (2, −3)   (c) (−1, 2)   (d) (0, −4)

Solution It is best to use quad-ruled graph paper to do these exercises. We darken a horizontal and vertical line for the *x*- and *y*-axes. For now, we will let the distance between adjacent lines represent a distance of 1 unit. To graph each point, we begin at the origin. To graph (3, 4), we move to the right (positive) 3 units along the *x*-axis. **From there** we turn and move up (positive) 4 units and make a dot. We label the location (3, 4). We follow a similar procedure for each point.

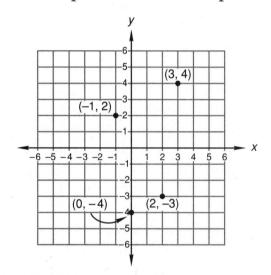

Example 2 Find the coordinates for points *A, B,* and *C* on this coordinate plane.

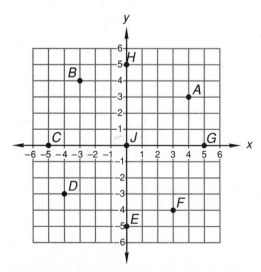

Solution We first find the point on the *x*-axis that is directly above or below the designated point. That is the first coordinate. Then we determine how many units above or below the *x*-axis the point is. That is the second coordinate.

> **Point A (4, 3)**
> **Point B (− 3, 4)**
> **Point C (− 5, 0)**

Practice Refer to the coordinate plane in Example 2 to find the coordinates of the following points.

a. Point D **b.** Point E

c. Point F **d.** Point J

Sketch a coordinate plane and graph the following points.

e. $(5, 0)$ **f.** $(3, -2)$

g. $(-1, -3)$ **h.** $(-2, 4)$

Problem set 119

1. What is the quotient when the product of 0.2 and 0.05 is divided by the sum of 0.2 and 0.05?

2. The table shows how many students earned certain scores on the last test. Find the (a) mode and (b) range of these scores.

CLASS TEST SCORES

SCORE	NUMBER OF STUDENTS
100	IIII
95	HHT I
90	HHT III
85	HHT II
80	III
75	I
70	I

3. Melissa finished a 3-mile race in 20 minutes. What was her average speed in miles per hour?

4. Use two unit multipliers to convert 1 km² to square meters.

5. Tim has $5 in quarters and $5 in dimes. What is the ratio of the number of quarters to the number of dimes?

Use a ratio box to solve Problems 6–8.

6. Jaime ran the first 3000 meters in 9 minutes. At that rate, how long will it take Jaime to run 5000 meters?

7. Sixty is 20 percent more than what number?

8. To attract customers, the merchant reduced all prices by 25 percent. What was the reduced price of an item that cost $36 before the price reduction?

9. Write an equation to solve this problem. Sixty is 150 percent of what number?

10. Draw a diagram of this statement. Then answer the questions that follow.

> Diane kept $\frac{2}{3}$ of her baseball cards and gave the remaining 234 cards to her brother.

   (a) How many cards did Diane have before she gave some to her brother?

   (b) How many baseball cards did Diane keep?

11. Compare: $a - b \bigcirc b - a$ if $a > b$

12. If a card is drawn from a normal deck of 52 cards (13 spades, 13 hearts, 13 diamonds, 13 clubs), what is the probability that the card is

   (a) a king?

   (b) a heart?

13. Find the area of this trapezoid. Dimensions are in centimeters.

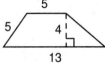

Find the volume of each solid. Dimensions are in inches.

14.

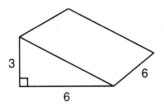

3

6

6

15.

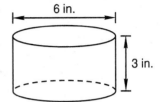

6 in.

3 in.

16. The skateboard cost $36. The tax rate is 6.5 percent.

    (a) What is the tax on the skateboard?

    (b) What is the total price, including tax?

17. Complete the table.

FRACTION	DECIMAL	PERCENT
(a)	0.15	(b)

18. What number is $66\frac{2}{3}$ percent of 48?

19. Multiply and write the product in scientific notation.

$$(6 \times 10^{-8})(8 \times 10^4)$$

20. Find the missing numbers in this function.

$y = \frac{2}{3}x - 1$

x	y
6	☐
0	☐
−3	☐

21. Use a ratio box to solve this problem. The ratio of the measures of the two acute angles of the right triangle is 7 to 8. What is the measure of the smallest angle of the triangle?

22. Graph the following points on a coordinate plane:

    (a) $(3, -2)$ (b) $(-5, 0)$ (c) $(-4, -3)$

23. Find the coordinates for points *M*, *N*, and *P* on this coordinate plane.

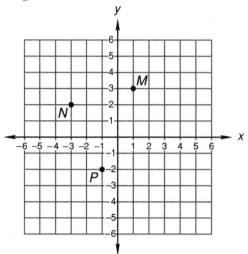

Solve:

24. $1\frac{1}{2}p + 6 = 36$

25. $7y - 1.3 = 5$

Add, subtract, multiply, or divide, as indicated:

26. $\sqrt{144} - \left(\sqrt{36}\right)\left(\sqrt{4}\right)$

27. $\dfrac{1 \text{ mi}}{4 \text{ min}} \cdot \dfrac{60 \text{ min}}{1 \text{ hr}}$

28. $\left(1\frac{5}{9}\right)(1.5) \div 2\frac{2}{3}$

29. $9.5 - \left(4\frac{1}{5} - 3.4\right)$

30. $\dfrac{(-18) + (-12) - (-6)(3)}{-3}$

LESSON 120

Estimating Angle Measures

We have practiced reading the measure of an angle from a protractor scale. The ability to measure an angle with a protractor is an important skill. The ability to **estimate** the measure of an angle is also a valuable skill. In this lesson we will learn a technique to help us estimate the measure of an angle. We will also practice using a protractor as we check our estimates.

To estimate a measurement, we need a mental image of the units to be used in the measurement. To estimate angle measure, we need a mental image of a degree scale—a mental protractor. We can "build" a mental image of a protractor from a mental image we already have—the face of a clock.

This clock face is a full circle. A full circle is 360°. A clock face is divided into 12 numbered divisions. From one numbered division to the next is $\frac{1}{12}$ of a full circle. One twelfth of 360° is 30°. Thus, the measure of the angle formed by the hands of the clock at 1 o'clock is 30°, at 2 o'clock is 60°, and at 3 o'clock is 90°. A clock face is further divided into 60 smaller divisions. From one small division to the next is $\frac{1}{60}$ of a circle. One sixtieth of a circle is 6°. Thus from one minute mark to the next on the face of a clock is 6°.

Here we have drawn an angle on the face of a clock. The vertex of the angle is at the center of the clock. One side of the angle is set at 12.

The other side of the angle is at "8 minutes after." Since each minute of separation represents 6°, the measure of this angle is 8 × 6°, which is 48°. With some practice we can usually estimate the measure of an angle within 5° of its actual measure.

Example (a) Record your estimate of the measure of ∠*BOC*.

(b) Use a protractor to find the measure of ∠*BOC*.

(c) By how many degrees did your estimate miss your measurement?

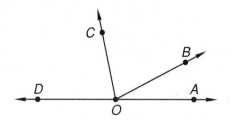

Solution (a) We use a mental image of a clock face on ∠*BOC* with one side of the angle set at 12. Mentally we may see that the other side is more than 10 minutes after. Perhaps it is 12 minutes after. Since 12 × 6° is 72°, we estimate that *m*∠*BOC* is **72°**.

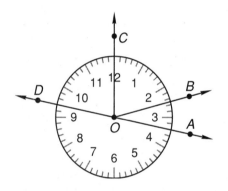

(b) We trace angle *BOC* on our paper and extend the sides so that we can use a protractor. We find that *m*∠*BOC* is **75°**.

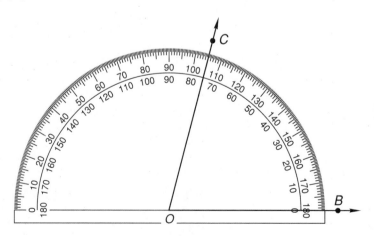

(c) Our estimate, 72°, misses our measurement, 75°, by **3°**.

Practice Find the measure of each angle shown on the clock face.

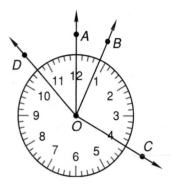

a. ∠AOB

b. ∠AOC

c. ∠AOD

In practice problems **d–g**, estimate the measure of each angle. Then use a protractor and measure each angle. By how many degrees did your estimate miss your measurement?

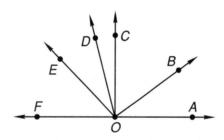

d. ∠BOC   e. ∠DOC   f. ∠FOE   g. ∠FOB

Problem set
120

1. In May the merchant bought 3 tons of beans at an average price of $280 per ton. In June the merchant bought 5 tons of beans at an average price of $240 per ton. What was the average price of all the beans bought by the merchant in May and June?

2. What is the quotient when 9 squared is divided by the square root of 9?

3. The Adams' car has a 16-gallon gas tank. How many tanks of gas will the car use on a 2000-mile trip if the car averages 25 miles per gallon?

4. In a triangular prism, what is the ratio of the number of vertices to the number of edges?

Use a ratio box to solve Problems 5–7.

5. If 58 dollars equals 100 marks, what is the cost in dollars of an item that sells for 250 marks?

6. Sixty is 20 percent less than what number?

7. The average number of customers increased 25 percent during the sale. If the average number of customers before the sale was 120 per day, what was the average number of customers per day during the sale?

8. Write an equation to solve this problem. Sixty is what percent of 50?

9. Use two unit multipliers to convert 1.2 m² to square centimeters.

10. Draw a diagram of this statement. Then answer the questions that follow.

   Twenty-four of the eggs were cracked. This was $\frac{1}{6}$ of the total number of eggs in the crate.

   (a) How many eggs were in the crate?

   (b) What percent of the eggs in the crate were not cracked?

11. Compare: $x + y$ ◯ $x - y$ if $y > 0$

12. Evaluate: $\dfrac{a + b}{c}$ if $a = -4$, $b = -3$, and $c = -2$

13. The perimeter of a certain square is 1 yard. Find the area of the square in square inches.

14. What is the probability of this spinner

   (a) stopping on 3?

   (b) not stopping on 3?

15. Find the volume of each solid. Dimensions are in centimeters.

(a)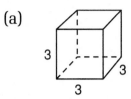

3

3

3

(b)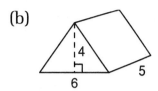

4

6

5

16. Find the total price, including 7 percent tax, of 20 square yards of carpeting priced at $14.50 per square yard.

17. Complete the table.

Fraction	Decimal	Percent
(a)	(b)	$3\frac{3}{4}\%$

18. What is $33\frac{1}{3}$ percent of $24?

19. Multiply and write the product in scientific notation.

$$(3 \times 10^3)(8 \times 10^{-8})$$

20. (a) Find the circumference of this circle.

(b) Find the area of this circle.

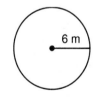

6 m

Leave π as π

21. Use the clock face to estimate the measure of each angle.

(a) $\angle BOC$

(b) $\angle COA$

(c) $\angle DOA$

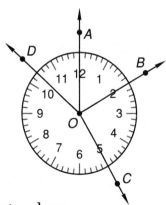

22. Graph each point on a coordinate plane.

(a) $(-3, 2)$ (b) $(0, -5)$ (c) $(3, -4)$

23. Find the coordinates for points R, S, and T on this coordinate plane.

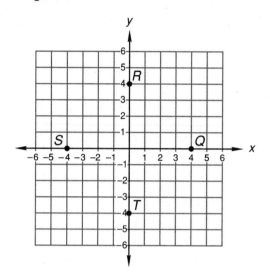

Solve:

24. $0.8m + 1.5 = 4.7$

25. $2\frac{1}{2}x - 7 = 13$

Add, subtract, multiply, or divide, as indicated:

26. $3^3 - \sqrt{49} + 3 \cdot 2^4$

27. $\begin{array}{r} 3 \text{ yd } 2 \text{ ft } 7\frac{1}{2} \text{ in.} \\ + \phantom{3 \text{ yd } 2 \text{ ft } } 5\frac{1}{2} \text{ in.} \\ \hline \end{array}$

28. $12.5 - \left(3\dfrac{3}{5} + 2.7\right)$

29. $2.7 \div \left(3 \div 1\dfrac{2}{3}\right)$

30. $\dfrac{(-4) - (-8)(-3)(-2)}{-2}$

**LESSON
121**

Similar Triangles

We often use "tick marks" to indicate that the measures of angles are equal.

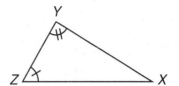

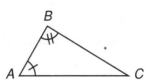

In these figures the single tick marks indicate that angles A and Z have equal measures. The double tick marks indicate that angles B and Y have equal measures.

If three angles in one triangle have the same measures as three angles in another triangle, the triangles are called *similar triangles*. Similar triangles look alike.

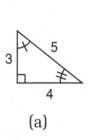

(a)

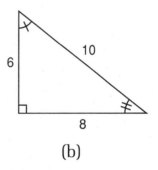

(b)

These triangles are not the same size, but they have the same shape because the angles of the same measures are in both triangles. These triangles are similar triangles. If an angle in one triangle has the same measure as an angle in a similar triangle, the angles are called **corresponding angles**. The sides opposite corresponding angles are called **corresponding sides**.

Example 1   List the pairs of corresponding angles and the pairs of corresponding sides for these triangles.

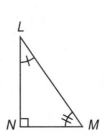

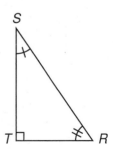

Solution The tick marks tell us that the angles in the triangle on the left have the same measures as the angles in the triangle on the right. This means that the triangles are similar triangles.

Corresponding Angles	Corresponding Sides
∠N and ∠T	$\overline{LM}$ and $\overline{SR}$
∠L and ∠S	$\overline{NL}$ and $\overline{TS}$
∠M and ∠R	$\overline{NM}$ and $\overline{TR}$

The lengths of corresponding sides in similar triangles are proportional. This means that the ratios formed by corresponding sides are equal.

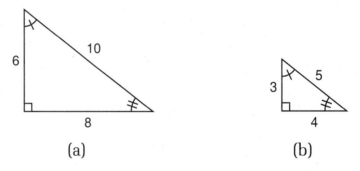

(a) (b)

Every side in triangle (a) is twice as long as the corresponding side in triangle (b). If we write the ratios of corresponding sides and put the sides for triangle (a) on top, we get

Shortest Sides	Middle Sides	Longest Sides
$\dfrac{6}{3}$	$\dfrac{8}{4}$	$\dfrac{10}{5}$

Each one of these ratios equals 2. If we put the sides of triangle (b) on top, we get

Shortest Sides	Middle Sides	Longest Sides
$\dfrac{3}{6}$	$\dfrac{4}{8}$	$\dfrac{5}{10}$

Each one of these ratios equals one half.

Example 2   Write the equal ratios of corresponding sides for these triangles.

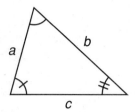

Solution   The angles are equal, so the triangles are similar. We decide to write the lengths of the sides of the left triangle on top.

$$\frac{a}{y} = \frac{b}{x} = \frac{c}{z}$$

Example 3   Find the length of side a.

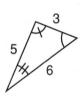

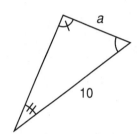

Solution   We will write a proportion and solve for a. We decide to write the ratios so that the sides from the left triangle are on top.

$$\frac{6}{10} \qquad \frac{3}{a}$$

Since these ratios are equal, we can connect them with an equals sign, cross multiply, and solve for a.

$$\frac{6}{10} = \frac{3}{a} \qquad \text{equal ratios}$$

$$6a = 30 \qquad \text{cross multiplied}$$

$$a = 5 \qquad \text{solved}$$

Practice   **a.**   Identify each pair of corresponding angles and each pair of corresponding sides in these two triangles.

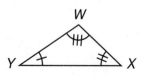

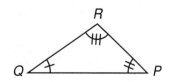

b. Find the length of side *x*.

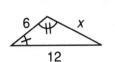 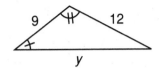

Problem set 121

1. Ginger gave the clerk $10 for a tape that cost $8.95 plus 6 percent tax. How much money should she get back?

2. Three hundred billion is how much less than two trillion? Write the answer in scientific notation.

3. During the second semester Joe's test scores were

   95, 90, 80, 85, 90, 100, 85, 90, 95, 80

   Find the (a) median, (b) mode, and (c) range of these scores.

Use a ratio box to solve Problems 4 and 5.

4. Coming down the long hill, Nelson averaged 24 miles per hour. If it took him 5 minutes to come down the hill, how long was the hill?

5. If Nelson traveled 3520 yards in 5 minutes, how far could he travel in 8 minutes at the same rate?

6. The points (3, 2), (3, −2), (−1, −2), and (−1, 2) are the vertices of a square. The area of the square is how many square units?

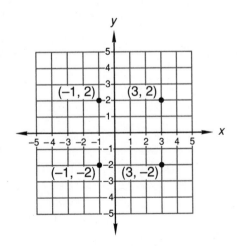

7. Three fourths of a yard is how many inches?

8. Use a ratio box to solve this problem. The ratio of leeks to radishes growing in the garden was 5 to 7. If 420 radishes were growing in the garden, how many leeks were there?

Write equations to solve Problems 9–11.

9. Forty is 250 percent of what number?

10. Forty is what percent of 60?

11. What decimal number is 40 percent of 6?

12. Use a ratio box to solve this problem. The tuition increased 10 percent this year. If the tuition this year is $2310, what was the tuition last year?

13. What is the average of the two numbers marked by arrows on this number line?

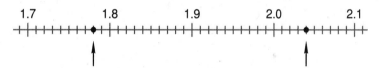

14. Complete the table.

FRACTION	DECIMAL	PERCENT
(a)	3.25	(b)

15. Compare: $x + y$ ◯ $x - y$ if x is positive and y is negative

16. Multiply and write the product in scientific notation.

$$(5.4 \times 10^8)(6 \times 10^{-4})$$

17. Find (a) the circumference and (b) the area of a circle with a radius of 10 millimeters.

18. Find the area of this trapezoid. Dimensions are in feet.

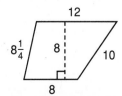

19. Find the volume of each of these solids. Dimensions are in meters.

(a)   (b)

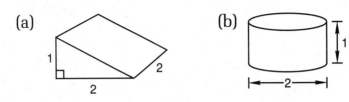

20. Refer to the figure shown. What are the measures of the following angles?

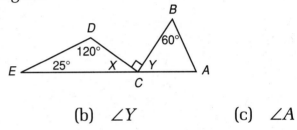

(a) ∠X   (b) ∠Y   (c) ∠A

21. The triangles are similar. Find *x*. Dimensions are in centimeters.

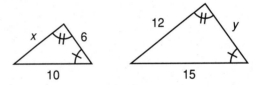

22. Use two unit multipliers to convert 10 yards to inches.

23. Estimate by first rounding each number to one nonzero digit:

$$\frac{(38{,}470)(607)}{79}$$

Solve:

24. $1.2m + 0.12 = 12$ **25.** $1\frac{3}{4}y - 2 = 12$

Add, subtract, multiply, or divide, as indicated:

26. $\sqrt{225} - \sqrt{121}$

27. $\dfrac{15\ \text{mi}}{1\ \text{hr}} \cdot \dfrac{1\ \text{hr}}{60\ \text{min}} \cdot \dfrac{1760\ \text{yd}}{1\ \text{mi}}$

28. $5\frac{1}{2} + \left(6\frac{1}{2} - 3.45\right)$ **29.** $3\frac{1}{3} \div \left(4.5 \div 1\frac{1}{8}\right)$

30. $\dfrac{(-2) - (+3) + (-4)(-3)}{(-2) + (+3) - (+4)}$

LESSON 122

Scale and Scale Factor

Scale In the preceding lesson we discussed similar triangles. Scale models and scale drawings are other examples of similar shapes. Scale models and scale drawings are reduced (or enlarged) renderings of actual objects. As is true of similar triangles, the lengths of corresponding parts of scale models and the objects they represent are proportional.

The **scale** of the model is stated as a ratio. For instance, if a model airplane is $\frac{1}{24}$ the size of the actual airplane, the scale is stated as $\frac{1}{24}$, or 1:24. We may use the given scale to write a proportion to find a measurement either on the model or on the actual object. A ratio box helps us put the numbers in the proper places.

Example 1 A model airplane is built on a scale of 1:24. If the wingspan of the model is 18 inches, the wingspan of the actual airplane is how many feet?

Solution We will construct a ratio box as we do with other ratio problems. In one column we write the ratio numbers which are the scale numbers. In the other column we write the measures. The first number of the scale refers to the model. The second number refers to the object. We can use the entries in the ratio box to write a proportion.

	Ratio scale	Measure
Model	1	18
Object	24	w

$\longrightarrow \quad \dfrac{1}{24} = \dfrac{18}{w}$

$w = 432$

The wingspan of the model was given in inches. Solving the proportion, we find that the full-size wingspan is 432 inches. We are asked for the wingspan in feet, so we convert units from inches to feet.

$$432 \text{ in.} \cdot \frac{1 \text{ ft}}{12 \text{ in.}} = 36 \text{ ft}$$

We find that the wingspan of the airplane is **36 feet**.

Example 2 Sarah is molding a model of a car from clay. The scale of the model is 1:36. If the height of the car is 4 feet 6 inches, what should be the height of the model in inches?

Solution First we convert 4 feet 6 inches to inches.

$$4 \text{ feet } 6 \text{ inches} = 54 \text{ inches}$$

Then we construct a ratio box using 1 and 36 as the ratio numbers, write the proportion, and solve.

	Ratio scale	Measure
Model	1	m
Object	36	54

$$\frac{1}{36} = \frac{m}{54}$$

$$36m = 54$$

$$m = \frac{54}{36}$$

$$m = 1\frac{1}{2}$$

The height of the model car should be $1\frac{1}{2}$ **inches**.

Scale factor We have solved proportions by using cross products. Sometimes a proportion can be solved more quickly by noting the scale factor. The scale factor is the number of times larger (or smaller) the terms of one ratio are when compared with the terms of the other ratio. The scale factor in the proportion below is 6 because the terms of the second ratio are 6 times the terms of the first ratio.

$$\overset{\times\ 6}{\frac{3}{4}} = \overset{}{\frac{18}{24}}$$
$$\underset{\times\ 6}{}$$

Example 3 Solve: $\dfrac{3}{7} = \dfrac{15}{n}$

Solution Instead of finding cross products, we note that 3 times 5 equals 15. Thus the scale factor is 5. We use this scale factor to find n.

$$\overset{\times\ 5}{\frac{3}{7}} = \frac{15}{35}$$
$$\underset{\times\ 5}{}$$

We find that n is **35**.

Practice

a. The blueprints were drawn to a scale of 1:24. If a length of a wall on the blueprint was 6 in., what was the length in feet of the wall in the house?

b. Bret is carving a model ship from balsa wood on a scale of 1:36. If the ship is 54 feet long, the model ship should be how many inches long?

Solve by using the scale factor.

c. $\dfrac{5}{7} = \dfrac{15}{w}$

d. $\dfrac{x}{3} = \dfrac{42}{21}$

Problem set 122

1. Use a ratio box to solve this problem. The regular price of the item was $45, but the item was on sale for 20 percent off. What was the sale price?

2. With one toss of a die, what is the probability of rolling

   (a) an odd number greater than 1?

   (b) an even number less than 2?

3. In her first 6 games, Ann averaged 10 points per game. In her next 9 games, Ann averaged 15 points per game. How many points per game did Ann average during her first 15 games?

4. Ingrid started her trip at 8:30 a.m. with a full tank of gas and an odometer reading of 43,764 miles. When she stopped for gas at 1:30 p.m., the odometer read 44,010 miles. If it took 12 gallons to fill the tank, her car averaged how many miles per gallon?

5. In Problem 4, Ingrid traveled at an average speed of how many miles per hour?

6. Use a ratio box to solve this problem. If 5 dollars equals 30 kronas, what is the cost in dollars of an item priced at 75 kronas?

7. Write an equation to solve this problem. Three fifths of Tom's favorite number is 60. What is Tom's favorite number?

8. On a coordinate plane graph the points $(-3, 2)$, $(3, 2)$, and $(-3, -2)$. If these points designate three of the vertices of a rectangle, what are the coordinates of the fourth vertex of the rectangle?

9. What is the ratio of counting numbers to integers in this set of numbers?

$$\{-3, -2, -1, 0, 1, 2\}$$

10. Find a^2 if $\sqrt{a} = 3$.

Write equations to solve Problems 11–13.

11. Forty is what percent of 250?

12. What is 60 percent of $40?

13. Forty percent of what number is 60?

14. Use a ratio box to solve this problem. The number of students in chorus increased 25 percent this year. If there are 20 more students in chorus this year than there were last year, how many students are in chorus this year?

15. Segment *BC* is how much longer than segment *AB*?

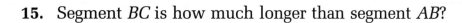

16. Graph on a number line: $x \leq 3$

17. Complete the table.

FRACTION	DECIMAL	PERCENT
(a)	(b)	1.4%

18. Multiply and write the product in scientific notation.

$$(1.4 \times 10^{-6})(5 \times 10^4)$$

19. Find the missing numbers in this function.

$y = -2x$

x	y
3	☐
0	☐
−2	☐

20. Find (a) the circumference and (b) the area of a circle that has a diameter of 2 feet.

21. Estimate the measure of ∠*ABC*. Then trace the angle, extend the sides, and measure the angle with a protractor.

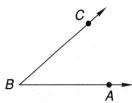

22. Find x. Then find the area of the triangle on the left. Dimensions are in inches.

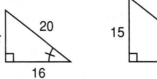

Solve:

23. $\frac{3}{5}m + 8 = 20$ **24.** $0.3x - 2.7 = 9$

25. $\frac{12}{53} = \frac{120}{n}$

Add, subtract, multiply, or divide, as indicated:

26. $\sqrt{5^3 - 5^2}$ **27.**   1 gal 1 qt
                                     $-$ 1 qt 1 pt
                                     _____

28. $(0.25)\left(1\frac{1}{4} - 1.2\right)$ **29.** $7\frac{1}{3} - \left(1\frac{3}{4} \div 3.5\right)$

30. $\dfrac{(-2)(3) - (3)(-4)}{(-2)(-3) - (4)}$

LESSON
123

Pythagorean Theorem

The longest side of a right triangle is called the **hypotenuse**. Every right triangle has a property that makes the right triangle a very important triangle in mathematics. **The area of the square drawn on the hypotenuse of a right triangle equals the sum of the areas of the squares drawn on the other two sides.**

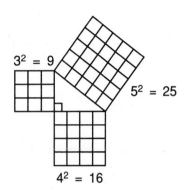

The triangle on the left is a right triangle. On the right we have drawn squares on the sides of the triangle. The areas of the squares drawn on the sides are 9, 16, and 25. Notice that the area of the largest square equals the sum of the areas of the other two squares.

$$25 = 16 + 9$$

This property of right triangles was known to the Egyptians as early as 2000 B.C., but it is named for a Greek who lived about 650 B.C. The Greek's name was Pythagoras, and the property is called the **Pythagorean theorem.** The Greeks are so proud of this mathematician that they have issued a postage stamp that illustrates the theorem. Here we show a reproduction of this stamp.

To solve problems that require the use of the Pythagorean theorem, we will sketch the right triangle and draw the squares on each side.

Example 1 Copy this triangle. Draw a square on each side. Find the area of each square. Then find c.

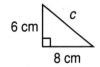

Solution We copy the triangle and sketch a square on each side of the triangle, using a side of the triangle for one side of each square.

We were given the lengths of the two shorter sides. The areas of the squares on these sides are 36 cm² and 64 cm².

$$36 + 64 = 100$$

The sum of 36 and 64 is 100, so the area of the largest square is 100. This means that a side of the largest square must be 10 because $10^2 = 100$. Thus

$$c = \textbf{10 cm}$$

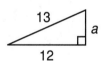

Example 2 In this triangle, find a. Dimensions are in inches.

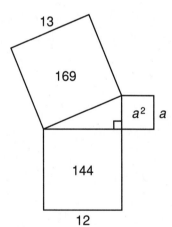

Solution We copy the triangle and draw a square on each side. The area of the largest square is 13×13, or 169 in.². The area of one of the smaller squares is 144 in.². So a^2 plus 144 must equal 169.

$$a^2 + 144 = 169$$

If we subtract 144 from both sides, we find that a^2 equals 25.

$$
\begin{array}{r}
a^2 + 144 = 169 \\
-144 \quad -144 \\
\hline
a^2 = 25
\end{array}
$$

This means that a equals 5 because 5 squared is 25.

$$a = \textbf{5 in.}$$

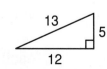

Example 3 Find the perimeter of this triangle. Dimensions are in centimeters.

Solution We can use the Pythagorean theorem to find side c. The areas of the two smaller squares are 16 and 9. The sum of these areas is 25, so the area of the largest square is 25. Thus the length of side c is 5. Now we add the lengths of the sides to find the perimeter.

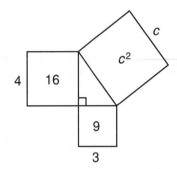

Perimeter $= 4$ cm $+ 3$ cm $+ 5$ cm $= $ **12 cm**

Practice a. Use the Pythagorean theorem to find side a.

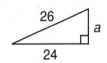

b. Use the Pythagorean theorem to find side b.

Problem set 123

1. The meal cost $15. Christie left a tip that was 15 percent of the cost of the meal. How much money did Christie leave for a tip?

2. Twenty-five ten-thousandths is how much greater than twenty millionths? Write the answer in scientific notation.

3. Find the (a) mean, (b) median, (c) mode, and (d) range of the number of days in the months of a leap year.

4. The 2-pound box cost $2.72. The 48-ounce box cost $3.60. The smaller box cost how much more per ounce than the larger box? (There are 16 ounces in a pound.)

5. Use a ratio box to solve this problem. If 80 pounds of seed cost $96, what would be the cost of 300 pounds of seed?

6. Five eighths of a pound is how many ounces?

7. Use a ratio box to solve this problem. The ratio of stalactites to stalagmites in the cavern was 9 to 5. If the total number of stalactites and stalagmites was 1260, how many stalagmites were in the cavern?

Write equations to solve Problems 8 and 9.

8. Ten percent of what number is 20?

9. Twenty is what percent of 60?

10. The ordered pairs (1, 0), (0, −1), and (3, 2) designate points that lie on the same line. Graph the points on a coordinate plane and draw the line.

11. The cost of a 10-minute call to Boise decreased by 20 percent. If the cost before the decrease was $3.40, what was the cost after the decrease? Use a ratio box to solve the problem.

12. What is the area of the shaded region of this rectangle?

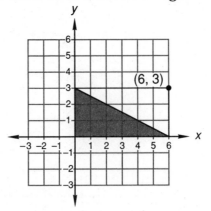

13. Use a ratio box to solve this problem. On a 1:60 scale model airplane, the wingspan is 8 inches. The wingspan of the actual airplane is how many inches? How many feet is this?

14. Complete the table.

Fraction	Decimal	Percent
$1\frac{1}{3}$	(a)	(b)

15. Divide 365 by 12 and write the quotient as a mixed number.

16. Multiply and write the product in scientific notation.

$$(8.1 \times 10^{-6})(9 \times 10^{10})$$

17. Evaluate: $\dfrac{x^2 - xy}{x}$ if $x = 8$ and $y = 7$

18. Use the Pythagorean theorem to find c.

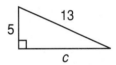

19. Find the volume of this solid. Dimensions are in centimeters.

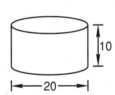

20. Refer to the figure shown. Find the measures of the following angles.

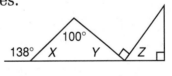

(a) $\angle X$ (b) $\angle Y$ (c) $\angle Z$

21. The triangles are similar. Find x. Dimensions are in inches.

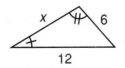

22. Estimate: $\dfrac{(41,392)(395)}{81}$

Solve:

23. $4n + 1.64 = 2$ **24.** $3\frac{1}{3}x - 1 = 49$

25. $\dfrac{17}{25} = \dfrac{m}{75}$

Add, subtract, multiply, or divide, as indicated:

26. $3^3 + 4^2 - \sqrt{225}$

27. $3\dfrac{1}{5} + 6.3 + 2\dfrac{1}{2}$

28. $\left(3\dfrac{1}{3}\right)(0.75)(40)$

29. $\dfrac{-12 - (6)(-3)}{(-12) - (-6) + (3)}$

30. $\dfrac{7 \text{ days}}{1 \text{ week}} \cdot \dfrac{24 \text{ hr}}{1 \text{ day}} \cdot \dfrac{60 \text{ min}}{1 \text{ hr}}$

LESSON 124

Estimating Square Roots • Special Angles

Estimating square roots

We review square roots by remembering that because 2 times 2 equals 4, we say that the square root of 4 is 2.

$$2 \times 2 = 4 \qquad \text{so} \qquad \sqrt{4} = 2$$

Because 3 times 3 equals 9, we say that the square root of 9 is 3.

$$3 \times 3 = 9 \qquad \text{so} \qquad \sqrt{9} = 3$$

Because 5 times 5 equals 25, we say that the square root of 25 is 5.

$$5 \times 5 = 25 \qquad \text{so} \qquad \sqrt{25} = 5$$

The number 30 does not have a whole number square root. We can estimate the square root of 30 by guessing and checking. If we know how to estimate square roots, then we will know if the answer we get for a square root on a calculator is reasonable. Let's estimate the square root of 30.

$$4 \times 4 = 16 \qquad \text{4 is too small}$$
$$5 \times 5 = 25 \qquad \text{5 is too small}$$
$$6 \times 6 = 36 \qquad \text{6 is too large}$$

Since 30 is between 25 and 36, we see that the square root of 30 is a number between 5 and 6. If we use the square root key on a calculator, we get

$$\sqrt{30} = 5.4772256$$

This result is reasonable because the square root of 30 is a number between 5 and 6.

Example The square root of 90 is between which two consecutive whole numbers?

Solution We begin by guessing.

6 × 6 = 36	6 is too small
7 × 7 = 49	7 is too small
8 × 8 = 64	8 is too small
9 × 9 = 81	9 is too small
10 × 10 = 100	10 is too large

Since 90 is between 81 and 100, we see that **the square root of 90 is a number between 9 and 10.**

Special angles If the sum of the measures of two angles is 90°, we say that the angles are **complementary angles**. If the sum of the measures of two angles is 180°, we say that the angles are **supplementary angles**.

Angles *A* and *B* are complementary angles because their combined measures total 90°. Angles *C* and *D* are supplementary angles because their combined measures total 180°. Some people confuse the meanings of these words. We can avoid confusion by remembering that the "c" in complementary resembles a right angle.

⟨ omplementary looks like complementary

The "s" in supplementary resembles two right angles, which total 180°.

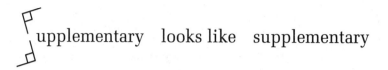

upplementary   looks like   supplementary

When two lines cross, four angles are formed. The pairs of opposite angles are called **vertical angles**. The measures of vertical angles are equal.

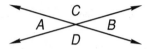

Angles *A* and *B* are a pair of vertical angles, and the measure of angle *A* equals the measure of angle *B*. Angles *C* and *D* are a pair of vertical angles, and the measure of angle *C* equals the measure of angle *D*.

It is easy to see why vertical angles are equal. Let's let angle *C* be a 150° angle.

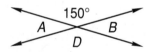

The 150° angle and angle *A* make a straight angle (the angles are supplementary), so angle *A* is a 30° angle. The 150° angle and angle *B* make a straight angle (the angles are supplementary), so angle *B* is a 30° angle.

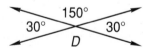

Now either 30° angle and angle *D* form a straight angle, so angle *D* is a 150° angle.

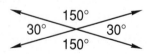

From this we can see why vertical angles are equal.

Practice Each square root below is between which two consecutive whole numbers?

a. $\sqrt{7}$ **b.** $\sqrt{70}$ **c.** $\sqrt{700}$

Describe each pair of angles in **d** and **e** as complementary or supplementary.

d.

e.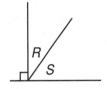

f. Find the measure of $\angle a$.

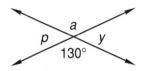

Problem set 124

1. Gabriel paid \$20 for $2\frac{1}{2}$ pounds of cheese that cost \$2.60 per pound and 2 boxes of crackers that cost \$1.49 each. How much money should he get back?

2. What is the probability that the spinner will not stop on a prime number?

3. What is the average of the first 10 counting numbers?

4. At an average speed of 50 miles per hour, how long would it take to complete a 375-mile trip?

5. Use a ratio box to solve this problem. The Johnsons traveled 300 kilometers in 4 hours. At that rate, how long will it take them to travel 500 kilometers? Write the answer in hours and minutes.

6. Three fourths of Bill's favorite number is 36. What number is one half of Bill's favorite number?

7. Use a ratio box to solve this problem. The ratio of winners to losers in the contest was 1 to 15. If there were 800 contestants, how many winners were there?

Write equations to solve Problems 8 – 10.

8. Three hundred is 6 percent of what number?

9. Twenty is what percent of 10?

10. What is 6.5 percent of $40?

11. The ordered pairs (0, 0), (−2, −4), and (2, 4) designate points that lie on the same line. Graph the points on a coordinate plane and draw the line.

12. Arrange in order of size from least to greatest:

$$2,\ 2^2,\ \sqrt{2},\ -2$$

13. Use a ratio box to solve this problem. The population of the colony decreased by 30 percent after the first winter. If the population after the first winter was 350, what was the population before the first winter?

14. Nathan used this graph to mold a scale model car from clay. The car was 4 feet high, and he used the graph to see that the model should be 2 inches high. If the length of the car's bumper is 5 feet, use the graph to find the proper length of the model's bumper.

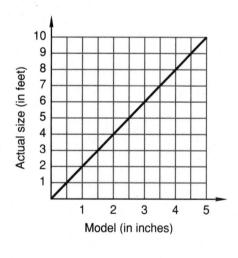

15. Compare: $xy \bigcirc \dfrac{x}{y}$ if x is positive and y is negative

16. Complete the table.

Fraction	Decimal	Percent
(a)	(b)	72%

17. Multiply and write the product in scientific notation.

$$(4.5 \times 10^6)(6 \times 10^3)$$

18. Each square root is between which two consecutive whole numbers?

(a) $\sqrt{40}$ (b) $\sqrt{20}$

19. Find (a) the circumference and (b) the area of a circle that has a radius of 7 inches. Use $\frac{22}{7}$ for π.

20. Find a.

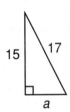

Find the volume of each solid. Dimensions are in centimeters.

21.

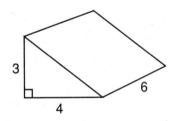

22.

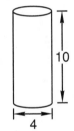

23. In this figure, find the measures of angles a, b, and c.

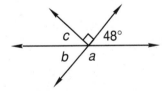

Solve:

24. $4\frac{1}{2}x + 7 = 70$ **25.** $\dfrac{15}{w} = \dfrac{45}{3.3}$

Add, subtract, multiply, or divide, as indicated:

26. $\sqrt{6^2 + 8^2}$ **27.** 1 ton − 350 lb   **28.** $3\frac{1}{3}\left(7.2 \div \frac{3}{5}\right)$

29. $8\frac{5}{6} - 2.5 - 1\frac{1}{3}$ **30.** $\dfrac{(18) - (2)(-3)}{(-3) + (-2) - (-4)}$

LESSON 125

Multiplying Three or More Signed Numbers • Powers of Negative Numbers

Multiplying three or more signed numbers

One way to multiply three or more signed numbers is to multiply the factors in order from left to right, keeping track of the signs with each step, as we show here.

$(-3)(-4)(+5)(-2)(+3)$	problem
$(+12)(+5)(-2)(+3)$	multiplied $(-3)(-4)$
$(+60)(-2)(+3)$	multiplied $(5)(12)$
$(-120)(+3)$	multiplied $(60)(-2)$
-360	multiplied $(-120)(3)$

Another way to keep track of the signs when multiplying signed numbers is to count the number of negative factors. Notice the pattern in the multiplications below.

$$-1 = -1 \qquad \text{odd}$$
$$(-1)(-1) = +1 \qquad \text{even}$$
$$(-1)(-1)(-1) = -1 \qquad \text{odd}$$
$$(-1)(-1)(-1)(-1) = +1 \qquad \text{even}$$
$$(-1)(-1)(-1)(-1)(-1) = -1 \qquad \text{odd}$$

When there is an even number of negative factors, the product is positive. When there is an odd number of negative factors, the product is negative.

Example 1   Find the product: $(+3)(+4)(-5)(-2)(-3)$

Solution   There are three negative factors (an odd number), so the product will be a negative number. We multiply and get

$$(+3)(+4)(-5)(-2)(-3) = -360$$

We did not consider the signs of the positive factors because positive factors do not affect the sign of the product.

Powers of negative numbers   We remember that the exponent of a power indicates how many times the base is used as a factor.

$$(-3)^4 \text{ means } (-3)(-3)(-3)(-3)$$

Example 2   Simplify: (a) $(-2)^4$ (b) $(-2)^5$

Solution   (a)   The expression $(-2)^4$ means $(-2)(-2)(-2)(-2)$. Since there is an even number of negative factors, the product is a positive number. Since 2^4 is 16, we find that $(-2)^4$ is **+16**.

(b)   The expression $(-2)^5$ means $(-2)(-2)(-2)(-2)(-2)$. This time there is an odd number of negative factors, so the product is a negative number. Since 2^5 equals 32, we find that $(-2)^5$ equals **−32**.

Practice   Simplify:

a. $(-5)(-4)(-3)(-2)(-1)$ b. $(+5)(-4)(+3)(-2)(+1)$

c. $(-2)^3$ d. $(-3)^4$ e. $(-9)^2$ f. $(-1)^5$

Problem set 125

1.   The dinner bill totaled $25. Mike left a 15 percent tip. How much money did Mike leave for a tip?

2.   When the square root of 9 is subtracted from 9 squared, the difference is how much greater than 9?

3.   The table shows a tally of the scores earned by students on a class test. Find (a) the mode and (b) the median of the 29 scores in the class.

CLASS TEST SCORES

SCORE	NUMBER OF STUDENTS
100				
95	⊬⊬	
90	⊬⊬	
85	⊬⊬ ||||
80					
70		

4. The plane completed the flight in $2\frac{1}{2}$ hours. If the flight covered 1280 kilometers, what was the plane's average speed in kilometers per hour?

5. Use a ratio box to solve this problem. Jeremy earned $25 for 4 hours of work. How much would he earn for 7 hours of work at the same rate?

6. Eight fifths of a kilometer is how many meters?

7. Use a ratio box to solve this problem. If 40 percent of the lights were on, what was the ratio of lights on to lights off?

8. Use the Pythagorean theorem to find the length of the longest side of a triangle whose vertices are $(3, 1)$, $(3, -2)$, and $(-1, -2)$.

9. Use a ratio box to solve this problem. Sam saved $25 buying the suit at a sale that offered 20 percent off. What was the regular price of the suit?

Write equations to solve Problems 10 and 11.

10. What decimal number is 1 percent of 150?

11. What percent of 25 is 20?

12. Use a ratio box to solve this problem. The merchant bought the item for $30 and sold it for 60 percent more. How much profit did the merchant make on the item?

13. Compare:

$$(-1)(-1)(-1)(-1) \bigcirc (-1)(-1)(-1)(-1)(-1)$$

14. Use a ratio box to solve this problem. The $\frac{1}{20}$ scale model of the rocket stood 54 inches high. What was the height of the actual rocket?

15. Graph on a number line: $x \geq 0$

16. Complete the table.

FRACTION	DECIMAL	PERCENT
(a)	1.75	(b)

17. Multiply and write the product in scientific notation.

$$(4.8 \times 10^4)(8 \times 10^{-8})$$

18. Quadrilateral *ABCD* is a rectangle. The measure of $\angle ACB$ is 34°. Find the measure of each of these angles.

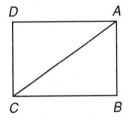

(a) $\angle CAB$ (b) $\angle CAD$

19. These two triangles are similar. Find *x*.

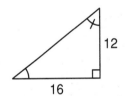

20. Find (a) the circumference and (b) the area of a circle with a diameter of 2 feet.

21. Estimate the measure of $\angle AOB$. Then use a protractor to measure the angle.

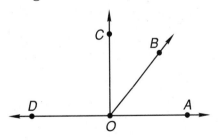

22. Which of these numbers is between 12 and 14?

    a. $\sqrt{13}$ **b.** $\sqrt{130}$ **c.** $\sqrt{150}$

23. Find the volume of this right prism.

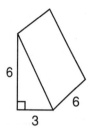

6

3

6

24. Find the volume of this right circular cylinder.

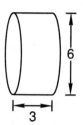

6

3

Solve:

25. $6\frac{2}{3}f - 5 = 5$

26. $\dfrac{12}{2.5} = \dfrac{m}{25}$

Add, subtract, multiply, or divide, as indicated:

27. $10\frac{1}{2} \cdot 1\frac{3}{7} \div 25$

28. $12.5 - 8\frac{1}{3} + 1\frac{1}{6}$

29. (a) $(-5)(+4)(-3)(+2)$ (b) $(-12)(+3)(-2)(-1)$

30. (a) $\dfrac{(-3)(-2)(-1)}{(-3)(+2)}$ (b) $\dfrac{(-6)(-3)(-2)}{(-4)(-1)}$

LESSON
126

Semicircles

A **semicircle** is half of a circle. Thus, the length of a semicircle is half the circumference of a whole circle. The area enclosed by a semicircle and a diameter is half the area of the full circle.

We will practice finding the lengths of semicircles and the areas they enclose by calculating the perimeters and areas of figures that contain semicircles. Unless directed otherwise, we will use 3.14 as the approximation for π when we perform the calculations. A calculator will be helpful.

Example 1 Find the perimeter of this figure. Dimensions are in meters.

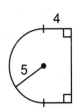

Solution The perimeter of the semicircle is half the perimeter of a circle whose diameter is 10.

$$\text{Perimeter of semicircle} = \frac{\pi(10 \text{ m})}{2} \approx 15.7 \text{ m}$$

Now we can write all the dimensions.

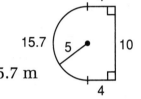

Perimeter $\approx$ 10 m + 4 m + 4 m + 15.7 m
$\approx$ **33.7 m**

Example 2 Find the area of this figure. Dimensions are in meters.

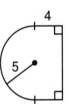

Solution We divide the figure into two parts, and then we find the area of each part. The area of the figure equals A_1 plus A_2.

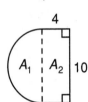

$$A_1 = \frac{\pi r^2}{2} \approx \frac{3.14(25 \text{ m}^2)}{2} \approx 39.25 \text{ m}^2$$

$$A_2 = L \times W = 4 \times 10 \quad = 40 \text{ m}^2$$

Total area $\approx$ **79.25 m²**

Practice Find (a) the perimeter and (b) the area of this figure. Dimensions are in centimeters.

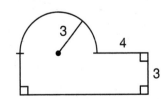

Problem set 126

1. The merchant sold the item for $12.50. If 40 percent of the selling price was profit, how much money did the merchant earn in profit?

2. With one toss of a die, what is the probability of rolling a prime number?

3. Bill's average score for 10 tests was 88. If his lowest score, 70, is not counted, what was his average for the remaining tests?

4. The 36-ounce container cost $3.42. The 3-pound container cost $3.84. The smaller container cost how much more per ounce than the larger container?

5. Sean read 18 pages in 30 minutes. If he had finished page 128, how many hours will it take him to finish his 308-page book if he reads at the same rate?

6. Matthew was thinking of a certain number. If $\frac{5}{6}$ of the number was 75, what was $\frac{3}{5}$ of the number?

7. Use a ratio box to solve this problem. The ratio of crawfish to tadpoles in the creek was 2 to 21. If there were 1932 tadpoles in the creek, how many crawfish were there?

Write equations to solve Problems 8 and 9.

8. What percent of $60 is $45?

9. What number is 45 percent of 60?

10. The ordered pairs $(-1, 0)$, $(-3, 4)$, and $(0, -2)$ designate points that lie on the same line. Graph the points on a coordinate system and draw the line.

Write equations to solve Problems 11 and 12.

11. What percent of 60 is 75?

12. What percent of 75 is 60?

13. Complete the table.

Fraction	Decimal	Percent
(a)	(b)	2.2%

14. Michelle used this graph to find how long it would take her to drive a certain distance. According to the graph, how long would it take her to drive 75 miles?

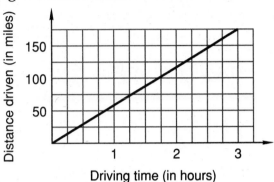

15. Compare: $ab \bigcirc a - b$ if a is positive and b is negative

16. Multiply and write the product in scientific notation.

$$(3.6 \times 10^{-4})(9 \times 10^{8})$$

17. Find the area of this figure. Dimensions are in centimeters.

18. Find the perimeter of the figure in Problem 17.

19. Find the volume of this solid in cubic inches. Dimensions are in feet.

20. What angle is formed by the hands of a clock at 5:00?

21. Find $m\angle x$.

22. The triangles are similar. Find x.

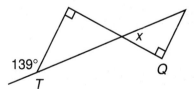

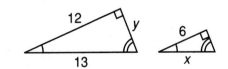

23. Use the Pythagorean theorem to find y in the triangle in Problem 22.

Solve:

24. $2\frac{3}{4}w + 4 = 48$

25. $2.4n - 0.12 = 4.8$

Add, subtract, multiply, or divide, as indicated:

26. $\sqrt{(3^2)(10^2)}$

27.   5 lb 7 oz
      − 2 lb 8 oz

28. $12.5 - \left(8\frac{1}{3} + 1\frac{1}{6}\right)$

29. $4\frac{1}{6} \div 3\frac{3}{4} \div 2.5$

30. $\dfrac{(-3)(4)}{-2} - \dfrac{(-3)(-4)}{-2}$

LESSON
127

Surface Area

The total area of the outside surfaces of a geometric solid is called the **surface area** of the solid.

This cube has six faces (surfaces). Each face has an area of 9 square inches. Thus the total surface area is

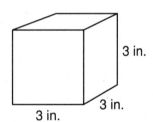

3 in.

3 in.

3 in.

$9 \text{ in.}^2 + 9 \text{ in.}^2 + 9 \text{ in.}^2 + 9 \text{ in.}^2 + 9 \text{ in.}^2 + 9 \text{ in.}^2 = 54 \text{ in.}^2$

Example   Find the surface area of this block. Dimensions are in centimeters.

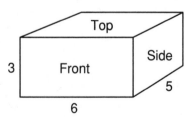

Top

Side

3   Front

6

5

Solution   The block has six rectangular faces. The areas of the top and bottom are equal. The areas of the front and back are equal, and the areas of the left and right sides are equal. We add the areas of these six faces to find the total surface area.

Area of top	= 5 cm	× 6 cm	= 30 cm²
Area of bottom	= 5 cm	× 6 cm	= 30 cm²
Area of front	= 3 cm	× 6 cm	= 18 cm²
Area of back	= 3 cm	× 6 cm	= 18 cm²
Area of side	= 3 cm	× 5 cm	= 15 cm²
Area of side	= 3 cm	× 5 cm	= 15 cm²
Total surface area			**= 126 cm²**

Practice Find the surface area of this rectangular solid. Dimensions are in meters.

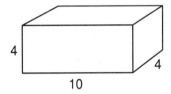

Problem set 127

1. Use a ratio box to solve this problem. The regular price of the dress was $30. The dress was on sale for 25 percent off. What was the sale price?

2. Twenty billion is how much greater than nine hundred million? Write the answer in scientific notation.

3. The mean of the following numbers is how much less than the median? Use a calculator.

   $$3.2, \ 4.28, \ 1.2, \ 3.1, \ 1.17$$

4. Evaluate: $\sqrt{a^2 - b^2}$ if $a = 10$ and $b = 8$

5. If Glenda is paid at a rate of $8.50 per hour, how much will she earn if she works $6\frac{1}{2}$ hours?

6. Use a ratio box to solve this problem. If 6 kilograms of flour costs $2.48, what is the cost of 45 kilograms of flour?

7. Five eighths of a mile is how many yards? (1 mi = 1760 yd)

8. Use a ratio box to solve this problem. The ratio of Whigs to Tories at the assembly was 7 to 3. If 210 party members had assembled, how many were Tories?

Write equations to solve Problems 9–11.

9. What percent of $60 is $3?

10. Sixty percent of what number is 6?

11. What fraction is 10 percent of 4?

12. Use a ratio box to solve this problem. The merchant sold the item at a 30 percent discount from the regular price. If the regular price was $60, what was the sale price?

13. The coordinates $(-2, -2)$, $(-2, 2)$, and $(1, -2)$ are the coordinates of the vertices of a right triangle. Find the length of the hypotenuse of this triangle.

14. Compare: $a^3 \bigcirc a^2$ if a is negative

15. Use a ratio box to solve this problem. Begin by converting 60 feet to inches. Brandon is making a model plane at a 1:36 scale. If the length of the actual plane is 60 feet, how many inches long should he make his model?

16. Complete the table.

Fraction	Decimal	Percent
(a)	1.7	(b)

17. Divide 1000 by 33 and write the quotient as a decimal with a bar over the repetend.

18. Multiply and write the product in scientific notation.

$$(8 \times 10^{-4})(3.2 \times 10^{-10})$$

19. Find the perimeter of this figure. Dimensions are in meters.

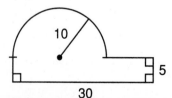

20. Find the missing numbers in this function.

$y = -2x - 1$

x	y
3	☐
-2	☐
0	☐

21. Find the surface area of this cube. Dimensions are in millimeters.

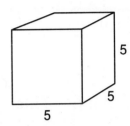

5
5
5

22. Find the volume of this right circular cylinder. Dimensions are in centimeters.

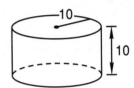

10
10

23. Find $m\angle b$.

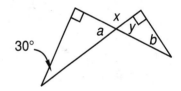

30°
a
x
y
b

24. The triangles are similar. Find x.

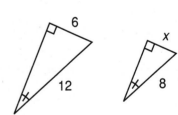

6
12
x
8

Solve:

25. $5\frac{1}{2}x + 4 = 48$

26. $\dfrac{3.9}{75} = \dfrac{c}{25}$

Add, subtract, multiply, or divide, as indicated:

27. $3.2 \div \left(2\frac{1}{2} \div \frac{5}{8}\right)$

28. $42\frac{5}{12} - \left(8.5 + 1\frac{1}{3}\right)$

29. 1 yd 2 ft 3 in. + 2 yd 1 ft 10 in.

30. $\dfrac{(-10)(-4) \ - \ (3)(-2)(-1)}{(-4) \ - \ (-2)}$

LESSON 128

Solving Literal Equations • Transforming Formulas

Solving literal equations A **literal equation** is an equation that contains letters instead of numbers. We can rearrange (transform) literal equations by using the rules we have learned.

Example 1 Solve for x: $x + a = b$

Solution We solve for x by getting x alone on one side of the equation. We do this by adding $-a$ to both sides of the equation.

$$
\begin{array}{ll}
x + a = b & \text{equation} \\
\underline{ -a \quad -a} & \text{add } -a \text{ to both sides} \\
\quad\quad x = b - a & \text{added}
\end{array}
$$

Example 2 Solve for x: $ax = b$

Solution To solve for x, we divide both sides of the equation by a.

$$
\begin{array}{ll}
ax = b & \text{equation} \\[4pt]
\dfrac{\cancel{a}x}{\cancel{a}} = \dfrac{b}{a} & \text{divided by } a \\[8pt]
x = \dfrac{b}{a} & \text{simplified}
\end{array}
$$

Transforming formulas **Formulas** are literal equations that we can use to solve certain kinds of problems. Often it is necessary to change the way a formula is written.

Example 3 Solve for W: $A = LW$

Solution   This is a formula for finding the area of a rectangle. We see that W is to the right of the equals sign and is multiplied by L. To undo the multiplication by L, we can divide both sides of the equation by L.

$$A = LW \qquad \text{equation}$$

$$\frac{A}{L} = \frac{\cancel{L}W}{\cancel{L}} \qquad \text{divided by } L$$

$$\frac{A}{L} = W \qquad \text{simplified}$$

Practice   **a.** Solve for x: $x - a = b$

b. Solve for w: $wx = y$

c. The formula for the area of a parallelogram is

$$A = bh$$

Solve this equation for b.

Problem set   **1.** Max paid $20 for 3 pairs of socks priced at $1.85 per pair
128   and a T-shirt priced at $8.95. The sales tax was 6 percent.
                 How much money should he get back?

2. If the spinner is spun once, what is
the probability that the spinner will
end up pointing to a one-digit prime
number?

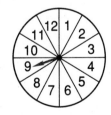

3. At $2.80 per pound, the cheddar cheese costs how many
cents per ounce?

4. Brenda's average score after 6 tests was 90. If her lowest
score, 75, is not counted, what was her average score on
the remaining 5 tests?

5. The ordered pairs (2, 3), (−2, −1), and (0, 1) designate points that lie on the same line. Graph the points on a coordinate system and draw the line.

6. Use a ratio box to solve this problem. Justin finished 3 problems in 4 minutes. At that rate, how long will it take him to finish the remaining 27 problems?

7. The price of the glove was $46. The tax was 6 percent. How much money was needed to pay for the glove?

8. Write an equation to solve this problem. What number is 225 percent of 40?

9. Arrange in order of size from least to greatest:

$$2, \frac{2}{2}, \sqrt{2}, 2^2$$

10. Use a ratio box to solve this problem. The ratio of residents to visitors in the community pool was 2 to 3. If there were 60 people in the pool, how many were visitors?

Write equations to solve Problems 11 and 12.

11. Sixty-six is $66\frac{2}{3}$ percent of what number?

12. Seventy-five percent of what number is 2.4?

13. Use a ratio box to solve this problem. The number of students enrolled in chemistry increased 25 percent this year. If there are 80 students enrolled in chemistry this year, how many were enrolled in chemistry last year?

14. Complete the table.

Fraction	Decimal	Percent
(a)	(b)	105%

15. Graph the positive integers that are less than 4 on a number line.

16. Divide 6.75 by 81 and write the quotient rounded to three decimal places.

17. Multiply and write the product in scientific notation.

$$(4.8 \times 10^{-10})(6 \times 10^{-6})$$

18. Evaluate: $x^2 + bx + c$ if $x = 3$, $b = -5$, and $c = 6$

19. Find the area of this figure. Dimensions are in millimeters.

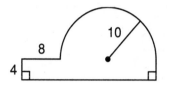

20. Find the surface area of this right triangular prism. Dimensions are in centimeters.

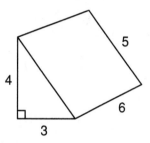

21. Find the volume of this cylinder. Dimensions are in inches.

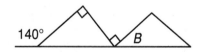

22. Find $m\angle B$.

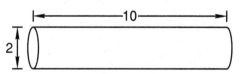

23. Solve for x: $x + c = d$

24. Solve for x: $ax = b$

Solve:

25. $12x + 8 = 14$

26. $\dfrac{15}{8} = \dfrac{m}{32}$

Add, subtract, multiply, or divide, as indicated:

27. $25 - [3^2 + 2(5 - 3)]$

28. 1 ton − 100 pounds

29. $3\dfrac{3}{4} - \left[\left(1\dfrac{1}{2}\right)\left(2\dfrac{2}{3}\right) - \dfrac{5}{6}\right]$

30. $(-3)(-2)(+4)(-1) + (-3) + (-4) - (-2)$

LESSON
129

Graphing Functions

In the table below we have recorded the values of y that are paired with x values of 2, 0, and −1 for the function $y = x + 2$.

$$y = x + 2$$

x	y
2	4
0	2
−1	1

The word "linear" comes from the word "line." The function $y = x + 2$ is called a **linear function** because the graphs of all pairs of x and y for this function lie on the same straight line. In the figure below, we graph the pairs of x and y from the table above and draw the line through these points.

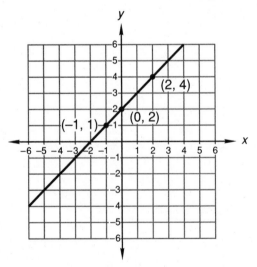

The graph of every pair of x and y that satisfies the equation $y = x + 2$ will be on this line.

Example   Make a table that shows three pairs of x and y values for the function $y = 2x - 1$. Graph these number pairs on a coordinate plane and draw a line through the points to show other number pairs of the function.

Solution   First we think of three numbers we would like to use in place of x. We decide to use 2, 0, and -2. We use small numbers so that the pairs of points can be graphed on a small graph.

$$y = 2x - 1$$

x	y
2	☐
0	☐
−2	☐

Next we use the equation to find the values of y, and we record these numbers in the table.

$$y = 2x - 1$$

x	y
2	3
0	−1
−2	−5

In the table we have recorded three x, y pairs. They are (2, 3), (0, −1), and (−2, −5). Now we graph the points and draw a line through the points.

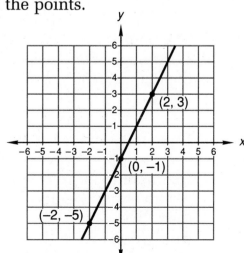

We note that, had we decided to use other numbers for the graph of x, the table would contain different x, y pairs. However, these pairs would also be on the line we have drawn.

Practice   Make a table for each of these linear functions. Find three pairs of x and y for each function. Then plot the pairs and draw the graphs of the functions.

a. $y = x + 2$

b. $y = -2x$

Problem set 129

1. The shirt regularly priced at $21 was on sale for $\frac{1}{3}$ off. What was the sale price?

2. Nine hundred seventy-five billion is how much less than one trillion? Use words to write the answer.

3. What is the (a) range and (b) mode of this set of numbers?

$$16, 6, 8, 17, 14, 16, 12$$

4. Use a ratio box to solve this problem. Riding her bike from home to the lake, Sonia averaged 18 miles per hour (per 60 minutes). If it took her 40 minutes to reach the lake, how far did she ride?

5. The points $(3, -2)$, $(-3, -2)$, and $(-3, 6)$ are the vertices of a triangle. Find the perimeter of the triangle. (*Hint:* Use the Pythagorean theorem to find the length of the hypotenuse.)

6. Five sixths of a yard is how many inches?

7. Use a ratio box to solve this problem. The ratio of earthworms to cutworms in the garden was 5 to 2. If there were 140 worms in the garden, how many earthworms were there?

Write equations to solve Problems 8–10.

8. Sixty is 125 percent of what number?

9. Sixty is what percent of 25?

10. What number is 60 percent of 125?

11. Use a ratio box to solve this problem. The average cost of a new car increased 8 percent in one year. Before the increase, the average cost of a new car was $16,550. What was the average cost of a new car after the increase?

12. In a can there are 30 red marbles, 40 green marbles, and 50 blue marbles. If a marble is drawn from the can, what is the probability that the marble will not be red?

13. Complete the table.

FRACTION	DECIMAL	PERCENT
$\frac{5}{6}$	(a)	(b)

14. Compare: $x + y \bigcirc x - y$ if x is is a whole number and y is an integer

15. Multiply and write the product in scientific notation.

$$(1.8 \times 10^{10})(9 \times 10^{-6})$$

16. Between which two consecutive whole numbers does $\sqrt{200}$ lie?

17. Find three pairs of x and y for the function $y = x + 1$. Graph these number pairs on a coordinate plane and draw a line through these points.

18. Find the area of this figure. Dimensions are in centimeters.

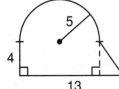

19. Find the surface area of this rectangular solid. Dimensions are in inches.

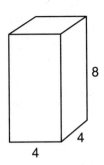

8

4

4

20. Find the volume of this right circular cylinder. Dimensions are in centimeters.

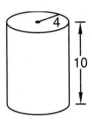

4

10

21. The polygon *ABCD* is a rectangle. Find *m∠x*.

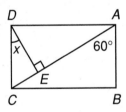

22. Use two unit multipliers to convert 2 m² to square centimeters.

23. Solve for x in each literal equation.

(a) $x - y = z$ (b) $w = xy$

Solve:

24. $\dfrac{a}{21} = \dfrac{1.5}{7}$ **25.** $6x + 5 = 7$

Add, subtract, multiply, or divide, as indicated:

26. $62 + 5\{20 - [4^2 + 3(2 - 1)]\}$

27. 2 yd − 4 ft 5 in.   **28.** $5\dfrac{1}{6} + 3.5 - \dfrac{1}{3}$

29. $(5.5)\left(3\dfrac{1}{2}\right)(0.2)$

30. $\dfrac{(5)(-3)(2)(-4) \; + \; (-2)(-3)}{-6}$

LESSON
130

Formulas and Substitution

A formula is a literal equation that describes a relationship between two or more variables. Formulas are used in mathematics, science, economics, the construction industry, food preparation, and wherever measurement is used.

To use a formula, we replace the letters in the formula with measures that are known. Then we solve the equation for the measure we wish to find.

Example 1 Use the formula $d = rt$ to find t when d is 36 and r is 9.

Solution This formula describes the relationship between distance (d), rate (r), and time (t). We replace d with 36 and r with 9 and then solve the equation for t.

$$d = rt \qquad \text{formula}$$

$$36 = 9t \qquad \text{substituted}$$

$$t = 4 \qquad \text{divided by 9}$$

Another way to find t is to first solve the formula for t.

$$d = rt \qquad \text{formula}$$

$$\frac{d}{r} = t \qquad \text{divided by } r$$

Then replace d and r with 36 and 9, respectively, and simplify.

$$\frac{36}{9} = t \qquad \text{substituted}$$

$$4 = t \qquad \text{divided}$$

Example 2   Use the formula $F = 1.8C + 32$ to find F when C is 37.

Solution   This formula is used to convert measurements of temperature from degrees Celsius to degrees Fahrenheit. We replace C with 37 and simplify.

$$F = 1.8C + 32 \qquad \text{formula}$$

$$F = 1.8\,(37) + 32 \qquad \text{substituted}$$

$$F = 66.6 + 32 \qquad \text{multiplied}$$

$$F = 98.6 \qquad \text{added}$$

Thus 37 degrees Celsius equals **98.6** degrees Fahrenheit.

Practice   **a.** Use the formula $A = bh$ to find b when A is 20 and h is 4.

b. Use the formula $A = \frac{1}{2}bh$ to find b when A is 20 and h is 4.

c. Use the formula $F = 1.8C + 32$ to find F when C is -40.

Problem set 130

1. The main course cost $8.35. The beverage cost $1.25. Dessert cost $2.40. Jason left a tip that was 15 percent of the total price of the meal. How much money did Jason leave for a tip?

2. Twelve hundred-thousandths is how much greater than twenty millionths? Write the answer in scientific notation.

3. Arrange the following numbers in order. Then find the median and the mode of the set of numbers.

    8, 12, 9, 15, 8, 10, 9, 8, 7, 4

4. One card is to be drawn from a normal deck of 52 cards. What is the probability of drawing a 5?

5. Use a ratio box to solve this problem. Milton can exchange $200 for 300 Swiss francs. At that rate, how many dollars would a 240-franc Swiss watch cost?

6. Three eighths of a ton is how many pounds?

7. Use a ratio box to solve this problem. The jar was filled with red beans and brown beans in the ratio of 5 to 7. If there were 175 red beans in the jar, what was the total number of beans in the jar?

Write equations to solve Problems 8–10.

8. What number is 2.5 percent of 800?

9. Ten percent of what number is 2500?

10. Fifty-six is what percent of 700?

11. Use a ratio box to solve this problem. During the off-season, the room rates at the resort were reduced by 35 percent. If the usual rates were $90 per day, what would be the cost of a 2-day stay during the off-season?

12. What is the area of the shaded region of this rectangle?

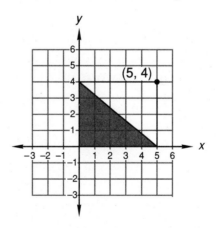

13. Use a ratio box to solve this problem. Liz is drawing a floor plan of her house. On the plan, 1 inch equals 2 feet. What is the floor area of a room that measures 6 inches by $7\frac{1}{2}$ inches on the plan?

14. Complete the table.

FRACTION	DECIMAL	PERCENT
$1\frac{1}{4}$	(a)	(b)

15. Multiply and write the product in scientific notation.

$$(2.8 \times 10^5)(8 \times 10^{-8})$$

16. Use the formula $c = 2.54n$ to find c when n is 12.

17. Make a table that shows three pairs of numbers for the function $y = 2x$. Then graph the number pairs on a coordinate plane and draw a line through the points.

18. Find the perimeter of this figure. Dimensions are in inches.

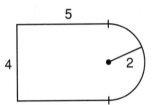

19. Find the surface area of this cube. Dimensions are in inches.

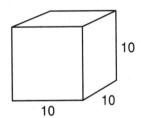

20. Find the volume of this right circular cylinder. Dimensions are in centimeters.

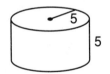

21. Find $m\angle x$.

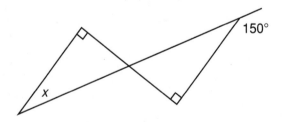

22. The triangles are similiar. Find y. Dimensions are in centimeters.

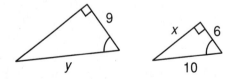

23. Use the Pythagorean theorem to find x in the triangle above.

Solve:

24. $\dfrac{d}{35} = \dfrac{1.6}{5}$ **25.** $1\dfrac{2}{3}x - 3 = 32$

Add, subtract, multiply, or divide, as indicated:

26. $28 - \{21 - 3[2 + 2(2)]\}$ **27.**

$$\begin{array}{r} 6 \text{ lb } 10 \text{ oz} \\ + \ 2 \text{ lb } \ \ 8 \text{ oz} \\ \hline \end{array}$$

28. $2.75 - \left(1\dfrac{1}{3} - 0.5\right)$ **29.** $3\dfrac{3}{4} \cdot 2\dfrac{2}{3} \div 10$

30. $(-6) - (7)(-4) + 5 + \dfrac{(-8)(-9)}{(-3)(-2)}$

LESSON
131

Simple Interest

When you put your money in a bank, the bank uses your money to make more money. The bank pays you to let them use your money. The amount of money you deposit is called the **principal**. The amount of money they pay you is called **interest**. The interest is a percentage of the money deposited. If you deposit $100 at 6 percent simple interest,* the bank will pay you $6 a year to use your money. If you take your money out after 3 years, the bank will pay you a total of $118.

$100.00	principal
$6.00	first year interest
$6.00	second year interest
$6.00	third year interest
$118.00	total

---

*With **simple interest,** interest is paid on the principal only and not on previous interest earnings. Most bank accounts pay compound interest, which pays interest on previous interest earnings.

Example   Roger deposits $700 in the bank at 8 percent simple interest for 3 years. How much money will he have in the bank at the end of 3 years?

Solution   First we calculate the interest for 1 year.

$$\begin{array}{r} \$700.00 \\ \times \quad 0.08 \\ \hline \$\ 56.00 \end{array}$$

The interest for 3 years will be

$$3 \times \$56.00 = \$168.00$$

At the end of 3 years, he will have

$$\begin{array}{ll} \$700.00 & \text{principal} \\ \$168.00 & \text{interest for 3 years} \\ \hline \mathbf{\$868.00} & \text{total} \end{array}$$

Practice   **a.** How much interest is earned in 3 years on a deposit of $3000 at 7 percent simple interest?

b. Jena deposited $5000 at 7 percent simple interest. Six months later she withdrew the deposit and the interest she had earned. How much money did she withdraw? (*Hint*: Consider six months as $\frac{6}{12}$ of a year.)

Problem set
131

1. Bill bought 3 paperback books for $5.95 each. The tax rate was 6 percent. If he pays for the purchase with a $20 bill, how much money should he get back?

2. When the sum of $\frac{5}{6}$ and $\frac{5}{9}$ is divided by the product of $\frac{5}{6}$ and $\frac{5}{9}$, what is the quotient?

3. What is the sum of the first five positive odd numbers?

4. George burned 100 calories running 1 mile. How many miles would he need to run to burn 350 calories?

5. If a dozen roses cost $4.90, what is the cost of 30 roses?

6. Three fourths of a day is how many hours?

7. The average of four numbers was 8. Three of the numbers were 2, 4, and 6. What was the fourth number?

Write equations to solve Problems 8–10.

8. One hundred fifty is what percent of 60?

9. What number is 60 percent of 60?

10. Sixty percent of what number is 150?

11. The points (3, 1), (−1, 1), and (−1, −2) are the vertices of a right triangle. What is the length of the hypotenuse of the right triangle?

Use a ratio box to solve Problems 12 and 13.

12. The price of the dress was reduced by 40 percent. If the sale price was $48, what was the regular price?

13. The car model was built on a 1:36 scale. If the length of the car is 180 inches, how many inches long is the model?

14. The square root of 80 is between which two consecutive whole numbers?

15. Complete the table.

FRACTION	DECIMAL	PERCENT
(a)	1.25	(b)

16. Graph $x \geq -2$ on a number line.

17. Multiply and write the product in scientific notation.

$$(6.3 \times 10^7)(9 \times 10^{-3})$$

18. The probability that it will rain is $\frac{2}{5}$. What is the probability that it will not rain?

19. How much interest is earned in 5 years on a deposit of $4000 at 9 percent simple interest?

20. Find the value of $\dfrac{mc}{x}$ if $m = 2$, $c = 10$, and $x = 5$.

21. These triangles are similar. Find x.

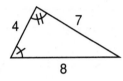 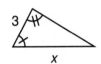

22. Find the volume of this triangular prism. Dimensions are in inches.

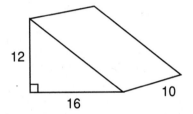

23. Find $m\angle x$.

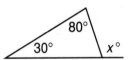

Solve:

24. $\dfrac{11}{42} = \dfrac{44}{r}$

25. $3\dfrac{1}{3}w - 4 = 36$

Add, subtract, multiply, or divide, as indicated:

26. $16 - \{27 - 3[8 - (3^2 - 2^3)]\}$

27. $\dfrac{60 \text{ mi}}{1 \text{ hr}} \cdot \dfrac{1 \text{ hr}}{60 \text{ min}}$

28. $3\dfrac{1}{3} + 1.5 + 4\dfrac{5}{6}$

29. $20 \div \left(3\dfrac{1}{3} \div 1\dfrac{1}{5}\right)$

30. $(-3)^2 + (-2)^3$

LESSON
132

Compound Probability

We know that the probability of getting heads on one toss of a coin is $\frac{1}{2}$. We can state this fact with the following equation in which $P(H)$ stands for "the probability of heads."

$$P(H) = \frac{1}{2}$$

The probability of getting two heads in a row is $\frac{1}{2}$ times $\frac{1}{2}$, or $\frac{1}{4}$. We can see this if we draw a diagram. If we toss a coin one time, we can get heads or tails.

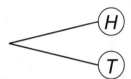

If the first toss comes up heads, the second toss could come up either heads or tails.

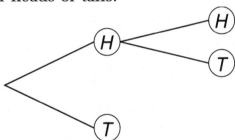

If the first toss came up tails, the second toss could come up either heads or tails.

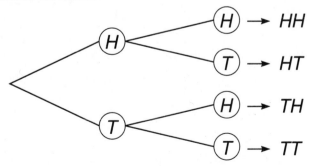

We see that there are 4 possible outcomes. They are

$$HH \qquad HT \qquad TH \qquad TT$$

The probability of each of these 4 outcomes is one fourth.

Thus, the probability of getting *HH* is $\frac{1}{4}$.

$$P(H, H) = \frac{1}{4}$$

The probability of independent events occurring in a specified order is the product of the probabilities of each event.

Thus $P(H, H, T)$ is $\frac{1}{2} \cdot \frac{1}{2} \cdot \frac{1}{2} = \frac{1}{8}$

so $P(H, T, T, H)$ is $\frac{1}{2} \cdot \frac{1}{2} \cdot \frac{1}{2} \cdot \frac{1}{2} = \frac{1}{16}$

and $P(T, H, T, H, T)$ is $\left(\frac{1}{2}\right)^5 = \frac{1}{32}$

Example 1   What is the probability of getting a 2 on the first spin and a 1 on the second spin?

Solution   The probability of getting a 2 is $\frac{1}{4}$. The probability of getting a 1 is $\frac{1}{4}$. The probability of independent events occurring in a specified order is the product of the individual probabilities.

$$P(2, 1) = \frac{1}{4} \cdot \frac{1}{4} = \mathbf{\frac{1}{16}}$$

Example 2   Jim tossed a coin once and it turned up heads. What is the probability that he will get heads on the next toss of the coin?

Solution   Past events do not affect the probability of future events. There are only 2 possible outcomes. The next toss of the coin will turn up either heads or tails. The probability that it will turn up heads is $\frac{1}{2}$.

Example 3   What is the probability of rolling a 4 and then a number above 3 in two rolls of a single die?

Solution $P(4) = \frac{1}{6}$ $P(>3) = \frac{1}{2}$

So $\qquad P(4, >3) = \dfrac{1}{6} \cdot \dfrac{1}{2} = \dfrac{1}{12}$

Practice Use this information to answer each question: The probability of a bird is $\frac{1}{2}$. The probability of a dog is $\frac{1}{4}$.

   a. What is the probability of getting bird, bird, dog in that order?

   b. What is the probability of getting bird, dog, bird, dog in that order?

Problem set 132

1. Sherman deposited $3000 in an account paying 8 percent simple interest yearly. He withdrew his money and interest 3 years later. How much money did he withdraw?

2. What is the square root of the sum of 3 squared and 4 squared?

3. Find (a) the median and (b) the mode of the following quiz scores.

CLASS QUIZ SCORES

SCORE	NUMBER OF STUDENTS
100	2
95	7
90	6
85	6
80	3
70	3

4. The trucker completed the 840-km haul in 10 hours 30 minutes. What was the trucker's average speed in kilometers per hour?

5. Use a ratio box to solve this problem. Barbara earned $28 for 6 hours of work. At that rate, how much would she earn for 9 hours of work?

6. A mile is about eight fifths of a kilometer. About how many meters is eight fifths of a kilometer?

7. If 60 percent of the students were boys, what was the ratio of boys to girls?

8. The points $(3, 11)$, $(-2, -1)$, and $(-2, 11)$ are the vertices of a right triangle. Use the Pythagorean theorem to find the length of the hypotenuse of this triangle.

9. Use a ratio box to solve this problem. Mike paid $48 for a jacket at 25 percent off of the regular price. What was the regular price of the jacket?

Write equations to solve Problems 10 and 11.

10. What decimal number is 2 percent of 360?

11. What percent of 2.5 is 2?

12. Use a ratio box to solve this problem. Troy bought a baseball card for $6 and sold it for 25 percent more than he paid for it. How much profit did he make on the sale?

13. What is the probability of having a coin turn up tails on 4 consecutive tosses of a coin?

14. How much interest is earned in 6 months on $4000 deposited at 9 percent simple interest?

15. Complete the table.

FRACTION	DECIMAL	PERCENT
$\frac{5}{8}$	(a)	(b)

16. Multiply and write the product in scientific notation.

$$(3 \times 10^{-4})(4 \times 10^8)$$

17. Convert 300 kg to grams.

18. Solve the formula $d = rt$ for t.

19. Make a table that shows three pairs of numbers for the function $y = -x$. Then graph the number pairs on a coordinate plane and draw a line through the points.

20. Find the perimeter of this figure. Dimensions are in centimeters.

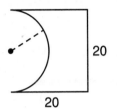

21. Find the surface area of this right triangular prism. Dimensions are in feet.

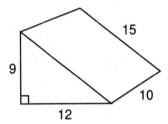

22. What fraction of 42 is 7?

23. These triangles are similar. Find x. (*Hint*: Mentally rotate one triangle so that it looks like the other triangle.)

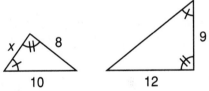

Solve:

24. $\dfrac{16}{2.5} = \dfrac{48}{f}$ **25.** $2\dfrac{2}{3}x - 3 = 21$

Add, subtract, multiply, or divide, as indicated:

26. $5^2 - [40 - 2(10 + 3^2)]$

27. $1 \text{ yd}^2 \cdot \dfrac{3 \text{ ft}}{1 \text{ yd}} \cdot \dfrac{3 \text{ ft}}{1 \text{ yd}}$

28. $2\dfrac{3}{4} - \left(1.5 - \dfrac{1}{6}\right)$ **29.** $3.5 \div 1\dfrac{2}{5} \div 3$

30. $(4) - (-3)(-2)(-1) + \dfrac{(-5)(4)(-3)(2)}{-1}$

LESSON 133

Volume of a Pyramid and a Cone

Pyramids A **pyramid** is a geometric solid that has three or more triangular faces and a base that is a polygon. Each of these figures is a pyramid.

The volume of a pyramid is $\frac{1}{3}$ the volume of a prism that has the same base and height. To find the volume of a pyramid, we will first find the volume of a prism that has the same base and height. Then we will divide the result by 3.

Example 1 The cube just contains the pyramid. Each edge of the cube is 6 centimeters.

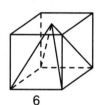

(a) Find the volume of the cube.

(b) Find the volume of the pyramid.

6

Solution (a) The volume of the cube equals the area of the base times the height.

$$\text{Area of base: } 6 \text{ cm} \times 6 \text{ cm} = 36 \text{ cm}^2$$

$$\text{Volume} = \text{base} \cdot \text{height}$$

$$= (36 \text{ cm}^2)(6 \text{ cm})$$

$$= \textbf{216 cm}^3$$

(b) The volume of the pyramid is $\frac{1}{3}$ the volume of the cube. Dividing by 3 (or multiplying by $\frac{1}{3}$), we find that the volume of the pyramid is

$$\frac{216 \text{ cm}^3}{3} = \textbf{72 cm}^3$$

Cones The volume of a cone is $\frac{1}{3}$ the volume of the cylinder with the same base and height.

Example 2   Find the volume of this circular cone. Dimensions are in centimeters.

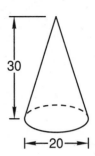

Solution   We first find the volume of a cylinder with the same base and height as the cone.

Volume of cylinder = area of circle · height

$$= (3.14) (10 \text{ cm})^2 \cdot 30 \text{ cm}$$

$$= 9420 \text{ cm}^3$$

Then we find $\frac{1}{3}$ of this volume.

Volume of cone $= \frac{1}{3}$ (volume of cylinder)

$$= \frac{1}{3} \cdot 9420 \text{ cm}^3$$

$$= \mathbf{3140 \text{ cm}^3}$$

Practice   Find the volume of each figure. Dimensions are in centimeters.

a.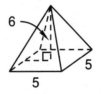

b.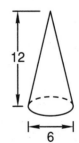

Problem set 133

1. Use a ratio box to solve this problem. The regular price of the item was $24, but it was on sale for 25 percent off. What is the item's sale price?

2. Ten billion is how much greater than nine hundred eighty million? Write the answer in scientific notation.

3. The median of the following numbers is how much less than the mean?

1.4, 0.5, 0.6, 0.75, 5.2

4. Nelda worked for 5 hours and earned $24. How much did Nelda earn per hour? Christy worked for 6 hours and earned $33. How much did Christy earn per hour? Christy earned how much more per hour than Nelda?

5. If 24 kilograms of seed costs $31, what is the cost of 42 kilograms of seed at the same rate? Use a ratio box to solve the problem.

6. A kilometer is about $\frac{5}{8}$ of a mile. A mile is 1760 yards. A kilometer is about how many yards?

7. A card was drawn from a deck of 52 playing cards. The card was then replaced. Another card was drawn. What was the probability that both cards were hearts?

Write equations to solve Problems 8 and 9.

8. What percent of $30 is $1.50?

9. Fifty percent of what number is $2\frac{1}{2}$?

10. Trinh left $5000 in an account that paid 8 percent simple interest annually. How much interest was earned in 3 years?

11. Use a ratio box to solve this problem. A merchant sold an item at a 20 percent discount from the regular price. If the regular price was $12, what was the sale price of the item?

12. The points (0, 4), (−3, 2), and (3, 2) are the vertices of a triangle. Find the area of the triangle.

13. Use two unit multipliers to convert 6 ft^2 to square inches.

14. Use a ratio box to solve this problem. Jessica sculptured a figurine from clay at $\frac{1}{24}$ of the actual size of the model. If the model was 6 feet tall, how many inches tall was the figurine?

15. Complete the table.

Fraction	Decimal	Percent
(a)	0.5	(b)

16. Multiply and write the product in scientific notation.

$$(6.3 \times 10^6)(7 \times 10^{-3})$$

17. Tim can get from point A to point B by staying on the sidewalk and turning left at the corner C, or he can take the shortcut and walk straight from point A to point B. How many yards does he save by taking the shortcut instead of staying on the sidewalk? Begin by using the Pythagorean theorem to find the length of the shortcut.

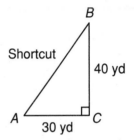

18. (a) Solve the formula $A = \frac{1}{2}bh$ for h.

(b) Use the formula $A = \frac{1}{2}bh$ to find h when $A = 16$ and $b = 8$.

19. Make a table that shows three pairs of numbers for the function $y = -2x + 1$. Then graph the number pairs on a coordinate plane and draw a line through the points to show other number pairs of the function.

20. Find the volume of the pyramid. Dimensions are in meters.

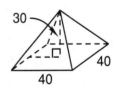

21. Find the volume of the cone. Dimensions are in centimeters.

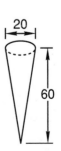

22. Refer to the figure to find the measure of each angle.

(a) ∠D

(b) ∠E

(c) ∠A

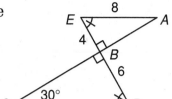

23. In the figure in Problem 22, *EB* is 4 cm, *BD* is 6 cm, and *EA* is 8 cm. Find *CD*.

Solve:

24. $\dfrac{7.5}{d} = \dfrac{25}{16}$

25. $1\dfrac{3}{5}w + 17 = 49$

Add, subtract, multiply, or divide, as indicated:

26. $5^2 - \{4^2 - [3^2 - (2^2 - 1^2)]\}$

27. $\dfrac{440 \text{ yd}}{1 \text{ min}} \cdot \dfrac{1 \text{ min}}{60 \text{ sec}} \cdot \dfrac{3 \text{ ft}}{1 \text{ yd}}$

28. $1\dfrac{3}{4} + 2\dfrac{2}{3} - 3\dfrac{5}{6}$

29. $\left(1\dfrac{3}{4}\right)\left(2\dfrac{2}{3}\right) \div 3\dfrac{5}{6}$

30. $(-7) - (-3) - (2)(-3) + (-4) - (-3)(-2)(-1)$

LESSON
134

Probability, Chance, and Odds

Probability, chance, and odds are different ways of expressing the likelihood that an event will occur. Recall that probability is the ratio of the number of ways a particular event can happen to the total number of equally likely possible outcomes.

$$\text{Probability} = \frac{\text{number of favorable outcomes}}{\text{number of possible outcomes}}$$

Thus, the probability that this spinner will end up in region A is $\frac{1}{4}$.

We can use the word **chance** and a percent to describe a probability. Since the fraction $\frac{1}{4}$ is equivalent to 25 percent, we can say that the chance of the spinner stopping in region A is 25 percent.

While probability is the ratio of the number of favorable outcomes to the number of possible outcomes, **odds** is the **ratio of the number of favorable outcomes to the number of unfavorable outcomes.** Using the spinner example, one outcome is A, and three outcomes are not A. Thus, the odds of the spinner ending up in region A are

1 to 3

Note that odds are usually expressed by using the word "to" and not by using a division line. However, odds are reduced as are other ratios.

Example 1 A 20 percent chance of rain was forecast.

(a) What is the probability that it will rain?

(b) What are the odds that it will rain?

Solution (a) To express chance as probability, we simply write the percent as a fraction and reduce.

$$20\% \; = \; \frac{20}{100} \; = \; \frac{1}{5}$$

The probability that it will rain is $\frac{1}{5}$.

(b) Since the probability of rain is $\frac{1}{5}$, the probability that it will not rain is $\frac{4}{5}$. Thus, for every favorable outcome there are 4 unfavorable outcomes. Therefore, the odds that it will rain are **1 to 4**.

Example 2   The odds that a marble drawn from a bag will be red is 3 to 2.

(a) What is the probability that a red marble will be drawn?

(b) What is the chance of drawing a red marble?

Solution   If the odds are 3 to 2, we mean

$$
\begin{array}{ll}
3 & \text{favorable outcomes} \\
\underline{2} & \text{unfavorable outcomes} \\
5 & \text{possible outcomes}
\end{array}
$$

(a) The probability of drawing a red marble is

$$\frac{\text{favorable}}{\text{possible}} = \frac{3}{5}$$

(b) The chance of drawing a red marble is

$$\frac{3}{5} = 60\%$$

Practice   Use this information to answer questions **a–d**.

In a bag there are 4 blue marbles, 3 red marbles, 2 green marbles, and 1 yellow marble.

a. What is the chance of drawing a green marble?

b. What are the odds of drawing a blue marble?

c. What is the chance of drawing a blue or yellow marble?

d. What are the odds of drawing a red or green marble?

Problem set 134

1. The regular price was \$72.50, but it was on sale for 20% off. What was the total sale price including 7% sales tax? Use a ratio box to find the sale price. Then find the sales tax and total price.

2. On his first 4 tests, Eric's average score was 87. What score does he need to average on his next 2 tests to have a 6-test average of 90?

3. In a bag there are 6 red marbles, 9 green marbles, and 12 blue marbles. One marble is to be drawn from the bag.

    (a) What is the probability that the marble will be blue?

    (b) What is the chance that the marble will be green?

    (c) What are the odds that the marble will not be red?

4. If a box of 12 dozen pencils costs $10.80, then what is the cost per pencil?

5. How much interest is earned on $5000 at 8 percent simple interest in 6 months?

6. One fourth of the students in the class earned an A. One third of the students earned a B. If 6 students earned an A, how many students earned a grade lower than a B?

Use a ratio box to solve Problems 7 and 8.

7. The ratio of cars to trucks passing by the checkpoint was 5 to 2. If 3500 cars and trucks passed by the checkpoint, how many were cars?

8. There were 20 percent more rainy days in April than there were in March. If there were 10 rainy days in March, then how many rainy days were there in April? (*Hint*: Let the number of rainy days in March equal 100 percent.)

Write equations to solve Problems 9 and 10.

9. What is 120 percent of $240?

10. Sixty is what percent of 150?

11. The points (3, 2), (6, −2), (−2, −2), and (−2, 2) are the vertices of a trapezoid.

    (a) Find the area of the trapezoid.

    (b) Find the perimeter of the trapezoid.

12. Arrange these numbers in order from least to greatest.

$$\sqrt{6}, \ 6^2, \ -6, \ 0.6$$

13. Complete the table.

FRACTION	DECIMAL	PERCENT
$1\frac{4}{5}$	(a)	(b)

14. Multiply and write the product in scientific notation.

$$(2 \times 10^{-6})(5 \times 10^{-9})$$

15. Divide 0.02 by 1.1 and write the answer with a bar over the repetend.

16. Convert 12 inches to centimeters. (1 in. = 2.54 cm)

17. (a) Solve the formula $C = \pi d$ for d.

  (b) Use the formula $C = \pi d$ to find d when C is 62.8 and π is 3.14.

18. Find three pairs of numbers for the function $y = 2x + 1$. Then graph the number pairs on a coordinate plane and draw a line through the points to show other number pairs of the function.

19. Find the perimeter of this figure. Dimensions are in centimeters.

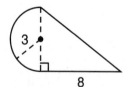

20. Find the surface area of this cube. Dimensions are in feet.

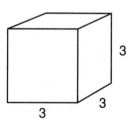

21. Find the volume of this cylinder. Dimensions are in meters.

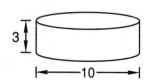

22. Refer to the figure to find (a)–(c).

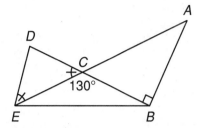

   (a) $m\angle ACB$

   (b) $m\angle CAB$

   (c) $m\angle CDE$

23. An aquarium that is 40 cm long, 10 cm wide, and 20 cm deep is filled with water. Find the volume of the water in the aquarium.

Solve:

24. $0.8m - 1.2 = 6$

25. $\dfrac{x}{3.2} = \dfrac{27}{24}$

Simplify:

26. $92 - \sqrt{81} + \sqrt{9}$

27. 1 kilogram − 50 grams

28. $(1.2)\left(3\dfrac{3}{4}\right) \div 4\dfrac{1}{2}$

29. $2\dfrac{3}{4} - 1.5 - \dfrac{1}{6}$

30. $(-3)(-2) - (2)(-3) - (-8) + (-2)(-3) - (-5)$

LESSON
135

Volume, Capacity, and Weight in the Metric System

Metric units of volume, capacity, and weight are related. The relationship describes the volume and the weights of a quantity of water under certain standard conditions. There are two commonly used references.

One milliliter of water has a volume of **1 cubic centimeter** and a weight of **1 gram**.

One cubic centimeter can contain 1 milliliter of water, which has a weight of 1 gram.

One liter of water has a volume of **1000 cubic centimeters** and a weight of **1 kilogram**.

One thousand cubic centimeters can contain 1 liter of water, which has a weight of 1 kilogram.

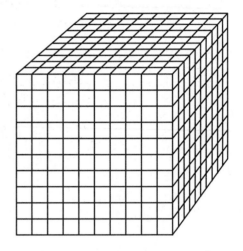

Example 1   Ray has a fish aquarium that is 50 cm long and 20 cm wide. If the aquarium is filled with water to a depth of 30 cm, (a) how many liters of water would be in the aquarium, and (b) what is the weight of the water in the aquarium?

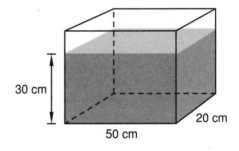

Solution   First we find the volume of the water in the aquarium.

$$(50 \text{ cm})(20 \text{ cm})(30 \text{ cm}) = 30{,}000 \text{ cm}^3$$

(a) Each cubic centimeter of water is 1 milliliter. Thirty thousand milliliters is **30 liters.**

(b) Each liter of water has a weight of 1 kg, so the weight of the water in the aquarium is **30 kg.**

Example 2   Jan wanted to find the volume of a vase. She filled a 1-liter beaker with water and then used all but 240 milliliters to fill the vase.

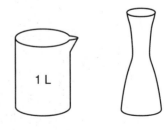

(a) What was the volume of the vase?

(b) If the weight of the vase was 640 grams, what was the weight of the vase filled with water?

Solution   (a) The 1-liter beaker contained 1,000 mL of water. Since Jan used 760 mL (1,000 mL − 240 mL), the volume of the inside of the vase was **760 cm³**.

(b) The weight of the water (760 g) plus the weight of the vase (640 g) is **1400 g**.

Practice   a. What is the weight of 2 liters of water?

b. What is the volume of 3 liters of water?

c. When the bottle was filled with water, the weight increased by 1 kilogram. How many milliliters of water were added?

d. A tank that is 25 cm long, 10 cm wide, and 8 cm deep can hold how many liters of water?

Problem set 135   1. How much interest is earned in 9 months on a deposit of $7000 at 8 percent? (*Hint*: 9 months is $\frac{9}{12}$ of a year.)

2. With two tosses of a coin,

(a) What is the probability of getting two heads?

(b) What is the chance of getting two tails?

(c) What are the odds of getting heads, then tails?

3. On the first 4 days of their trip the Schmidts averaged 410 miles per day. On the fifth day they traveled 600 miles. How many miles per day did they average for the first 5 days of their trip?

4. The 18-ounce container cost $2.16. The 1-quart container cost $3.36. The smaller container cost how much more per ounce than the larger container?

Use a ratio box to solve Problems 5 and 6.

5. Adam typed 160 words in 5 minutes on his typing test. At that rate, how long would it take him to type an 800-word essay?

6. The ratio of guinea pigs to rats running the maze was 7 to 5. Of the 120 guinea pigs and rats running the maze, how many were guinea pigs?

7. Kelly was thinking of a certain number. If $\frac{3}{4}$ of the number was 48, then what was $\frac{5}{8}$ of the number?

Write an equation to solve Problems 8 and 9.

8. What percent of $60 is $20?

9. What fraction is 50 percent of $\frac{3}{4}$?

10. The points $(-3, 4)$, $(5, -2)$, and $(-3, -2)$ are the vertices of a triangle.

    (a) Find the area of the triangle.

    (b) Find the perimeter of the triangle.

11. A glass aquarium with dimensions as shown has a mass of 5 kg when empty. What is the mass of the aquarium when it is half full of water?

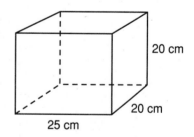

20 cm

20 cm

25 cm

12. Complete the table.

Fraction	Decimal	Percent
(a)	0.875	(b)

13. Compare: $a \div b \bigcirc a - b$ if a is positive and b is negative

14. Multiply and write the product in scientific notation.

$$(6.4 \times 10^6)(8 \times 10^{-8})$$

15. Convert 36 inches to centimeters. (1 in. = 2.54 cm)

16. (a) Solve the formula $A = \frac{1}{2}bh$ for b.

(b) Use the formula $A = \frac{1}{2}bh$ to find b when A is 24 and h is 6.

17. Find three pairs of numbers for the function $y = -2x$. Then graph the number pairs on a coordinate plane and draw a line through the points to show other number pairs of the function.

18. Find the area of this figure. Dimensions are in millimeters.

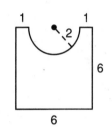

19. Find the surface area of the cube. Dimensions are in centimeters.

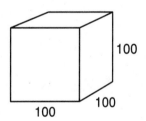

20. Find the volume of the cylinder. Dimensions are in inches.

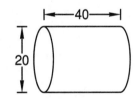

21. Refer to the figure to find (a)–(c).

(a) $m\angle YXZ$

(b) $m\angle WXV$

(c) $m\angle WVX$

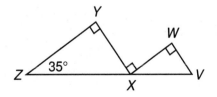

22. In the figure in Problem 21, ZX is 21 cm, YX is 12 cm, and XV is 14 cm. Write a proportion to find WV.

23. A pyramid is cut out of a cube of plastic with dimensions as shown. What is the volume of the pyramid?

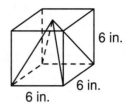

6 in.

6 in.

6 in.

Solve:

24. $0.4n + 5.2 = 12$

25. $\dfrac{18}{y} = \dfrac{36}{28}$

Add, subtract, multiply, or divide, as indicated:

26. $\sqrt{5^2 - 3^2} + \sqrt{5^2 - 4^2}$

27. 3 yd − 2 ft 1 in.

28. $3.5 \div \left(1\dfrac{2}{5} \div 3\right)$

29. $12.75 + 3\dfrac{1}{2} + \dfrac{7}{8}$

30. $\dfrac{(3)(-2)(4)}{(-6)(2)} + (-8) + (-4)(+5) - (2)(-3)$

Appendix

Supplemental Practice Problems for Selected Lessons

This appendix contains additional practice problems for concepts presented in selected lessons. It is very important that no problems in the regular problem sets be omitted to make room for these problems. This book is designed to produce long-term retention of concepts, and long-term practice of all the concepts is necessary. The practice problems in the problem sets provide enough initial exposure to concepts for most students. If a student continues to have difficulty with certain concepts, some of these problems can be assigned as remedial exercises.

Supplemental Practice for Lesson 5

List the whole numbers from 1–10 that are factors of:

1. 36   **2.** 3600   **3.** 350

4. 1326   **5.** 4320   **6.** 950

7. 12,000   **8.** 35,420   **9.** 36,270

10. 123,450   **11.** 1,000,000   **12.** 2520

Supplemental Practice for Lesson 16

Reduce each fraction to lowest terms.

1. $\dfrac{15}{20}$ **2.** $\dfrac{8}{24}$ **3.** $\dfrac{9}{24}$ **4.** $\dfrac{12}{18}$

5. $\dfrac{24}{30}$ **6.** $\dfrac{16}{32}$ **7.** $\dfrac{24}{36}$ **8.** $\dfrac{28}{35}$

9. $3\dfrac{15}{18}$ **10.** $6\dfrac{18}{24}$ **11.** $8\dfrac{9}{15}$ **12.** $4\dfrac{18}{32}$

Supplemental Practice for Lesson 22

Write the prime factorization of each of these numbers.

1. 81   **2.** 300   **3.** 2000

4. 625   **5.** 450   **6.** 1200

7. 440   **8.** 750   **9.** 10,000

10. 128   **11.** 780   **12.** 1540

Supplemental Practice for Lesson 23

Simplify:

1. $\dfrac{15}{6}$ **2.** $\dfrac{28}{8}$ **3.** $\dfrac{30}{12}$ **4.** $\dfrac{36}{10}$

5. $\dfrac{40}{6}$ **6.** $\dfrac{48}{15}$ **7.** $\dfrac{50}{12}$ **8.** $\dfrac{36}{8}$

9. $3\dfrac{15}{9}$ **10.** $4\dfrac{16}{12}$ **11.** $9\dfrac{15}{6}$ **12.** $8\dfrac{24}{10}$

Supplemental Practice for Lesson 26

Add or subtract, as indicated:

1. $5\frac{3}{5} + 2\frac{4}{5}$ **2.** $7\frac{3}{8} + 1\frac{3}{8}$ **3.** $2\frac{3}{7} + 3\frac{4}{7}$

4. $5\frac{3}{4} + 3\frac{3}{4}$ **5.** $6\frac{5}{8} + 5\frac{7}{8}$ **6.** $8\frac{5}{9} + 2\frac{7}{9}$

7. $6\frac{7}{8} - 2\frac{1}{8}$ **8.** $5 - 3\frac{1}{4}$ **9.** $6 - 2\frac{3}{5}$

10. $5\frac{1}{3} - 1\frac{2}{3}$ **11.** $4\frac{2}{5} - 1\frac{4}{5}$ **12.** $6\frac{1}{6} - 2\frac{5}{6}$

Supplemental Practice for Lesson 30

Multiply or divide, as indicated:

1. $3\frac{3}{4} \times \frac{2}{5}$ **2.** $2\frac{1}{3} \times 3$ **3.** $1\frac{4}{5} \times 3\frac{1}{3}$

4. $7 \times 2\frac{2}{3}$ **5.** $\frac{5}{8} \times 3\frac{1}{5}$ **6.** $2\frac{1}{4} \times 1\frac{3}{5}$

7. $3\frac{1}{2} \div 3$ **8.** $2\frac{3}{4} \div \frac{3}{4}$ **9.** $1\frac{1}{2} \div 2\frac{2}{3}$

10. $3\frac{1}{3} \div 1\frac{3}{4}$ **11.** $6 \div 3\frac{3}{5}$ **12.** $\frac{5}{8} \div 3\frac{1}{2}$

Supplemental Practice for Lesson 35

Add or subtract, as indicated:

1. $\frac{3}{5} + \frac{3}{10}$ **2.** $2\frac{5}{6} + 1\frac{1}{2}$ **3.** $\frac{3}{4} + \frac{1}{2} + \frac{3}{8}$

4. $\frac{5}{6} + \frac{3}{4}$ **5.** $3\frac{3}{5} + 2\frac{2}{3}$ **6.** $\frac{5}{6} + \frac{3}{8} + \frac{7}{12}$

7. $\frac{5}{8} - \frac{1}{2}$ **8.** $3\frac{5}{6} - 1\frac{1}{2}$ **9.** $4\frac{3}{4} - 1\frac{1}{3}$

10. $\frac{8}{12} - \frac{2}{3}$ **11.** $6\frac{3}{5} - 3\frac{1}{3}$ **12.** $5\frac{1}{4} - 1\frac{5}{6}$

Supplemental Practice for Lesson 37

Name these decimal numbers.

1. 16.125

2. 5.03

3. 105.105

4. 0.001

5. 160.166

6. 4000.321

Write each of these as decimal numerals.

7. One hundred twenty-three thousandths

8. One hundred and twenty-three thousandths

9. One hundred twenty and three thousandths

10. Five hundredths

11. Twenty and nine hundredths

12. Twenty-nine and five tenths

13. One thousand and two hundred twelve thousandths

14. One thousand two hundred and twelve thousandths

Supplemental Practice for Lesson 38

Round to the nearest whole number.

1. 23.459

2. 164.089

3. 86.6427

Round to two decimal places.

4. 12.83333

5. 6.0166

6. 0.1084

Round to the nearest thousandth.

7. 0.08333

8. 0.45454

9. 3.14159

10. Round 283.567 to the nearest hundred.

11. Round 283.567 to the nearest hundredth.

12. Round 126.59 to the nearest ten.

Supplemental Practice for Lesson 40   Add or subtract, as indicated:

1. 45.3 + 2.64 + 3   **2.** 0.4 + 0.5 + 0.6 + 0.7

3. 3.6 + 2.75 + 0.194 + 3   **4.** 12.8 + 6.32 + 15

5. 278.4 + 3.26 + 1.475   **6.** 10 + 1.0 + 0.1 + 0.01

7. 14.327 − 6.5   **8.** 10.8 − 9.67   **9.** 6.5 − 4.321

10. 10 − 4.76   **11.** 0.1 − 0.019   **12.** 5 − 4.937

Supplemental Practice for Lesson 42   Find the perimeter of each of these polygons. Dimensions are in centimeters.

1.
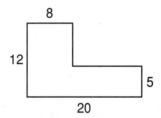

2.
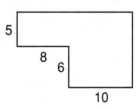

3.
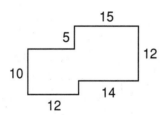

4.
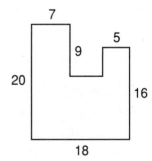

Supplemental Practice for Lesson 48   Change each of these numbers to a reduced fraction or mixed number.

1. 0.48   **2.** 3.75   **3.** 0.125

4. 12.6   **5.** 0.025   **6.** 1.08

Change each of these numbers to a decimal number.

7. $\dfrac{5}{8}$ **8.** $\dfrac{1}{3}$ **9.** $2\dfrac{2}{5}$

10. $6\frac{1}{6}$ **11.** $\frac{11}{20}$ **12.** $5\frac{5}{9}$

Supplemental Practice for Lesson 50

Complete each division.

1. $0.15 \div 0.5$ **2.** $14.4 \div 0.06$ **3.** $18 \div 0.4$

4. $5 \div 0.8$ **5.** $12.5 \div 0.04$ **6.** $288 \div 1.2$

7. $4.3 \div 0.01$ **8.** $1.5 \div 0.12$ **9.** $9 \div 1.8$

10. $4.5 \div 2.5$ **11.** $8 \div 0.04$ **12.** $12.5 \div 0.5$

Supplemental Practice for Lesson 52

Simplify:

1. 8^2 **2.** 2^6 **3.** 3^3 **4.** 10^5

5. $3^2 + 2^3$ **6.** $5^2 - 4^2$ **7.** 4^3 **8.** 15^2

9. $\frac{10^4}{10^3}$ **10.** $\frac{8^2}{2^3}$ **11.** 25^2 **12.** $5^4 - 5^3$

Supplemental Practice for Lesson 58

Change:

1. 40 inches to feet and inches

2. 200 seconds to minutes and seconds

Simplify:

3. 3 ft 21 in. **4.** 2 hr 90 min

Add and simplify:

5.   3 yd 2 ft 7 in.
    + 1 yd 1 ft 8 in.

6.   5 hr 18 min 23 sec
    + 2 hr 45 min 48 sec

Supplemental Practice for Lesson 60

Find the area of each figure. Dimensions are in centimeters.

1.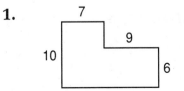

2.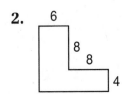

3.

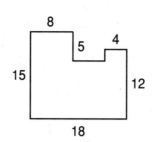

4.

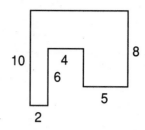

Subtract:

1. 5 ft 7 in. − 3 ft 10 in.

2. 10 min 13 sec − 3 min 28 sec

3. 4 yd 6 in. − 2 ft 8 in.

4. 1 hr 10 min − 25 min 40 sec

5. 8 yd 2 ft 4 in.
  − 1 yd 2 ft 9 in.

6. 3 hr 17 min 30 sec
  − 2 hr 48 min 43 sec

Copy and complete the table.

FRACTION	DECIMAL	PERCENT
$\frac{5}{6}$	(1)	(2)
(3)	1.2	(4)
(5)	(6)	8%
$1\frac{3}{5}$	(7)	(8)
(9)	0.075	(10)
(11)	(12)	125%

Evaluate:

1. $ab - bc + abc$ if $a = 5$, $b = 4$, and $c = 2$

2. $xy + \dfrac{x}{y} - 5$ if $x = 8$ and $y = 4$

3. $abc - ab - \dfrac{a}{c}$ if $a = 6$, $b = 4$, and $c = 3$

4. $m - mn$ if $m = \dfrac{3}{4}$ and $n = \dfrac{1}{2}$

5. $wx + xz - z$ if $w = 1.2$, $x = 0.5$, and $z = 0.1$

6. $ab - ac - \dfrac{ab}{c}$ if $a = 4$, $b = 3$, and $c = 2$

Supplemental Practice for Lesson 78

Find each sum.

1. $(-36) + (+54)$

2. $(-15) + (-26)$

3. $(-6) + (-12) + (+15)$

4. $(+4) + (-12) + (+21)$

5. $(-6) + (-8) + (-7) + (-2)$

6. $(-9) + (-15) + (+50)$

7. $(+42) + (-23) + (-19)$

8. $(-54) + (+76) + (-17)$

9. $\left(-3\dfrac{1}{2}\right) + \left(-2\dfrac{1}{4}\right)$

10. $\left(-1\dfrac{1}{3}\right) + \left(+2\dfrac{5}{6}\right)$

11. $(-4.3) + (+2.63)$

12. $(-1.7) + (-3.2) + (-1.8)$

Supplemental Practice for Lesson 79

Find the area of each triangle. Dimensions are in centimeters.

1.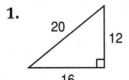

2.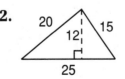

3.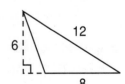

4.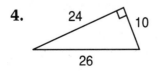

5.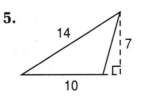

6.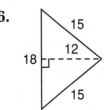

Supplemental Practice for Lesson 80

Change each percent to a fraction or decimal number before multiplying.

1. What is 50% of 250?   2. What is 5% of 40?

3. What is 25% of 48?   4. What is 90% of 30?

5. What is 100% of 65?   6. What is 1% of 5000?

7. What is 8% of 48?   8. What is 75% of 64?

9. What is 80% of 21?   10. What is 2% of 600?

11. What is 70% of 80?   12. What is 80% of 70?

Supplemental Practice for Lesson 85

Find the circumference of each circle. Dimensions are in centimeters.

1.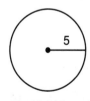

Use 3.14 for π

2.

Use $\frac{22}{7}$ for π

3.

Leave π as π

4.

Use 3.14 for π

5.

Use $\frac{22}{7}$ for π

6.

Leave π as π

Supplemental Practice for Lesson 86

Rewrite each subtraction or addition. Then find the sum.

1. $(-3) - (-8)$ 2. $(-12) + (+20)$

3. $(+8) - (-15)$ 4. $(+6) - (18)$

5. $(-3) + (-4) - (-5)$ 6. $(+3) - (-4) - (+5)$

7. $(-2) - (-3) - (-4)$ 8. $(+2) - (3) - (-4)$

9. $(-6) - (-7) + (8)$ **10.** $(+8) - (+9) - (-12)$

11. $(-3) - (-1) - (-8) - (2)$ **12.** $(-9) - (10) - (-11)$

Supplemental Practice for Lesson 89

Express in the customary form of scientific notation.

1. 0.15×10^7 **2.** 48×10^{-8}

3. 20×10^5 **4.** 0.72×10^{-4}

5. 0.125×10^{12} **6.** 22.5×10^{-6}

7. 17.5×10^{10} **8.** 0.375×10^{-8}

Supplemental Practice for Lesson 96

Find each area. Dimensions are in centimeters.

1. 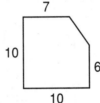   **2.**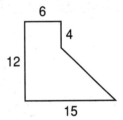

3. 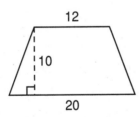   **4.**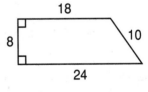

Supplemental Practice for Lesson 97

Simplify:

1. $\dfrac{62\frac{1}{2}}{100}$ **2.** $\dfrac{12}{\frac{3}{4}}$ **3.** $\dfrac{16\frac{2}{3}}{100}$ **4.** $\dfrac{10}{1\frac{2}{3}}$

Change each percent to a fraction.

5. $6\frac{2}{3}\%$ **6.** $87\frac{1}{2}\%$ **7.** $83\frac{1}{3}\%$ **8.** $3\frac{1}{3}\%$

Find "what percent."

9. Sixteen is what percent of 40?

10. Eight is what percent of 12?

11. What percent of 30 is 6?

12. What percent of 30 is 5?

Supplemental Practice for Lesson 103

Find the area of each circle. Dimensions are in centimeters.

1.

Use 3.14 for π

2.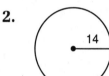

Use $\frac{22}{7}$ for π

3.

8

Leave π as π

4.

20

Use 3.14 for π

5.

14

Use $\frac{22}{7}$ for π

6.

8

Leave π as π

Supplemental Practice for Lesson 104

Write each product in scientific notation.

1. $(1.2 \times 10^5)(3 \times 10^6)$ **2.** $(3 \times 10^6)(6 \times 10^3)$

3. $(2.5 \times 10^5)(4 \times 10^7)$ **4.** $(4.2 \times 10^8)(2.5 \times 10^{12})$

5. $(4 \times 10^{-3})(2 \times 10^{-8})$ **6.** $(6 \times 10^{-7})(4 \times 10^{-5})$

7. $(2 \times 10^{-4})(6.5 \times 10^{-8})$ **8.** $(1.6 \times 10^{-5})(7 \times 10^{-7})$

9. $(6 \times 10^{-4})(4 \times 10^8)$ **10.** $(7 \times 10^{-9})(3 \times 10^5)$

11. $(1.4 \times 10^7)(8 \times 10^{-5})$ **12.** $(7.5 \times 10^{-8})(4 \times 10^6)$

Supplemental Practice for Lesson 110

Write each quotient in scientific notation.

1. $\dfrac{8 \times 10^8}{4 \times 10^4}$

2. $\dfrac{6 \times 10^3}{3 \times 10^6}$

3. $\dfrac{3.6 \times 10^6}{2 \times 10^{12}}$

4. $\dfrac{1.2 \times 10^8}{3 \times 10^4}$

5. $\dfrac{2.4 \times 10^{12}}{8 \times 10^7}$

6. $\dfrac{3 \times 10^7}{4 \times 10^5}$

7. $\dfrac{4.2 \times 10^6}{7 \times 10^9}$

8. $\dfrac{1 \times 10^8}{2 \times 10^{12}}$

9. $\dfrac{1.8 \times 10^7}{6 \times 10^{11}}$

10. $\dfrac{7.5 \times 10^{12}}{5 \times 10^7}$

11. $\dfrac{6.3 \times 10^8}{9 \times 10^4}$

12. $\dfrac{4 \times 10^6}{5 \times 10^{10}}$

Supplemental Practice for Lesson 119

Find the coordinates of each of these points of this coordinate plane.

1. Point A

2. Point B

3. Point C

4. Point D

5. Point E

6. Point F

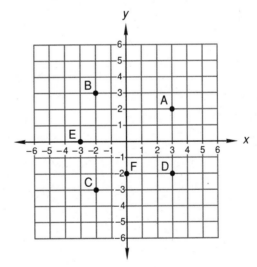

On a piece of graph paper draw an x-axis and a y-axis. Then graph and label the following points.

7. $(4, -3)$

8. $(-3, 4)$

9. $(-3, -4)$

10. $(3, 4)$

11. $(3, 0)$

12. $(0, 3)$

Glossary

absolute value The quality of a number that equals the distance of the graph of the number from the origin. Since the graphs of −3 and +3 are both 3 units from the origin, the absolute value of both numbers is 3.

acute angle An angle whose degree measure is between 0° and 90°.

acute triangle A triangle in which all three angles are acute angles.

addend One of two or more numbers that are to be added to find a sum.

adjacent angles Two angles that have a common side and a common vertex. The angles lie on opposite sides of their common side.

algebraic addition The combining of positive and/or negative numbers to form a sum.

algorithm A particular process for solving a certain type of problem. Often the process is repetitive, as in the long division algorithm.

altitude of a triangle The perpendicular distance from the base of a triangle to the opposite vertex; also called the *height* of the triangle.

angle In geometry, the figure formed by two rays that have a common endpoint.

area The number of square units of a certain size needed to cover the surface of a figure.

average   The sum of a group of numbers divided by the number of numbers in the group.

base   (1) A designated side (or a face) of a geometric figure. (2) The lower number in an exponential expression. In the exponential expression 2^5, the number 2 is the base and the number 5 is the exponent.

centimeter   One hundredth of a meter.

chance   A way of expressing the likelihood of an event; the probability of an event expressed as a percent.

circumference   The perimeter of a circle.

common factors   Identical factors of two or more indicated products.

complementary angles   Two angles whose sum is 90°.

composite number   A counting number that is the product of two counting numbers, neither of which is the number 1.

congruent polygons   Two polygons in which the corresponding sides have equal lengths and the corresponding angles have equal measures.

coordinate(s)   The number associated with a point on a number line; also the ordered pair of numbers associated with a point in the Cartesian plane.

corresponding sides   Sides of similar polygons that occupy corresponding positions. Corresponding sides are always opposite angles whose measures are equal.

counting numbers   Sometimes called the natural numbers, these numbers are 1, 2, 3, 4, 5, ...

decagon   A 10-sided polygon.

decimal fraction   A decimal number.

decimal number   A base 10 numeral that contains a decimal point.

decimal point   A dot placed in a decimal number to use as a place value reference point. The place to the left of the decimal point is always the units' (ones') place.

denominate number   A combination of a number and a descriptor that designates units. Examples are 4 ft, 16 tons, 42 miles per hour.

denominator   The bottom number in a fraction.

diameter   The distance across a circle through its center.

difference   The result of subtraction.

digit   In the base 10 system, any of the symbols 0, 1, 2, 3, 4, 5, 6, 7, 8, or 9.

directed numbers   Another name for signed numbers.

dividend   The number to be divided. In the expression $10 \div 2$, the dividend is 10 and the divisor is 2.

divisible   If one whole number is divided by another whole number and the quotient is a whole number (the remainder is zero), we say that the first whole number is divisible by the second whole number: 10 is divisible by 2.

divisor   The number by which another number is divided. In the expression $10 \div 2$, the divisor is 2 and the dividend is 10. Also, a factor of a number. Both 2 and 5 are divisors of 10.

edge   A line segment of a polyhedron where two faces intersect.

equation   A statement that two quantities are equal.

equilateral triangle   A triangle whose sides all have the same length.

equivalent fractions Fractions that have the same value.

estimate To determine an approximate value.

expanded form A way of writing a number as the sum of the products of the digits and the place values of the digits.

exponent The upper number in an exponential expression. A number that tells how many times another number is to be used as a factor. In the expression 2^5, 5 is the exponent and 2 is the base.

exponential expression An expression that indicates that one number is to be used as a factor a given number of times. The expression 4^3 tells us that 4 is to be used as a factor 3 times. The value of 4^3 is 64.

face A flat surface of a geometric solid.

factor (1) Noun. One of two or more numbers that are to be multiplied. In the expression $4xy$, the factors are 4, x, and y. (2) Verb. To write as a product of factors. We can factor the number 6 by writing it as 2×3.

fraction A part of a whole or the indicated division of two numbers, such as $\frac{4}{5}$.

fraction line The line segment that separates the numerator and the denominator of a fraction.

function A set of number pairs related by a certain rule so that for every number to which the rule may be applied, there is exactly one resulting number.

geometric solid A three-dimensional geometric figure. Spheres, cones, and prisms are examples of geometric solids.

gram A basic unit of mass in the metric system.

greater than One number is said to be greater than a second number if the graph of the number is to the right of the graph of the second number.

height   *See* altitude of a triangle.

heptagon   A seven-sided polygon.

hexagon   A six-sided polygon.

hypotenuse   The side of a right triangle that is opposite the right angle.

improper fraction   A fraction whose numerator is equal to or greater than the denominator; thus, a fraction equal to or greater than 1.

independent events   Two events are said to be independent if the outcome of one event does not affect the probability that the other event will happen. If a dime is tossed twice, the outcome (heads or tails) of the first toss does not affect the probability of getting heads or tails on the second toss.

integers   The whole numbers and the opposites of the positive whole numbers. The members of the set . . . , -2, -1, 0, 1, 2, ...

intersect   To share a common point or points. Lines that intersect meet at a common point.

inverse operation   Two operations are inverse operations if one operation will "undo" the other operation. If we begin with 3 and multiply by 2, the product is 6. If we divide 6 by 2, we will undo the multiplication by 2, and the answer will be 3, the original number.

invert   When said of a fraction, to interchange the numerator and denominator.

isosceles triangle   A triangle with at least two sides of equal length.

kilogram   One thousand grams.

kilometer   One thousand meters.

least common denominator (LCD)   Of two or more fractions, a denominator that is the least common multiple of the denominators of the fractions.

least common multiple (LCM)   The smallest whole number that every member of a set of whole numbers will divide evenly.

less than   One number is less than a second number if the graph of the number on a number line is to the left of the graph of the second number.

line   A straight collection of points extending without end.

line segment   A part of a line.

liter   The basic unit of capacity in the metric system.

lowest terms   In reference to a fraction, when the numerator and denominator contain no common factors.

mean   Of a set of numbers, the average of the set of numbers.

median   The middle number when a set of numbers is arranged in order from the least to the greatest.

meter   The basic unit of length in the metric system.

milliliter   One thousandth of a liter.

millimeter   One thousandth of a meter.

mixed number   A numerical expression composed of a whole number and a fraction, such as $2\frac{1}{2}$.

mode   The number in a set of numbers that appears the most often.

multiple   A product of a selected counting number and any other counting number. Multiples of 3 include 3, 6, 9, and 12.

multiplier   One of two numbers that are to be multiplied; a factor.

negative numbers   Numbers to the left of zero on the number line.

nonagon   A nine-sided polygon.

numeral   Symbol or groups of symbols used to represent a number.

numerator   The top number of a fraction.

obtuse angle   An angle whose measure is greater than 90° and less than 180°.

obtuse triangle   A triangle that contains an obtuse angle.

octagon   An eight-sided polygon.

odds   A way of describing the likelihood of an event; the ratio of favorable outcomes to unfavorable outcomes.

opposites   Two numbers whose sum is 0. Thus, a positive number and a negative number whose absolute values are equal. The numbers −3 and +3 are a pair of opposite numbers.

origin   The point on a number line with which the number zero is associated.

parallel lines   Lines in the same plane that do not intersect.

parallelogram   A quadrilateral that has two pairs of parallel sides.

pentagon   A five-sided polygon.

percent   (1) Per hundred. Forty percent is 40 per hundred. (2) Hundredth. Forty percent is forty hundredths.

perimeter   Of a plane geometric figure, the distance around the figure.

perpendicular lines   Two lines that intersect and form right angles.

pi (π)   The number of diameters equal to the circumference of a circle. Approximate values of pi are 3.14 and $\frac{22}{7}$.

plane   In mathematics, a flat surface that has no boundaries.

point   A location on a line, on a plane, or in space with no size.

polygon   A closed, plane geometric figure whose sides are line segments.

polyhedron   A geometric solid whose faces are polygons.

positive numbers   Numbers to the right of zero on the number line.

power   The value of an exponential expression. The expression 2^4 is read as 2 to the fourth power and has a value of 16. Thus, 16 is the fourth power of 2. The word *power* is also used to describe the exponent.

prime factorization   The expression of a composite number as a product of its prime factors.

prime factors   The factors of a number that are prime numbers.

prime number   A whole number greater than 1 whose only whole number divisors are 1 and the number itself.

prism   A polyhedron with two congruent parallel bases.

probability   A way of describing the likelihood of an event; the ratio of favorable outcomes to all possible outcomes.

product   The result obtained when numbers are multiplied.

proper fraction   A fraction whose numerator is less than the denominator.

proportion   Two equivalent ratios.

Pythagorean theorem   A description of a property of right triangles that states that the area of a square constructed on the longest side of a right triangle is equal to the areas of the squares constructed on the other two sides of the triangle.

quadrant   Any one of the four sectors of a rectangular coordinate system, which is formed by two perpendicular number lines that intersect at the origins of both number lines.

quadrilateral   A four-sided polygon.

quotient   The result of division.

radical expression   An expression that contains radical signs, such as $\sqrt{x}$, $\sqrt[3]{16}$, and $\sqrt[4]{xy}$, which indicate roots of a number.

radius   The distance from the center of a circle to a point on the circle.

range   The difference between the largest and smallest numbers in a set of numbers.

rate   A ratio of two measures.

ratio   A comparison of two numbers by division. The ratio of a to b is written $\frac{a}{b}$.

ray   A part of a line that begins at a point called the *origin* and continues without end.

reciprocals   Two numbers whose product is 1. The reciprocal of $\frac{4}{3}$ is $\frac{3}{4}$ since $\frac{4}{3} \times \frac{3}{4} = 1$.

rectangle   A parallelogram that has four right angles.

regular polygon   A polygon in which all sides have equal lengths and all angles have equal measures.

repetend   The repeating digits of a decimal number often indicated by a bar. In the number $0.08\overline{3}$ the repetend is 3.

rhombus   A parallelogram that has four sides whose lengths are equal.

right angle   One of the angles formed at the intersection of two perpendicular lines. A right angle has a measure of 90°.

right triangle   A triangle that contains a right angle.

root   The solution to an equation; also, the value of a radical expression.

scale factor   The number that relates corresponding sides of similar geometric figures.

scalene triangle   A triangle whose three sides are of different lengths.

scientific notation   A method of writing a number as a product of a decimal number and a power of 10.

semicircle   A half circle.

sequence   An ordered list of numbers arranged according to a certain rule.

signed numbers   Numbers that are either positive numbers or negative numbers.

similar triangles   Two triangles that have the same shape but may not be the same size. The corresponding angles of similar triangles are equal in measure and the lengths of the corresponding sides are proportional.

square   A rectangle with sides of equal length.

square root   A number which, when multiplied by itself, equals the given number. A square root of 49 is 7 because $7 \cdot 7 = 49$.

straight angle   An angle whose measure is 180°.

sum   The result of addition.

supplementary angles   Two angles the sum of whose measures is 180°.

surface area   The total area of the surface of a geometric solid.

trapezoid   A quadrilateral with exactly one pair of parallel sides.

triangle   A three-sided polygon.

unit conversion   The process of changing a denominate number to an equivalent denominate number that has different units.

unit multiplier   A fraction of denominate numbers whose value is 1.

unit price   The price of one unit of measure of a product.

variable   A letter used to represent a number that has not been designated.

vertex   A point of an angle, polygon, or polyhedron where two or more lines, rays, or segments meet.

volume   The number of cubic units of a certain size that equals the space occupied by a geometric solid.

whole number   The numbers 0, 1, 2, 3, 4, ...

Index

Absolute value
  in addition of signed numbers,
    347–349
  definition of, 325
  of signed numbers, 325
Acute angles, 88, 456
Acute triangles, 339
Addends, definition of, 1
Addition
  algebraic, 384–385
  aligning decimals for, 8, 184
  of decimal numbers, 184
  of exponents to multiply
    powers of 10, 473–474
  finding averages with, 151–152
  of fractions, 41, 162–163
  of integers on number lines,
    324–326
  missing numbers in, 11–12
  of mixed measures, 263
  of mixed numbers, 121–123
  on number lines, 20–21,
    325–326
  order of operations for, 279–280
  part-part-whole pattern of,
    63–66
  rule for equations, 393–396
  sequence of numbers in, 320
  of signed numbers, 325–326,
    347–349
  some and some more pattern of,
    44–47
  of whole numbers, 3
Algebra
  addition, 384–385
  division, 334
  minus sign in, 384
  multiplication, 334
  operations of arithmetic in,
    334–335
  symbols in, 334

Angles
  charts of degrees of, 456
  definition of, 88
  estimating measures of,
    554–556
  measuring, 456–458, 504–505,
    554–556
  naming, 88
  sides of (*see* Rays)
  sum of, in triangles, 504–505
  symbol for, 88 (*see also*
    Symbols)
  types of, acute, 88, 456
    complementary, 579
    obtuse, 88, 456
    right, 88, 456
    straight, 88, 456, 505–506
    supplementary, 579–580
    vertical, 580
  vertex of, 88
Area
  conversion of units of, 499–500
  formulas for, 596–597
  of geometric figures: circles,
    467–469
    complex shapes, 271–272,
      432–433
    parallelograms, 312–313, 467
    rectangles, 244–246, 467
    semicircles, 588–589
    trapezoids, 433–434
    triangles, 352–353, 467
  surface, of geometric solids,
    592–593
Arithmetic, operations of. *See*
  Operations of arithmetic
Arrowheads
  on number lines, 15, 20–21
  with rounding numbers,
    156–157, 176–177
  as symbol for lines, 32

Asterisk, as symbol for
multiplication, 2
Averages, 151–152, 292–293

Bar, as symbol for repeating
decimals, 215–217
Bar graphs, 198
Bases, 237
in areas of parallelograms,
312–313, 354
definition of, 313
and exponents, 236–237
in volume of solids, 410–411
Binary operations, definition of, 1
Borrowing, in subtraction of
mixed measures, 296–297
Braces, 343. *See also* Symbols;
Symbols of inclusion
Brackets, 343. *See also* Symbols;
Symbols of inclusion

Canceling, 131–132, 283. *See also*
Reducing fractions
Capacity, 299–301, 627–628
Cent sign, 6, 7. *See also* Money
Center, of circles, 375
Centimeters, 83, 409
Chance. *See* Probability
Charts
of characteristics of
quadrilaterals, 308
of degrees in angles, 456
of fraction-percent equivalents,
495
of operations of arithmetic in
algebra, 334
of place values, 23, 167, 240
Circle graphs, 199
Circles
area of, 467–469
circumference of, 375
definition of, 375
parts of, 375
as symbols, 156–157, 176–177,
448
Circumference, 375
definition of, 375
finding with pi, 380–381
Commas, in large numbers, 23–24

Common denominators, 160–161.
See also Least common
multiples
Common multiples, 143–144. *See
also* Common denominators;
Least common multiples;
Multiples
Comparison
of decimals, 172–173
in insufficient information
problems, 452
symbols of (*see* Symbols of
Comparison; Symbols)
Complementary angles, 579
Complex shapes, area of, 271–272,
432–433
Composite numbers
definition of, 105
divisors of, 105–106
factoring, 106–108
Compound probability, 613–615
Cones, 366, 619
Conversion
of decimals to fractions, 220
of equivalent measures,
300–301
of fractions to decimals,
220–222
of fractions to numbers, 69–70
of numbers to fractions,
114–115
of numbers to percents, 257
of percents to numbers, 258
with unit multipliers, 284–285,
371–372, 498–499
of units of area, 499–500
of weights, 371–372
Coordinates, rectangular, 548–551
Corresponding angles, 561–563
Corresponding sides, 561–563
Counting numbers, 3, 488. *See
also* Composite numbers;
Prime numbers
Cubes, 366, 367, 409–410
Cubic measurements, 409–410
Cylinders, 366, 367

Decagons, 92
Decimal-part-of-a-number
problems, 330–331

Decimal numbers
  adding, 184
  charts of, 167
  comparing, 172–173
  dividing by, 228–230
  dividing of, by whole numbers,
    211–213
  fraction-percent equivalents of,
    316–317
  as fractions, 166–168, 220–222
  multiplying, 207–208
    by powers of 10, 241
  on number lines, 180–181
  and operations with fractions,
    388–389
  place value of, 167–168
  placing of zeros in, 172–173,
    211–212
  repeating, 215–217
  rounding of, 176–177
  subtracting, 185
  and terminal zeros, 172–173,
    177
  in words, 171–172
Decimal points
  aligning, for addition, 8, 184
    for division, 9, 211–212,
      229–230
    for multiplication, 207–208,
      241
    for subtraction, 8, 185
  placing with money, 6–7
  placing with powers of 10, 241,
    275–276, 304–305
  in scientific notation, 275–276,
    304–305
Degrees
  definition of, 456
  measuring, in angles, 456–458
    in straight angles, 505–506
    in triangles, 504–505
Denominators
  common (see Common
    denominators)
  in decimal fractions, 166–167
  definition of, 36,
  in dividing fractions, 135–137,
    437–439
  least common (see Least
    common multiple)

Denominators (*Cont.*)
  ordinal numbers in, 36
  in reducing fractions, 77–79,
    131–132
Diagrams
  of fraction-of-a-group problems,
    117–119
  of number families, 490
Diameter, definition of, 375
Difference, definition of, 1
Digits
  place value of, 23, 167
  repeating, 215–217
Directed numbers, 325. *See also*
    Negative numbers; Positive
    numbers; Signed numbers
Dividend, definition of, 1
Divisibility
  definition of, 28
  tests for, 29
Division
  algebraic, 334
  aligning decimals for, 9,
    211–212, 229–230
  in converting improper
    fractions, 69
  by decimal numbers, 228–230
  of decimal numbers, 211–212
  definition of, 1
  finding averages with, 151–152
  of fractions, 135–137, 437–438
  line, as symbol of inclusion,
    343–344
  of mixed numbers, 140
  by 100 to find percents, 258
  order of operations for, 279–280
  in reducing fractions, 77–79
  remainders in, 211–212,
    225–226
  repeated method of, in
    factoring, 108
  rules for, 404–406, 428 (*see
    also* Rules)
  of signed numbers, 427–429
  symbols for, bars, 2
    boxes, 2
    lines, 36 (*see also* Symbols)
  in unit price problems, 233
  of whole numbers, 4

Divisors
  definition of, 1
  as factors, 28–29
  inverting, 136–137, 437–439
  of numbers, 105–106
Dollar signs, 6, 7. *See also* Money
Dots, as symbol for multiplication, 2

Edges, of polyhedrons, 366, 367
Ellipsis, definition of, 3
Endpoint, 32
Equal fractions. *See* Equivalent
    fractions
Equal groups problems, 59–61
Equalities/inequalities. *See* Greater
    than/less than; Inequalities;
    Symbols; Symbols of
    comparison)
Equals sign, 16, 447. *See also*
    Symbols; Symbols of
    comparison
Equations
  addition rule for, 393–396
  definition of, 100
  division rule for, 404–406
  to find percents, 442–443
  literal, 596
  with mixed numbers, 510–511
  multiplication rule for, 403–406
  solving, 100–102
Equilateral triangles, 339
Equivalent fractions
  definition of, 73
  multiplying to find, 73–74
Equivalent measures. *See also*
    Linear measures
  conversion of, 300–301
  tables of, 82, 83, 300
Equivalents, fraction-decimal-
    percent, 316–317
Estimating
  measures of angles, 554–556
  with rounded numbers,
      157–158
  square roots, 578–579
Evaluation of expressions
  with positive numbers, 334–335
  with signed numbers, 514–515

Expanded notations, 241. *See also*
    Powers of 10
Exponential expressions, 236–237.
    See also Exponents
Exponents
  adding, in multiplication of
      powers of 10, 473–474
  definition of, 236
  expressions with, 237
  to show repeated
      multiplication, 236–237

Faces, of polyhedrons, 366, 367
Factoring
  with repeated division method,
      108
  with tree method, 107
Factors
  definition of, 1
  as divisors, 28–29
Feet, 82, 409
Formulas
  for areas, 595–597
  definition of, 596, 605
  substituting letters in, 605–606
  transforming, 596–597
Fraction lines, 2
Fractional-part-of-a-number
    problems, 330–331, 422–424
Fraction-of-a-group problems,
    117–119, 414
Fractions
  adding, 41, 162–163
  canceling, 131–132, 283
  common denominators in,
      160–161
  conversion of to numbers,
      69–70
  decimal-percent equivalents,
      316–317
  as decimals, 166–168, 220–222
  definition of, 36
  denominators (*see*
      Denominators)
  dividing, 135–137, 437–439
  equivalent, 69, 73–74
  improper (*see* Improper
      fractions)
  lowest terms of, 78

Fractions (*Cont.*)
    mixed numbers as, 114–115
    multiplication of, 41
    on number lines, 37–38
    numerators (*see* Numerators)
    and operations with decimals,
        388–389
    percent equivalents of, 494–495
    proper, 110–111
    reciprocals of, 127–128,
        135–137 (*see also*
        Reciprocals)
    reducing (*see* Reducing
        fractions)
    simplifying, 110–111, 482–483
    subtracting, 41, 162–163
    terms of, 78
    whole numbers as, 115
Functions, 321, 483–485. *See also*
    Linear functions

g. *See* Grams
Gal. *See* Gallons
Gallons, 82, 299–301
Geometric solids, 366–367,
    409–410, 592–593
Grams, 370–372
Graphs, 197–199
    of inequalities, 448
    of linear functions, 600–602
    on number lines, 447–448
    of rectangular coordinates,
        548–551
    types of, bar, 198
        circle, 199
        line, 199
        picture, 197–198
Greater than/less than, 16,
    447–448. *See also*
    Inequalities; Symbols;
    Symbols of comparison

Height
    in parallelograms, 313
    in triangles, 353
    in volume of solids, 410
Heptagons, 92
Hexagons, 92, 93, 96

Hyphens, for two-digit numbers,
    24
Hypotenuse
    definition of, 572
    and Pythagorean theorem,
        572–575

Improper fractions, 69–70
    conversion of, 69
    definition of, 69
    mixed numbers as, 114–115
    simplifying, 110–111
    whole numbers as, 115
Inches, 82, 410
Inequalities
    definition of, 448
    graphing of, 448
    symbols for, 447 (*see also*
        Greater than/less than;
        Symbols of comparison)
Insufficient information problems,
    451–452
Integers
    adding on number lines,
        324–326
    definition of, 489
Interest, 609–610
International system (SI, or
    Système International). *See*
    Metric system
Intersecting lines, definition of, 87
Isosceles triangle, 339, 340

kg. *See* Kilograms
Kilograms, 82, 370–372
Kilometers, 82–83
km. *See* Kilometers

L. *See* Liters
Larger-smaller-difference word
    problems, 54–56
lb. *See* Pounds
Least common multiple (LCM),
    144. *See also* Common
    multiple; Denominators;
    Multiples

Length
  measures of, in metric system,
      83–84
    in U.S. system, 82
  table of units of, 82, 83
Letters
  naming angles with, 88
  naming lines with, 32–34
  naming polygons with, 93
  in writing equations, 100, 596
Linear functions
  definition of, 600
  graphing of, 600–602
Linear measures
  metric, 82, 83
  tables of, 82, 83
  U.S. system, 82
  using, 82–84
Line graphs, 199
Lines
  intersecting, 87, 549
  naming with letters, 32–34
  number (*see* Number lines)
  parallel, 87
  perpendicular, 87
  symbols for, 32, 33 (*see also*
      Symbols)
Line segments. *See* Segments
Literal equations, 596
Liters, 82, 299–301. *See also*
    Volume

m. *See* Meters
Mass, 371
Mean, 477–479
Median, 464, 477–479
Measurement. *See also* Rates;
    Ratios
  equivalent (*see* Equivalent
      measures)
  of liquids, 299–301 (*see also*
      Volume)
  mixed (*see* Mixed measures)
  unit conversion problems of,
      284–285, 371–372,
      499–500
Meters, 82–84, 409
Metric system
  definition of, 82

Metric System (*Cont.*)
  length measures in, 83–84
  square units in, 245
  tables of, 83, 300
  units of capacity in, 299–301
  units of volume in, 628–629
  weight measures in, 370
mg. See Milligrams
Milligrams, 370–372
Milliliters, 299–301
Millimeters, 83
Minuend, definition of, 1
Minus sign, in algebraic addition,
    384
Missing numbers
  in addition, 11–12
  finding in functions, 321,
      483–485
  in multiplication, 12–13
  in some and some more
        problems, 44–47
  in subtraction, 12
Mixed measures
  adding, 263
  converting, 262–263 (*see also*
      Conversion; Unit
      multipliers)
  subtracting, 296–297
Mixed numbers. *See also*
    Fractions
  adding, 121–124
  definition of, 37
  dividing, 140
  equations with, 510–511
  as improper fractions, 114–115
  multiplying, 140
  on number lines, 37–38
  simplifying, 110–111
  subtracting, 122–124
mL. See Milliliters
mm. See Millimeters
Mode, 464, 477–479
Money
  forms of writing, 6–7
  and interest problems, 609–610
  operations of arithmetic with,
      8–9
  placing decimal points with,
      6–7
  in unit price problems, 233–234

Multiples
    common, 143–144
    definition, 143
    least common (LCM), 144
Multiplication
    of decimal numbers, 207–208
    definition of, 1
    in division of fractions, 136
    equal groups pattern, 59–61
    to find equivalent fractions,
        73–74
    of fractions, 41
    missing numbers in, 12–13
    of mixed numbers, 140
    of numbers in scientific
        notation, 473–474
    by 100 percent, 257
    operations of, in algebra, 334
    order of operations for, 279–280
    by powers of 10, 241
    of powers of 10, 472–474
    of rates, 266–268
    repeated, with exponents,
        236–237
    rules for, 403–406, 428 (see also
        Rules; specific operations)
    sequence of numbers in, 321
    of signed numbers, 427–429,
        584
    symbols for, 2, 334 (see also
        Symbols)
    with unit multipliers, 284–285
    of variables, 334–335
    of whole numbers, 4

Negative numbers. See also Signed
        numbers
    absolute value of, 325
    adding with positive numbers,
        325–326, 347–349
    definition of, 16
    in evaluation of expressions,
        514–515
    powers of, 585
    as signed numbers, 324–325
    and zeros, 16, 304
Negative sign, 16
Newtons, 371
Nonagons, 92

Notations. See Symbols
Number lines
    adding on, 20–21, 324–326
        (see also Signed numbers)
    decimal numbers on, 180–181
    definition of, 15
    fractions on, 37–38
    graphs on, 447–448
    mixed numbers on, 37–38
    as model for addition, 325–326
    reading, 16, 38
    rounding with, 155–156
    subtracting on, 20–21
    zeros as origin on, 15
Number pairs, 321
Numbers
    averages of, 151–152, 292–293
    commas in, 23–24
    comparing of, 16–17 (see also
        Symbols of comparison)
    composite, 104–108 (see also
        Composite numbers)
    conversion of fractions to,
        69–70
    conversion of percents to, 258,
        357–358
    conversion of, to percents, 257
    counting, 3, 488 (see also
        Composite numbers; Prime
        numbers)
    decimal (see Decimal numbers)
    decimals in (see Decimal
        points)
    directed, 325 (see also Signed
        numbers)
    families of, 488–491
    missing (see Missing numbers)
    mixed (see Mixed numbers)
    negative (see Negative numbers)
    opposite, 16, 384
    ordering of, 16
    ordinals, in denominators, 36
    as percents, 257–258, 442–443
    percents of, 357–358
    prime (see Prime numbers)
    positive (see Positive numbers)
    rational, 489–490
    in scientific notation, 275–276,
        304–305, 399–400,
        473–474

Numbers (*Cont.*)
  sequences of, 320–321
  signed (*see* Signed numbers)
  whole (*see* Whole numbers)
  in words, 24
Numerators
  definition of, 36
  in dividing fractions, 135–137,
    437–439
  in reducing fractions, 77–79,
    131–132

Obtuse angles, 88, 456
Obtuse triangles, 339
Octagons, 92
Odds. *See* Probability
One (1)
  hidden, 437
  fractions equivalent to, 69, 73
Operations of arithmetic, 1–2
  in algebra, 334–335
  with fractions and decimals,
    388–389
  with money, 8–9
  order of, 279–280, 482–483
  symbols of, 334 (*see also*
    Symbols)
  within symbols of inclusion,
    343–344
  with whole numbers, 3–4
Opposites, of the opposite, 384
Order of operations, 279–280,
    482–483. *See also*
    Operations of arithmetic
Origin
  as zeros on number lines, 15
  as point of intersection of *x* axis
    and *y* axis, 549
Ounces, 299–301, 371–372
Outlines
  for equal groups problems, 59
  for part-part-whole pattern, 64
  for some and some more
    pattern, 64
Overbar, as symbol for segments,
    33
oz. *See* Ounces

Parallel lines, 87. *See also*
  Symbols
Parallelograms, 307–308
  area of, 312–313, 467
  rectangles as, 307
  squares as, 307
Parentheses
  and signed numbers, 514–515
  simplifying numbers within,
    2–3
  as symbol of inclusion, 2 (*see
    also* Symbols)
  using in evaluation of variables,
    334–335
Part-part-whole word problems,
    63–66
Pentagons, 92, 93
Percents, 442–443
  common fraction equivalents,
    494–495
  conversion of, to numbers, 258,
    357–358
    numbers to, 257, 358
  definition of, 256
  fraction-decimal equivalents,
    316–317
  and probability, 623–624
  symbol for, 257 (*see also*
    Symbols)
  using proportions with
    problems of, 461–463
  zeros with, 257
Perimeters
  definition of, 96
  of polygons, 96–97, 193–194
  of semicircles, 588–589
  word problems with, 97–98
Perpendicular lines, 87. *See also*
  Symbols
Pi (π)
  and area of circles, 468–469
  and circumference, 380–381
  definition of, 380
  symbol for, 380 (*see also*
    Symbols)
Picture graphs, 197–198
Pints, 299–301
Place values
  charts of, of decimal numbers,
    167

Place values, charts of (*Cont.*)
    with powers of 10, 240
    of whole numbers, 23
  in decimal fractions, 166
  of decimals, 167–168
  definition of, 23
  through hundred trillions, 23
Planes, definition of, 87
Polygons
  definition of, 92
  irregular, 93
  naming of, 92
  regular, 93
  perimeters of, 96–98, 193–194
  vertex of, 93
Polyhedrons, 366–367
Positive numbers. *See also* Signed
    numbers
  absolute value of, 325
  adding with negative numbers,
    325–326, 347–349
  definition of, 16
  evaluation of expressions with,
    334–335
  and scientific notation, 304
  as signed numbers, 324–325
  and zeros, 16, 304
Pounds, 82, 371–372
Powers, 237–238
  of negative numbers, 585
  of 10 (*see* Powers of 10)
Powers of 10
  chart of place values of, 240
  and expanded notations, 241
  multiplying, 472–474
  multiplying decimal numbers
    by, 241
  placing decimals with,
    275–276, 304–305
  in scientific notation, 275–276,
    399–400, 472–474
  use of zeros in writing, 240
Prime factorization, 107, 131
Prime numbers
  definition of, 104
  divisors of, 105–106
  factoring, 106–108
Principal, 609–610
Prisms, 365–367, 409
Probability, 538–540, 613–615,
    622–624

Product, definition of, 1
Proper fractions, simplifying of,
    110–111
Proportions, 203–204
  with percent problems,
    461–463
  with ratio problems, 288–289
Protractors, measuring angles
    with, 457–458, 554–556
pt. *See* Pints
Pyramids, 366, 618–619
Pythagorean theorem, 572–575

qt. *See* Quarts
Quadrants, 549
Quadrilaterals, 92, 93
  charts of characteristics of, 308
  classifying, 307–309
Quarts, 299–301
Quotient, 1

Radical sign, as symbol for square
    root, 249
Radius (Radii), 375–376
Range, 464, 477–479
Rates
  definition of, 253
  multiplying, 266–268
  problems of, 254, 417–419
  types of, exchange, 253
    mileage, 253
    speed, 253
Ratio boxes, use of
  with implied ratio problems,
    418–419
  with problems of scale,
    567–568
  with proportion and percent
    problems, 461–463
  with ratio word problems
    288–289, 361–362
Rational numbers, 489–490
Ratios, 188–190. *See also*
    Proportions
  cross products of, 203–204
  definition of, 188
  implied, 417–419
  as proportions, 203

Ratios (*Cont.*)
    as rate, 253
    in word problems, 189–190,
        288–289, 361–362,
        461–463
Rays
    of angles, 88
    naming, 32
    symbol for, 32 (*see also*
        Symbols)
Reciprocals, 127–128
    definition of, 127
    in division of fractions,
        135–137
    property of, 127
Rectangles
    area of, 244–246, 467
    in complex shapes, 272–273
    perimeter of, 96
    as quadrilaterals, 307–308
Rectangular coordinates, graphing,
    548–551
Rectangular prisms, 366, 367
Reducing fractions
    before multiplying, 131–132,
        283 (*see also* Canceling)
    numerators in, 77–79, 131–132
    with prime factorization, 131
Remainders, 211–212, 225–226
Repeated division method, of
    factoring, 108
Repetend, definition of, 215
Right angles, 88, 456
Right triangles
    definition of, 339–340
    and Pythagorean theorem,
        572–575
Rounding
    of decimal numbers, 176–177
    in estimating answers, 157–158
    with number lines, 155–156
    of whole numbers, 155–158
Rules
    for equations: addition,
        393–396
        division, 404–406
        multiplication, 403–406
    for signed numbers: division,
        428
        multiplication, 428

Scale, 567–568
Scale factor, 568–569
Scalene triangles, 339
Scientific notation
    decimals in, 275–276, 304–305
    large numbers in, 275–276
    multiplying numbers in,
        473–474
    and powers of 10, 275–276,
        399–400, 472–474
    small numbers in, 304–305
Segments. *See also* Lines
    definition of, 32
    length of, 33
    naming of, 32–34
    symbols for, 32, 33 (*see also*
        Symbols)
Semicircles, 588–589
Sequence
    missing numbers in, 321
    of numbers, 320–321
Signed numbers
    absolute value of, 325, 347–348
    adding, 325–326, 347–349
    definition of, 324–325
    as directed numbers, 325
    dividing, 427–429
    evaluating expressions with,
        514–515
    multiplying, 427–429, 584
    on number lines, 325–326
    order of operations with,
        482–483
    and parentheses, 514–515
    rules for, 428
    zeros with, 347–348
Similar triangles, 561–563
Simplifying, 2–3, 110–111,
    482–483
Slanted fraction line, 2
Solids. *See* Geometric solids
Some and some more word
    problems, 44–47
Some went away word problems,
    49–52
Spheres, 366, 367
Square roots, 249–250
    definition of, 249
    estimating, 578–579
    symbol for, 249 (*see also*
        Symbols)

Squares
  perimeter of, 97
  as quadrilaterals, 308
Square units, in area, 245
Straight angles, 88, 456, 505–506
Substitution, in formulas, 605–606
Subtraction
  aligning decimals for, 8, 185
  of decimal numbers, 185
  definition of, 1
  of fractions, 41, 162–163
  larger-smaller-difference pattern
    of, 54–56
  missing numbers in, 12
  of mixed measures, 296–297
  of mixed numbers, 122–124
  on number lines, 20–21
  order of operations for, 279–280
  some went away pattern of,
    49–52
  of whole numbers, 3
Subtrahend, definition of, 1
Sums
  in adding signed numbers, 347
  of angles measures in triangles,
    504–505
  definition, 1
Supplementary angles, 579–580
Surface area, of geometric solids,
    592–593
Symbols
  ∠ (angles), 88
  arrowheads, 32
  * (asterisk), 6
  ‾ (bars), 215–217
  ¢ (cent sign), 6
  circles, 156–157, 176–177, 448
  of comparison, = (equals sign),
    16, 447
    > < (greater than/less than),
      16, 447
  / (division), 2
  — (division bar), 2
  ‾⟌ (division box), 2
  $ (dollar sign), 6
  dot (multiplication), 2
  of inclusion, [ ] (brackets), 343
    { } (braces), 343
    — (division line), 344
    ( ) (parentheses), 2, 343

Symbols (*Cont.*)
  for lines, 32, 33
  ×, * (multiplication), 2
  – (negative), 16
  ‾ (overbar), 33
  ‖ (parallel lines), 87
  % (percent), 257
  ⊥ (perpendicular lines), 87
  π (pi), 380
  ⇌ (rays), 32
  √ (square root), 249
Symbols of comparison. *See*
    Symbols; *specific symbols*
Symbols of inclusion. *See also*
    Symbols; *specific symbols*
  definition of, 343
  operations of arithmetic within,
    343–344

t. See Tons
Tables. *See also* Metric system;
    U.S. system
  of equivalent measures, 300
  of fraction-decimal-percent
    equivalents, 317
  of units of length, 82, 83
Terms, definition of, 78
Theorem, Pythagorean, 572–575
Time, problems of, 55–56
Tons, 371
Totals, in ratio problems, 362. *See
    also* Sums
Transformation, of formulas,
    596–597
Trapeziums, 307, 308
Trapezoids, 307–309, 433–434
Tree method of factoring, 107
Triangles, 338–340
  area of, 352–353, 467
  classifying, 339
  measuring angles in, 504–505
  180° in, 504
  as polygons, 92, 93
  types of, acute, 339
    equilateral, 339
    isosceles, 339, 340
    obtuse, 339
    right, 339, 340, 572–575
    scalene, 339
    similar, 561–563

Triangular prisms, 366, 367
Two-step word problems, 147–148

Undecagons, 92
Unit multipliers, 283–285
  with conversion problems,
      284–285, 372,
      499–500
  definition of, 284
  multiple, 498–499
Unit price problems, 233–234
U.S. Customary system
  definition of, 82
  measuring length in, 82–84
  measuring weight in, 371–372
  square units in, 245
  tables of 82, 300
  units of capacity in, 299–301

Variables
  definition of, 334
  multiplying, 334–335
  with parentheses, 334–335
Vertex (Vertices)
  of angles, 88
  of polygons, 93
  of polyhedrons, 366, 367
Volume, 299–301
  of cones, 619
  of cubes, 410
  definition of, 409
  of prisms, 409
  of pyramids, 618–619
  of solids, 410
  units of capacity in, 299–300,
      627–628
  using cubes in measurement of,
      409–410

Weight, 370–372, 627–628
Whole numbers
  adding, 3
  averages of, 151–152
  definition of, 3, 488–489
  dividing, 4
  dividing decimal numbers by,
      211–212
  as fractions, 115
  multiplying, 4

Whole numbers (*Cont.*)
  operations of arithmetic with,
      3–4
  place values of, 23
  reading and writing, 23–24
  rounding, 155–158
  subtracting, 3
Word problems
  with perimeters, 97–98
  with probability, 613–615,
      623–624
  with rates, 253–254, 417–419
  with ratios, 189–190, 288–289,
      361–362, 461–463
  with scale, 567–568
  with statistics, 477–479
  types of, decimal part of a
      number, 330–331
    equal groups, 59–61
    fraction-of-a-group, 117–119,
        414
    fractional-part-of-a-number,
        330–331, 422–424
    larger-smaller-difference,
        54–56
    part-part-whole, 63–66
    quantitative comparison,
        451–452
    some and some more, 44–47
    some went away, 49–52
    time, 55–56
    two-step, 147–148
  with unit prices, 233–234

x-axis, 549

y-axis, 549

Zeros
  in adding signed numbers,
      347–348
  with cent sign, 7
  with decimal numbers,
      172–173, 211–212
  with dollar sign, 7
  and negative numbers, 16, 304
  as origin on number lines, 15
  with percents, 257
  and positive numbers, 16, 304
  terminal, 172–173, 177
  in writing powers of 10, 240